Micro/Nano Manufacturing

Micro/Nano Manufacturing

Special Issue Editors

André Zimmermann
Stefan Dimov

MDPI • Basel • Beijing • Wuhan • Barcelona • Belgrade

Special Issue Editors

André Zimmermann
University of Stuttgart and Hahn-Schickard,
Germany

Stefan Dimov
University of Birmingham,
UK

Editorial Office
MDPI
St. Alban-Anlage 66
4052 Basel, Switzerland

This is a reprint of articles from the Special Issue published online in the open access journal *Applied Sciences* (ISSN 2076-3417) from 2018 to 2019 (available at: https://www.mdpi.com/journal/applsci/special_issues/Micro-Nano_Manufacturing)

For citation purposes, cite each article independently as indicated on the article page online and as indicated below:

LastName, A.A.; LastName, B.B.; LastName, C.C. Article Title. *Journal Name* **Year**, *Article Number*, Page Range.

ISBN 978-3-03921-169-2 (Pbk)
ISBN 978-3-03921-170-8 (PDF)

Cover image courtesy of André Zimmermann.

Contents

About the Special Issue Editors

André Zimmermann, Univ.-Prof. Dr.-Ing., was born in Schweinfurt, Germany in 1971. He studied chemistry and crystallography at Julius-Maximilians-Universität Würzburg as well as materials science with a specialization in mechanical engineering at Technische Universität Darmstadt. After several stays in the USA at NIST and the University of Washington, he received his Ph.D. in 1999 at Technische Universität Darmstadt. He held positions as group manager at the Max-Planck-Institute for Metals Research, Stuttgart and as senior manager for electronic packaging within the corporate research and development sector of Robert Bosch GmbH in Waiblingen. Since January 2015, he has worked as a Professor of Micro Technology at the Institute for Micro Integration (IFM) of the University of Stuttgart. He also serves as Head of the Institute for Micro Assembly Technology at Hahn-Schickard in Stuttgart.

Stefan Dimov, Dipl. Eng., Ph.D., D.Sc., FIMechE, is Professor of Micro Manufacturing and Head of the Advanced Manufacturing Technology Centre at the Department of Mechanical Engineering, University of Birmingham. He obtained his Diploma Engineer and Doctoral degrees from Moscow State University of Technology and a Doctor of Science degree from Cardiff University. His research interests encompass the broad area of advanced manufacturing with a special focus on micro and nano manufacturing, additive manufacturing, and hybrid manufacturing technologies. His academic output includes more than 250 technical papers and 13 books. He has supervised over 30 Ph.D. theses to completion. He has won in excess of £30M in external research grants and contracts. He is Associate Editor of the *ASME Journal of Micro- and Nano-Manufacturing and Precision Engineering Journal*. In addition to pursuing and leading research, he has also been very active with knowledge transfer to industry, applying the results of his work to help multinational companies and SMEs to create wealth and safeguard jobs. He established the Multi-Material Micro Manufacture (4M) Community in 2004 through the FP6 4M NoE programme and currently is Executive Officer of the self-sustained 4M Association (www.4m-association.org). He was the recipient of the Thomas Stephens Group Prize in 2000 and 2003 and the Joseph Whitworth Prize in 2017, awarded by the Institution of Mechanical Engineers (IMechE).

Editorial

Special Issue on "Micro/Nano Manufacturing"

André Zimmermann [1,2] and Stefan Dimov [3,*

[1]   University of Stuttgart, Institute for Micro Integration (IFM), Allmandring 9 b, 70569 Stuttgart, Germany; zimmermann@ifm.uni-stuttgart.de
[2]   Hahn-Schickard, Allmandring 9 b, 70569 Stuttgart, Germany
[3]   Department of Mechanical Engineering, School of Engineering, University of Birmingham, Edgbaston, Birmingham B15 2TT, UK
*   Correspondence: S.S.Dimov@bham.ac.uk

Received: 5 June 2019; Accepted: 10 June 2019; Published: 11 June 2019

Keywords: micro and nano manufacturing; micro-fluidics; micro-optics; micro and nano additive manufacturing; micro-assembly; surface engineering and interface nanotechnology; micro factories; micro reactors; micro sensors; micro actuators

---

1. Introduction

Micro manufacturing is dealing with the fabrication of structures in the order of 0.1 to 1000 μm. The scope of nano manufacturing extends the size range of manufactured features to even smaller length scales: below 100 nm. A strict borderline between micro and nano manufacturing can hardly be drawn, such that both domains are treated as complementary and mutually beneficial within a closely interconnected scientific community. Both micro and nano manufacturing can be considered as important enablers for high-end products. Especially, such products are enabled by micro and nano features and structures to incorporate special optical, electronic, mechanical, fluidic or biological functions into existing and new emerging products, and thus lead to unique selling points. Application fields include, but are not restricted to, precision instrumentation, sensors, metrology, energy harvesting, mechatronic systems, transport, medical technologies and life sciences. This Special Issue is dedicated to recent advances in research and development within the field of micro and nano manufacturing. The included papers report recent findings and advances in manufacturing technologies for producing products with micro and nano scale features and structures as well as applications underpinned by the advances in these technologies. In particular, the Special Issue covers the following topics:

- Micro fabrication technologies, process chains and process characterization;
- Novel product designs, micro-assembly technologies and micro-handling;
- Surface engineering and interface nanotechnology;
- Process modeling and simulation;
- Processing and characterization of nanomaterials;
- Micro and nano additive manufacturing technologies;
- Micro and desktop factory concepts, systems, components and modules;
- On-line monitoring and inspection systems/methods;
- Applications of micro and nano technologies.

2. History of the World Congress on Micro and Nano Manufacturing

This Special Issue is predominantly based on selected papers presented at the World Congress on Micro and Nano Manufacturing (WCMNM 2018) held in September 2018 in Portoroz, Slovenia [1], which was organized jointly by 4M [2], I2M2 [3] and the International Forum on Micro Manufacturing (IFMM) [4]. Further contributions were collected via a call for papers used by the Applied Sciences journal. Although the gathering of the global scientific community in micro and nano manufacturing in

Portoroz was merely the second edition within the WCMNM series after WCMNM 2017 in Kaohsiung, Taiwan, the tradition of international conferences on micro manufacturing dates back to the year 2005. In this year, 4M organized the First International Conference on Multi-Material Micro Manufacture held in Karlsruhe, Germany [5]. Only one year later, I2M2 initiated its series of International Conferences on Micro Manufacturing (ICOMM) with a first meeting at Urbana-Champaign, USA, in 2006. Additionally, the first IFMM took place in Gifu, Japan, in 2010. These coinciding activities and joint efforts in America, Asia and Europe laid the foundation for WCMNM as a world-wide platform dedicated to micro and nano manufacturing.

3. Content of the Special Issue

The Special Issue contains 13 papers covering the topics listed above [6–18]. The treated micro fabrication technologies range from established processes like elliptical vibration cutting to novel process chains such as the formation of nanoparticle arrays by hot embossing and sputtering. Even for well-known processes, modeling can contribute to their better understanding, and thus, lead to improved process control and quality of produced products. Therefore, it is not surprising that a number of contributions in this Special Issue report modeling results, which can enhance the knowledge and in depth understanding of processes underpinning micro and nano manufacturing solutions and their respective applications. Since additive manufacturing is currently attracting a significant interest from industry and research in general, it is again not surprising that additive manufacturing contributions also feature prominently in in this issue. For example, advances in selective laser melting are reported together with investigations into achievable surface quality of additively manufactured microstructures, which represents a major challenge for the broader use of this technology for micro manufacturing. Last but not the least, this Special Issue reports numerous applications of micro and nano manufacturing technologies with a special focus on micro-optics and micro-fluidics.

The Special Issue is the culmination of the efforts and dedication of many people to make it reality. We would like to thank the WCMNM 2018 sessions' chairs for selecting conference contributions that defined the skeleton of this Special Issue. Furthermore, the work of the reviewers involved in preparing this Special Issue needs to be acknowledged.

Acknowledgments: We would like to thank the associate editor for the efforts in making this Special Issue reality.

Conflicts of Interest: The authors declare no conflict of interest.

References

1. Valentincic, J.; Dimov, S.; Byung-Guk Jun, M.; Dohda, K.P. Proceedings of the World Congress on Micro and Nano Manufacturing, Portoroz, Slovenia, 18–20 September 2018; Valentinčič, J., Byung-Guk Jun, M., Dohda, K., Dimov, S., Eds.; Research Publishing: Singapore, 2018; 5. Available online: http://rpsonline.com.sg/proceedings/9789811127281/html/preface.html (accessed on 6 June 2019).
2. 4M Association. Available online: http://www.4m-association.org/ (accessed on 3 June 2019).
3. International Institution for Micro Manufacturing. Available online: https://sites.google.com/i2m2.org/home/ (accessed on 3 June 2019).
4. International Forum on Micro Manufacturing. Available online: http://www3.u-toyama.ac.jp/ifmm/ (accessed on 3 June 2019).
5. Proceedings of the Institution of Mechanical Engineers, Part C: Journal of Mechanical Engineering Science, 2006; 220, pp. 1609–1705. Available online: ttps://journals.sagepub (accessed on 6 June 2019).
6. Han, Y.; Zhang, L.; Fan, C.; Zhu, W.; Beaucamp, A. Theoretical Study of Path Adaptability Based on Surface Form Error Distribution in Fluid Jet Polishing. *Appl. Sci.* **2018**, *8*, 1814. [CrossRef]
7. Lu, M.; Chen, B.; Zhao, D.; Zhou, J.; Lin, J.; Yi, A.; Wang, H. Chatter Identification of Three-Dimensional Elliptical vibration Cutting Process Based on Empirical Mode Decomposition and Feature Extraction. *Appl. Sci.* **2019**, *9*, 21. [CrossRef]
8. Li, K.; Xu, G.; Huang, X.; Xie, Z.; Gong, F. Manufacturing of Micro-Lens Array Using Contactless Micro-Embossing with an EDM-Mold. *Appl. Sci.* **2019**, *9*, 85. [CrossRef]

9. Kim, N.; Bhalerao, I.; Han, D.; Yang, C.; Lee, H. Improving Surface Roughness of Additively Manufactured Parts Using a Photopolymerization Model and Multi-Objective Particle Swarm Optimization. *Appl. Sci.* **2019**, *9*, 151. [CrossRef]

10. Yasuda, R.; Adachi, S.; Okonogi, A.; Anzai, Y.; Kamiyama, T.; Katano, K.; Hoshi, N.; Natsume, T.; Mogi, K. Fabrication of a Novel Culture Dish Adapter with a Small Recess Structure for Flow Control in a Closed Environment. *Appl. Sci.* **2019**, *9*, 269. [CrossRef]

11. Seybold, J.; Bülau, A.; Fritz, K.; Frank, A.; Scherjon, C.; Burghartz, J.; Zimmermann, A. Miniaturized Optical Encoder with Micro Structured Encoder Disc. *Appl. Sci.* **2019**, *9*, 452. [CrossRef]

12. Xie, S.; Wan, X.; Wei, X. Fabrication of Multiscale-Structure Wafer-Level Microlens Array Mold. *Appl. Sci.* **2019**, *9*, 487. [CrossRef]

13. Kim, N.; Yang, C.; Lee, H.; Aluru, N. Spatial Uncertainty Modeling for Surface Roughness of Additively Manufactured Microstructures via Image Segmentation. *Appl. Sci.* **2019**, *9*, 1093. [CrossRef]

14. Charles, A.; Elkaseer, A.; Thijs, L.; Hagenmeyer, V.; Scholz, S. Effect of Process Parameters on the Generated Surface Roughness of Down-Facing Surfaces in Selective Laser Melting. *Appl. Sci.* **2019**, *9*, 1256. [CrossRef]

15. Huang, C.; Tsai, M. Tunable Silver Nanoparticle Arrays by Hot Embossing and Sputter Deposition for Surface-Enhanced Raman Scattering. *Appl. Sci.* **2019**, *9*, 1636. [CrossRef]

16. Aizawa, T.; Morita, H.; Wasa, K. Low-Temperature Plasma Nitriding of Mini-/Micro-Tools and Parts by Table-Top System. *Appl. Sci.* **2019**, *9*, 1667. [CrossRef]

17. Gielisch, C.; Fritz, K.; Noack, A.; Zimmermann, A. A Product Development Approach in The Field of Micro-Assembly with Emphasis on Conceptual Design. *Appl. Sci.* **2019**, *9*, 1920. [CrossRef]

18. Papatzani, S.; Paine, K. From Nanostructural Characterization of Nanoparticles to Performance Assessment of Low Clinker Fiber-Cement Nanohybrids. *Appl. Sci.* **2019**, *9*, 1938. [CrossRef]

Article

From Nanostructural Characterization of Nanoparticles to Performance Assessment of Low Clinker Fiber–Cement Nanohybrids

Styliani Papatzani [1,2,*] and Kevin Paine [2]

[1] Hellenic Ministry of Culture and Sports, Directorate of Restoration of Medieval & Post-medieval Monuments, Tzireon 8-10, Athens 11742, Greece

[2] BRE Centre for Innovative Construction Materials, University of Bath, Bath BA2 7AY, UK; k.paine@bath.ac.uk

* Correspondence: spapatzani@gmail.com

Received: 20 April 2019; Accepted: 7 May 2019; Published: 11 May 2019

Abstract: With the current paper three nano-Montmorillonites (nMt) are applied in cement nanohybrids: an organomodified nMt dispersion, nC2; an inorganic nMt dispersion, nC3; and an organomodified powder, nC4. nC4 is fully characterized in this paper (X-ray diffraction, scanning electron microscopy/X-ray energy dispersive spectroscopy and thermal gravimetric analysis/differential thermogravimetry. Consecutively a ternary non pozzolanic combination of fiber–cement nanohybrids (60% Portland cement (PC) and 40% limestone (LS)) was investigated in terms of flexural strength, thermal properties, density, porosity, and water impermeability. Flexural strength was improved after day 28, particularly with the addition of the inorganic nMt dispersion. There was no change in density or enhancement in pozzolanic reactions for the powder nMt. Mercury intrusion porosimetry showed that the pore related parameters were increased. This can be attributed to mixing effects and the presence of fibers. Water impermeability tests yielded ambiguous results. Clearly, novel manufacturing processes of cement nanohybrids must be developed to eliminate mixing issues recorded in this research.

Keywords: low PC clinker; Portland limestone ternary fiber–cement nanohybrids; flexural strength; TGA/dTG; XRD; MIP; water impermeability tests

---

1. Introduction

Soil constitutes of clay minerals, which are classified according to their internal structure as 1:1 (one tetrahedral silicate sheet bonded to one octahedral hydroxide sheet) or 2:1 (one octahedral hydroxide sheet sandwiched between two tetrahedral silicate sheets). The latter group consists of illites, Kaolins, smectites, and others. Montmorillonite (MMT) is a dioctahedral smectite (one SiO_2 platelet/layer sandwiched between two Al_2O_3 platelets/layers). It can be found in pure form or in the form of bentonite, in which 50%–80% MMT is intermixed with chlorite, muscovite, illite, and kaolinite. If bentonite undergoes a purification process with the use of sodium ions, "sodium bentonite" is produced, which practically is pure MMT. Montmorillonite is; therefore, a two-dimensional naturally hydrophilic stack of platelets of nanosized thickness [1]. The lateral dimensions vary from a few nanometers to a few hundreds of micrometers [2].

Given that the European Commission in 2011 has specified that "nanomaterial" means a natural, incidental, or manufactured material containing particles, in an unbound state or as an aggregate or as an agglomerate and where, for 50% or more of the particles in the number size distribution, one or more external dimensions is in the size range 1–100 nm [3]. It is only natural that an extensive number of scientists are involved into the separation of the platelets in order to produce individually reactive platelets of nanosized thickness, generally termed as "nanoclay." If an organic surface modifier is inserted in the MMT layers to separate them, then an organomodified MMT (OMMT) is produced and is considered to be nano-Montmorillonite (nMt). Then, nanoclays or nMt's can be dispersed in ceramic, metallic, biopolymeric, or polymeric matrixes or in water. The preferred manufacturing and fabrication method depends on the purpose of use of the nanoclays or nMt's. Ultrasonic sonication, planetary centrifugal mixing, magnetic dispersion, roll milling, and combinations of them have been investigated. For example, a significant number of papers is published on the effect of the nanoclay dispersion methods on the mechanical behavior of E-glass/epoxy nanocomposites [4,5] or clay-polymer nanocomposites [6–8].

The use of nanoclays is widespread in environmental science; for example, for removal of contaminants in water [9], remediation of contaminated soils [10] or biodegradation of polyolefins [11], in medicine for drug delivery [12], and in other fields. The use of nanoclays in cement science is limited, notwithstanding the size compatibility between the main hydration product of cement, C–S–H, which is nanosized [13] and nMt, which is in the same order of the nanocrystalline structure of C–S–H [14]. In fact, most studies investigate the effect of the addition of thermally- and mechanically-treated (calcined at temperatures above 700 °C) nanoclays, predominantly kaolinites [15], MMT's or micro-sized OMMT [16], despite the environmental cost due to the calcination process [17]. Others, report results on the use of industrial nMt's (in liquid form purchased from J Nano Technology Co., Ltd.) used as an admixture without providing any characterization of the nanoparticles [14].

In recent years a number of studies are suggesting the replacement of Portland cement by an amount of nanoclays or nMt's in composite cements in an effort to produce "green" cementitious nanohybrids (i.e., nanohybrids that contain minimum amount of Portland cement clinker), which is accountable for almost 8% of the global manmade CO_2 emissions. As summarized in Figure 1, binary cements comprising PC and LS should not contain less than 65% PC. The same is valid for composite cements containing PC and/or LS/Fly ash/blast furnace slag/pozzolanas/silica fume, according to the Eurocodes [18]. Any other combination beyond these restrictions can potentially incur a series of issues, such as prolonged setting times, reduced compressive strengths, non-homogeneous microstructure.

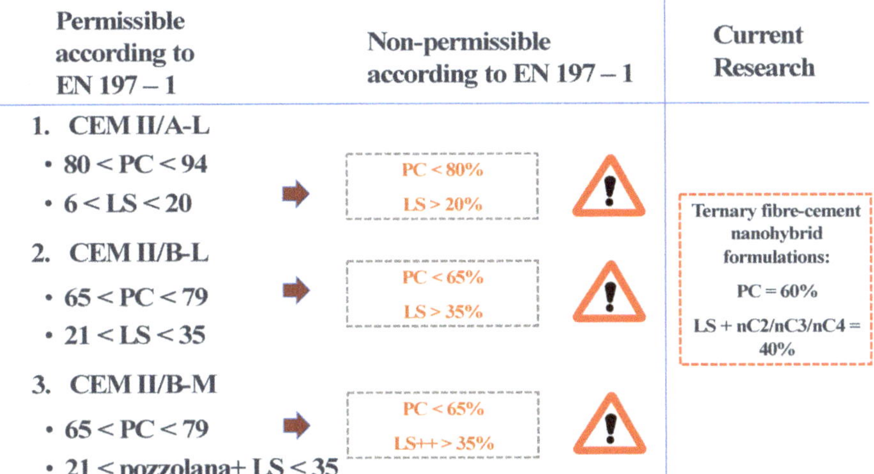

Figure 1. Permissible and non-permissible limits of clinker substitution/SCM (Supplementary Cementitious Materials) addition according to EN197-1 and challenge of current research. (PC stands for Portland cement and LS for limestone)

One of the key challenges in this paper was to produce fiber–cement nanohybrids with lower PC content than the one allowed by the current design codes (Figure 1) in an effort to design a "greener" formulation. It has been suggested that the direction of the platelets in intercalated or exfoliated nMt's may affect the load path and crack propagation within nanomodified pastes [19]. Moreover, it has been argued that nMt particles may further exfoliate within the hydrating cement matrix, tending to disperse in one direction under stirring [20], rendering them possibly more suitable for flexural than compressive strength enhancement. Adding to this, a recent study on 1% nanoclay addition to CEMI mortars showed an enhancement in the rate of strength gain for flexure [21] with respect to rate of strength gain for compression. However, although comparison for compression was made at day seven, 28, and 91, the comparison for flexure was made between seven and 28 days only and the matrix was binary, containing solely PC and nanoclay. What is important, though, is that nanoclay performed better in flexure, attributing to better adhesion of the specific nanoparticles in the matrix.

As shown in Table 1, the way the nanoparticles are produced affects the properties of the cementitious nanocomposites they will form. In fact, although a significant diversity of manufacturing and fabrication methods of nanoclays and nMt's exist for various applications as discussed earlier, as reviewed in Table 1, for cementitious nanocomposites the most common methods include ultrasonic sonication, ball milling, and electrical mixing. Still, in the reported papers, all formulations are binary (PC + nMt) except for one [22], hence there is a broad field for the investigation of higher order formulations, containing more than two constituents.

Table 1. Manufacturing and fabrication method of nano-montmorillonites/nanoclays suitable for cement nanocomposites according to literature.

Manufacturing and Fabrication Method	Cement Mix	Characterization Technique	Selected Results	Reference
Ultrasonic sonication	Ordinary Portland cement (OPC) type 42.5R and commercially available nano-kaolinite clay (NKC) powder was added as Portland cement (PC) replacement at 0%, 1%, 3%, and 5% by weight. The NKC was first dispersed in water using an ultrasonic dispersion method.	X-ray diffraction (XRD) analysis and transmission electron microscopy (TEM) techniques were carried out on the clay powder.	The samples with 5% NKC exhibited the highest compressive strength, chloride diffusion resistivity, relative dynamic modulus of elasticity, and the most electrical resistivity after 125 freeze–thaw cycles.	[23]
Hobart electric mixer	OPC is partially substituted by 1%, 2%, or 3% nanoclay by weight of OPC. The OPC and nanoclay were first dry mixed for 5 min in Hobart mixer at a low speed and then mixed for another 10 min at high speed until homogeneity was achieved. Cloisite30B is a natural montmorillonite modified with a quaternary ammonium salt, which was supplied by Southern Clay Products, USA.	X-ray diffraction (XRD) analysis.	The nanoclay behaves not only as a filler to improve microstructure, but also as an activator to promote pozzolanic reaction.	[24]
Electric mixer	OPC mortars where produced by partial substitution with NMK at 2%, 4%, 6%, and 8% by weight of cement. The nanoclay used in this investigation is kaolin clay supplied by Middle East Mining Investments Company (MEMCO), Cairo, Egypt. The nano-kaolin was heated for 2 h at 750 °C to give active amorphous NMK. The ingredients were homogenized on an electric mixer to assure complete homogeneity.	TEM imaging of NMK.	The compressive strength and the tensile strength of the cement mortars with NMK were higher than the reference mortar at the same water to binder (w/b) ratio. The enhancement in tensile strength reached 49%, whereas the enhancement in compressive was 7% at 8% NMK.	[25]
Electric mixer	A total of 1% or 2% Nano-montmorillonite modified foamed paste was produced with a high volume (70%) fly ash binder. The nMt used was produced by Zhejiang FengHong New Materials Co., Ltd., which has a purity of 99.5% (Technical Grade) The foaming agents and stabilizing agent had been dissolved in the water and used to soak the nMt for 24 h to facilitate its dispersion. Solids were mixed first and then mixed with water. Lastly, with the water containing nMt and foaming agents.	Not applicable.	Mix constituting of 70% FA, 1% alpha-olen sulfonate (AOS), 2% alcohol ethoxylate (AEO), 0.75% Na_3PO_4 and 1% nMt) exhibited the lowest thermal conductivity (0.071 W (m1 K1)) and reasonably high strength (3.23 MPa) at 28 days.	[22]

Table 1. *Cont.*

Manufacturing and Fabrication Method	Cement Mix	Characterization Technique	Selected Results	Reference
Electric mixer and ball mill	The blended cement paste samples were prepared by partial replacement of cement with 2%, 4%, 6%, 8%, 10%, 12%, and 14% nano-metakaolin (NMK) by weight of cement. All pastes were prepared with the same water to cement (W/C) ratio 0.3. The ingredients of the blended cement pastes were homogenized on an electric mixer to assure complete homogeneity. The NMK was thermally treated at 750 °C for 2 h to assure complete decomposition and to get active amorphous nano-metakaolin (NMK). The ingredients were homogenized on a roller in a porcelain ball mill with four balls for 1 h to assure complete homogeneity.	Differential thermal analysis (DTA) was performed for the nano-kaolin to specify the decomposition/calcinations temperature. XRD to confirm that the kaolinite phase transformed into amorphous phase TEM imaging confirmed the thermal activation, capturing reduction in grain size with ill-defined edges, which suggests some amorphous character and results in increasing the pozzolanic reactivity of NMK.	The optimum replacement was found to be 10% with which an enhancement of compressive strength by about 50% and flexure strength by 36% was achieved. Microstructure was also enhanced.	[26]
Distilled water and stirring for the OMMT, then electric mixing of the cementitious nanocomposites	Ordinary Portland cement type CEM II/A-LL 42.5 N (OPC) and OMMT at 1% cement replacement and at water to solid ratio (w/s) equal to 0.27. Sodium MMT (cation exchange capacity (CEC) 105 meq/100 g), was used for OMMT synthesis in the laboratory by applying the ion exchange method. Quaternary ammonium salt (QAS) methylbenzyl di-hydrogenated tallow ammonium chloride (Noramium MB2HT = 640 g/mol) was selected as a modifier to produce three OMT denoted as 06, 08, and 1 M, respectively (corresponding to 0.6, 0.8 and 1.0 cation exchange degree). Each OMMT was mixed with distilled water and stirred rigorously for one day at 20 °C, to ensure the formation of well-dispersed suspension. After stirring time terminated, the OMMT–water suspension was mixed with the required amount of cement.	The pozzolanic activity of the modified clays was determined by the Frattini test.	The properties of OMMT modified cement paste vary according to the cation exchange degree of the OMMT. OMMT of lower cation exchange degree (0.6 M) shows pozzolanic behaviour after 14 days, while OMMT of higher cation exchange degree (0.8 and 1 M) displays the activity later only after 28 days.	[27]

The work presented in this paper was part of a much broader research project, namely FIBCEM, (Nanotechnology-enhanced Extruded Fibre Reinforced For Cement-based Environmentally Friendly Sandwich Material for Building Applications) supported by the EU, involving industrial and academic partners throughout Europe, to investigate nanotechnologically enhanced cements. A number of studies have been conducted and published by the authors on ternary, quaternary, or quinary pastes, which did not include fibers or superplasticizers [28–32]. Of all nanoparticles employed in these studies, the nMt's, which are promising materials for a number of reasons, including the fact that they are naturally occurring, abundant, and cost effective, provided ambiguous results. It was found that their addition at significant concentrations can, indeed, cause a reduction in compressive strength of samples and this was attributed to a number of issues; primarily the modifier and dispersion agent used and the nanostructure of each dispersion itself [19,33], the quantity of nMt addition, and the absence of superplasticizer [1,28]. In summary, results suggest that in ternary (Portland limestone nanocomposites) [32] and quaternary combinations [1] 1% of nMt addition by total weight of the binder is optimal and can offer strength improvements in pastes.

Although a number of papers have been reported on the flexural performance of binary nanosilica [34], nanotube [35], or nanoclay [36] enhanced cements, no research has been presented in ternary blends reinforced with fibers. It is acknowledged that nMt nanoparticles can improve flexural strength and assign damping properties [37] to pastes they will be added to [19]. Moreover, as stated above, nMt particles may further exfoliate within the hydrating cement matrix, dispersing in one direction under stirring [20], rendering them possibly more suitable for flexural than compressive strength enhancement. For this, a new research program was implemented in order to assess the potentials of nMt in ternary fiber–cementitious nanohybrids, when tested in flexure. For the first time, the effect of differently manufactured nMt's, namely; one organically modified dispersion (nC2), one inorganic dispersion (nC3), and one organically modified nano-montmorillonite in powder form, industrial product, (nC4) was evaluated. In the present study, the optimal nMt content, (1% addition of nC2 or nC3 or nC4 by total mass of binder) as defined by previous studies, was investigated. Furthermore, Polyvinylalcohol (PVA) fibers were used as a low-cost type of fiber to provide resistance to cracking and to crack propagation [38]. To the best knowledge of the authors, such an extensive investigation of the nMt properties and the properties of the low-PC fiber reinforced nanohybrids has never been presented before.

2. Materials and Methods

2.1. Materials and Nanomaterials

The materials used were:

- Portland limestone cement CEMII/A-L42.5, with a limestone content of 14%, conforming to EN 197-1. The supplier gave the following clinker composition: 70% C_3S, 4% C_2S, 9% C_3A, 12% C_4AF. CEM II/A-L42.5. In mix proportioning the Portland cement (PC) content (86% by mass) was considered separately from the limestone (LS) content (14% by mass)
- Limestone (additional LS), conforming to EN 197-1. The total LS content of each paste was the sum of that contained in the Portland limestone cement and this additional *LS*.
- Fly ash (FA), conforming to EN 450. The oxide composition provided by the material data sheet was: 53.5% SiO_2, 34.3% Al_2O_3, 3,6% Fe_2O_3, 4.4% CaO.
- Organomodified nano-montmorillonite (nMt), nC2 dispersed in water with the help of an alkyl aryl sulfonate surfactant, containing about 15% by mass of nMt particles.
- Inorganic nano-montmorillonite (nMt), nC3 dispersed in water with the help of sodium tripolyphosphate, containing about 15% by mass of nMt particles.
- Organomodified nano-montmorillonite (nMt), nC4, an industrial product by Sigma-Aldrich, non-dispersed—in powder form. It consists of Montmorillonite K-10 (70–75 wt%) surface modified with 25–30 wt% methyl dihydroxyethyl hydrogenated tallow ammonium * Nanomer® I.34 MN. The supplier's data sheet gives the following additional information: 6.5 < pH < 7.0 and density = 1.7 g/cm^3.
- PVA fibers, kuralon H-1, 4 mm, added at 2% by weight.
- Superplasticizer viscocrete 20 HE, denoted as SP.

2.2. Methods

2.2.1. Formulation of Fiber–Cement Nanohybrids

To assess this hypothesis of the better flexural performance and to investigate the possibility of creating lower Portland cement (i.e., carbon footprint) formulations, in the present paper a ternary formulation comprising a non-pozzolanic reference paste containing 60% PC, 40% LS, 3% PVA fibers and 2% superplasticizer, denoted as F.PC60LS40PVA3SP2 (F is for flexure),

The general formula of the matrix of the ternary cementitious nanohybrids was:

$$PC(60) + LS\,(40 - x) + PVA(3) + SP(2) \tag{1}$$

where x = % of nMt solids and water to solids (W/S) ratio at 0.3.

The PC content and the water to solids (W/S) ratio was kept constant and the content of nMt solids was deducted from the LS content. This was done in order to keep the $Ca(OH)_2$ production during PC hydration comparable in all pastes, so as to detect possible pozzolanic reactivity of the nanoparticles in composite cement formulations, as shown in Table 2.

Table 2. Composition of ternary fiber cement nanohybrids—proportions % by total mass of solids.

Sample	PC (%)	LS (%)	nMt (%solids)	SP (%)	PVA fibres (%)	W/S
F.PC60LS40PVA3SP2+0%nC	60	40	0	2	3	0.3
F.PC60LS39PVA3SP2+1%nC2	60	39	1	2	3	0.3
F.PC60LS39PVA3SP2+1%nC3	60	39	1	2	3	0.3
F.PC60LS39PVA3SP2+1%nC4	60	39	1	2	3	0.3

Where, PC stands for Portland cement, LS for limestone, nMt for nanomontmorillonite, SP for superplasticizer, PVA for Polyvinyl alcohol and W/S for water to solids ratio.

The mixing procedure was standardized as in the case of research on nanosilica particles presented by the authors [31]. In specific:

- Dry mixing of all powder components was firstly carried out with a spatula by hand. For the powder nMt, mixing was carried out together with PC and LS.
- For formulations containing nMt in dispersion, the nC2 or nC3 dispersion was poured in a separate container together with water, stirred with the use of a magnetic stir bar for 1 min, and then added to the mixed powders.
- The PVA fibers were added last after they had been manually further separated.
- With the addition of water (and nC2/nC3 where applicable), the paste was mixed employing a dual shaft mixer at 1150 rpm for a duration of up to four min.

2.2.2. Analytical Testing

Characterization of nMt

The characterization of the powder nC4 followed the methodology presented in literature [19]; nc4 was, therefore, characterized in terms of basal spacing via X-ray diffraction (XRD), in terms of shape, size, and basal spacing via transmission electron microscopy (TEM) and elemental composition via scanning electron microscopy/ X-ray energy dispersive spectroscopy (SEM/EDX). The thermal properties of nC4 were examined via thermal gravimetric analyses (TGA). In greater detail:

X-ray Diffraction: XRD measurements were performed using a D8 ADVANCE X-ray diffractometer with CuKα radiation. Spectra were obtained in the range $4° < 2\theta < 20°$ at an angular step-size of $0.016°$ 2θ. The state and extent of dispersion–exfoliation of the NMt can be examined by XRD and TEM analysis with monitoring changes in basal spacing [19].

Transmission Electron Microscopy (TEM): Suspension of 10 ng/mL was prepared using nC4 and distilled water. Small drops of the diluted solutions were then deposited on copper mesh grids coated with a thin carbon film. Grids were dried at 25 °C prior to the insertion in the instrument. Samples were examined at a voltage of 120 kV with a GATAN Jeol JEM 1200 mkII. Images were recorded on a Gatan Dual View camera.

As explained in [19], when a polycrystalline structure, with randomly oriented grains, forming rings is detected, having the camera constant known, the calculation of the lattice spacing is allowed according to the following adapted formula [39]:

$$\frac{r}{2L} = \frac{\lambda}{d} \rightarrow d = \frac{2\lambda L}{r} \tag{2}$$

where L: the camera length and λ: the electron wavelength, which are independent of the specimen and constant for the TEM instrument and $d = d$-value and $r =$ distance from the diffraction center. Since d is inversely proportional to r, the largest d value is obtained by the innermost ring. For the specific TEM diffraction analysis, $\lambda L = 1$. These results can be compared with d value measured by XRD. The only variable measured with the help of the TEM software was r, manually taken to be equal to the distance between the center of the diffraction and the center of each individual ring. As an effect, greater error is expected when comparing TEM and XRD d-value measurements.

Scanning electron microscopy/ X-ray energy dispersive spectroscopy (SEM/EDX): For SEM/EDX elemental composition analyses nC4 was placed uncoated on a sheet of molybdenum, an element absent from the LnS dispersions for unbiased elemental analyses. A matrix of 5×5 spectra was acquired and the median of the elemental composition was presented. The Si/Al ratio was also calculated.

Thermal gravimetric analyses (TGA) and derivative thermogravimetry (dTG) were carried out using a Setaram TGA92 instrument. Approximately 20 mg of nC4 were placed in an alumina crucible and heated at a rate of 10 °C/min from 20 to 1000 °C under 100 mL/min flow of inert nitrogen gas. The differential thermogravimetric curve (DTG) was derived by the TG curve. The first derivative curve was produced for the various samples tested and was used for comparisons instead of the mass loss curve, as it yields sharp distinctive peaks.

Characterization of fiber–cement nanohybrids

Flexural (three-point bending) strength tests were carried out in accordance with BS EN 12467. Mean strength values of three specimens were calculated, as well as standard deviation at day 7, 28, 56, and 90.

For the paste characterization, arrest of hydration was performed following two different methodologies: oven drying and solvent exchange, as described by Calabria-Holley et al. [40]. For TGA/dTG, the oven drying technique was adopted. For the MIP investigation, solvent exchange was the technique employed for the arrest of hydration. Isopropanol was selected as the most appropriate solvent according to literature [41,42].

Thermogravimetric analysis (TGA) and derivative thermogravimetry (dTG) were carried out using Setaram TGA92. Each powder sample was placed in an alumina crucible and heated at a rate of 10 °C/min from 20 to 1000 °C in nitrogen atmosphere, as explained above for the characterization of nC4 at day 28, 56, and 90.

X-ray diffraction (XRD) measurements were performed using a D8 ADVANCE x-ray diffractometer with CuKα radiation controlled by a Dell PC. Spectra were obtained in the range $4° < 2\theta < 60°$. Analysis of peaks and *d*-spacing [according to Bragg's law ($n\lambda = 2d\sin\theta$) was carried out using EVA software, at day 28].

For the late age relative density measurements, BS EN 12390-7:2009 [18] was selected as a basis and the exact procedure followed is covered in literature [31]. Mean density values of three specimens were calculated, as well as standard deviation.

Water impermeability test was modified to account for the much smaller specimen used in this research (slabs 120×40 mm and 10 mm thickness) and were carried out in accordance with BS EN 492:2012 at day 7, 28, and 56. A transparent tube of 250 mm length was used as water column with an internal bore of 29 mm diameter. A control water column was also adopted to ensure zero water evaporation in the laboratory testing environment.

Mercury intrusion porosimetry (MIP) was executed with an Autopore III—Model unit 9420 supplied by Micromeritics, Hexton, Herts, UK. Stems of 3 mL capacity were filled with solids of the arrested hydration paste and tested at day 28.

3. Results and Discussion

3.1. Characterization of nMt Powder (nC4)

A thorough methodology on the characterization of organomodified and inorganic nano-montmorillonite dispersions that are suitable as supplementary cementitious materials in cement nanohybrids was presented by the authors [19]. It was postulated that the anionic surfactant used in nC2 mostly kept the platelets apart, notwithstanding scarce areas of re-agglomeration. On the contrary, dispersion nC3 was fully exfoliated. The elemental composition of nC2 and nC3 was also given in the related graph, as well as the Si/Al ratio as a measure of comparison between dispersions nC2 and nC3 and the powder nC4. Lastly, the thermal properties of nC4 were investigated in comparison with nC2 and nC3.

The XRD analysis of nC4 yielded a basal spacing (*d*-value) of 1.8 nm (Figure 2), which is more expected in unmodified nMt [43]. The reflection at 4.892° 2θ corresponded to the 001 plane for this nMt and, as explained in literature [19], the intercalated platelets, being periodically stacked, were traceable by the XRD, whereas the exfoliated platelets, being fully disordered, remained XRD silent [44].

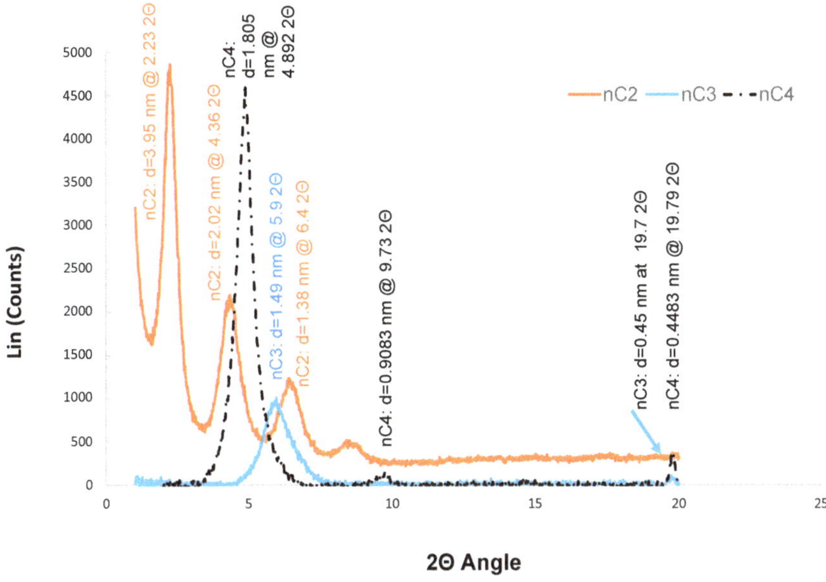

Figure 2. Comparison of the X-Ray diffraction pattern of the organomodified montmorillonite (MMT) dispersion (nC2), inorganic MMT dispersion (nC3) [19], and organomodified MMT powder (nC4).

Although nC4 was expected to bear resemblance to nC2, both being organomodified, as shown in Figure 2, the powder OMMT, nC4, in fact, bears greater resemblance to the inorganic dispersion nC3. With respect to the XRD analysis of nC2, the reflection at 2.23° 2θ corresponds to the 001 plane. The reflection at 4.36° 2θ corresponds to the 002 plane, while the reflection at 6.4° 2θ corresponds to the 003 plane. Higher order reflections such as 002 and 003, are typical of regular stacking of the modifier chains in the interlayer space [19]. However, the width at medium height of the peak corresponding to the 001 lane provides an estimate of the level of crystallinity of the nMT; therefore, the production process for nC4 produced less crystalline structures [45].

Lastly the typical montmorillonite reflection at 19.7° 2θ, which can act as a measure for the level of exfoliation, was non-existent for nC2, marginally traceable for nC3 (which is expected since nC3 is inorganic), and slightly more traceable for nC4 (which was not expected since it was organomodified). Therefore, nC2 seemed to have both intercalated (presence of higher order reflection, which are typical of regular stackings) and exfoliated (disappearance of peak at 19.7° 2θ) platelets, whereas nC3 seemed to be the better exfoliated of all three samples by XRD [19]. The condition of the nanostructure of the nMt's is expected to affect the mechanical, chemical, and thermal performance of the samples the nMt's will be added to significantly according to previous studies [19,32,46]

TEM micrographs showed good intercalation and possible exfoliation of nC4. Soft edges can be observed in Figure 3A,B. Moreover, nC4 agglomerated less than nC2 or nC3 [19] in water dispersion and showed significant differences in terms of particle shape and size, with platelets reaching the size of 300 nm.

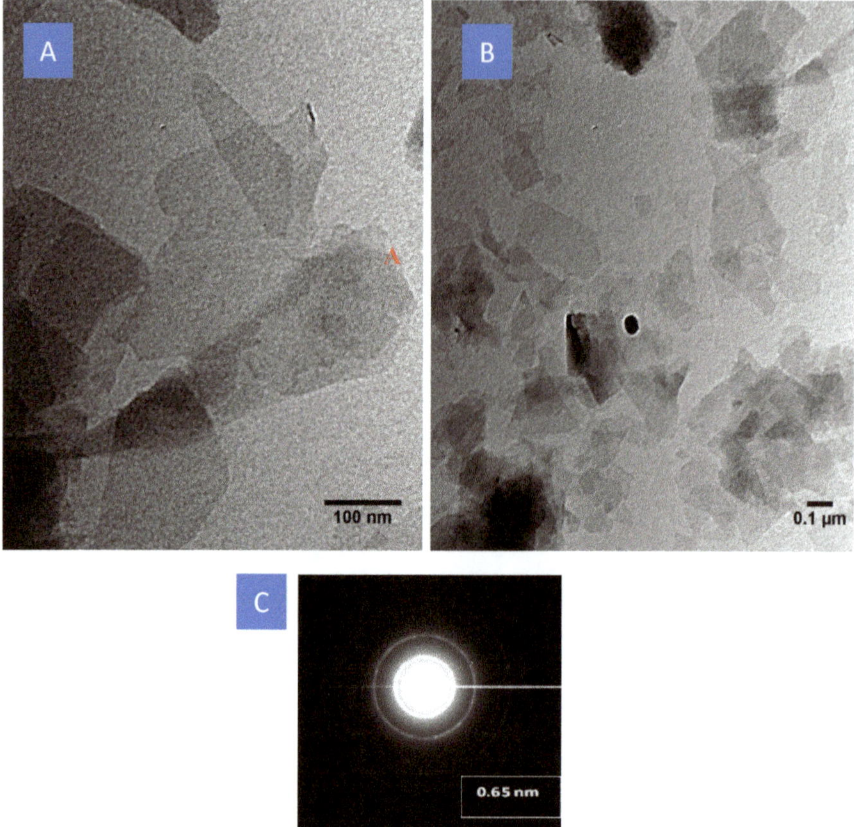

Figure 3. Transmission Electron Microscopy (TEM) micrograph of nC4 at (**A**) 150,000 times magnification, (**B**) 50,000 times magnification and, (**C**) TEM-diffraction pattern of nC4.

The diffraction analysis was carried out according to Equation (2) for the various d-values. The d-values obtained for the diffraction pattern of nC4 (Figure 3B) were: $d_1 = 2 \times 1/0.96 = 2.08$ nm, $d_2 = 2 \times 1/1.83 = 1.09$ nm, $d_3 = 2 \times 1/2.63 = 0.76$ nm, and $d_4 = 2 \times 1/3.17 = 0.63$ nm.

It is known that the study of diffraction patterns via TEM can clarify the extent of dispersion or exfoliation of the nMt [19]. These values clearly indicate that some platelets were intercalated and some exfoliated. The different d-spacing (1.8 nm) given by analysis XRD can be attributed to the fact that diffraction analysis is carried out on a single crystal, whereas XRD in the bulk of the powdered sample. Moreover, a greater error is expected on the TEM results, inherent to the technique adopted to extract the r parameter in Equation (2).

TEM micrographs and diffraction patterns and analyses for nC2 and nC3 can be found in literature [19].

TEM imaging and diffraction analysis of nC4 was followed by SEM/EDS analysis. A matrix of 5×5 randomly-selected spectra was collected (Table 3). The elemental composition of nC4 (Figure 4) showed that nC4 contained lower quantities of C, Si, and Al compared to the dispersed OMMT, nC2 and higher quantities of Fe and O. Furthermore, it was comprised of fewer elements and no Na was traced; therefore, nC4 is not a sodium-MMT originally. The Si/Al ratio of nC4 was found to be between 3.31 and 3.67, with a median of 3.48 and a standard deviation of 0.10.

Table 3. Scanning electron microscopy/ X-ray energy dispersive spectroscopy (SEM/EDX) counts summary (% atomic) of nC4.

Initially Undispersed nC—All Results in Atomic % Normalized by 100%						
Spectrum	C	O	Mg	Al	Si	Fe
(1, 1)	38.63	49.13	0.55	3.29	8.01	0.40
(2, 1)	39.31	45.63	0.62	3.94	9.95	0.54
(3, 1)	37.94	48.55	0.60	3.70	8.74	0.47
(4, 1)	39.87	48.95	0.47	2.86	7.40	0.44
(5, 1)	38.78	47.83	0.55	3.54	8.83	0.47
Total mean	41.01	45.56	0.51	3.37	8.30	1.25
Standard Deviation	2.95	2.95	0.08	0.52	1.47	0.99

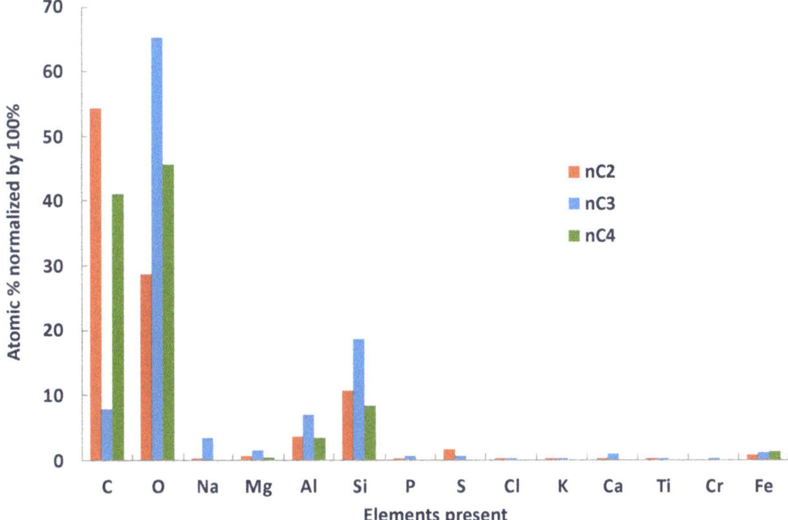

Figure 4. Comparison of the elemental composition of the organomodified nanomontmorillonite (MMT) dispersion (nC2), inorganic MMT dispersion (nC3) [19], and organomodified MMT powder (nC4).

The comparative graphs of the elemental composition and the Si/Al ratios of the organomodified MMT dispersion (nC2), the inorganic MMT dispersion (nC3), and the organomodified MMT powder (nC4) are presented in Figure 5. Values for nC2 and nC3 were found in literature [19]. In summary, the two figures allow for the following conclusions to be drawn:

- nC2 compared to nC4 has higher potential to form additional C–S–H, due to the higher quantities of silica and limited exfoliation. At the same time, the greater amount of carbon present is expected to cause reduction in compressive strengths [19].
- The inorganic nMt, nC3, exhibited the highest net amounts of Si and Al, essential for C–S–H and C–Al–H.
- Traces of Na were found in the nC2 dispersion, because an anionic surfactant containing Na was used.
- The pronounced presence of Na in nC3 proves that nC3 was formed by a natural sodium-MMT, whereas the absence of Na in sample nC4 implies that nC4 was not a sodium-MMT originally.
- The Ca content in nC3 could potentially take part in the C–S–H forming hydration reactions or acting as a seeding agent [13].
- Compared to nC2, the commercially available nC4 was better exfoliated, but showed marginally greater variation in Si/Al and more polycrystalline phases.

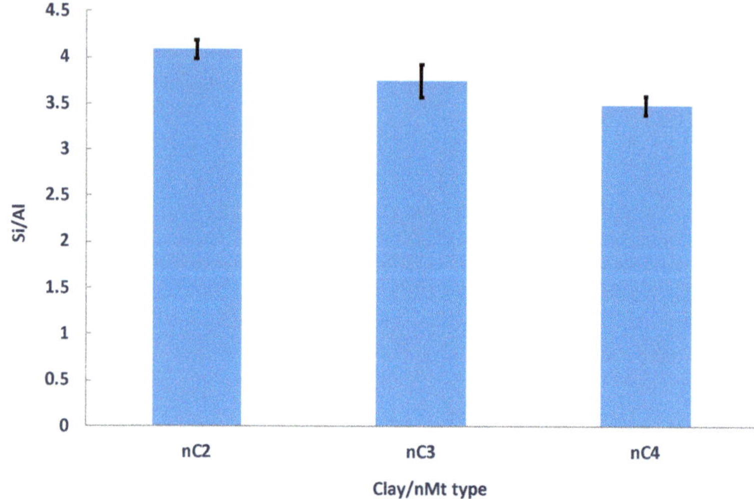

Figure 5. Comparison of the Si/Al ratios of the organomodified nanomontmorillonite (MMT) dispersion (nC2), the inorganic MMT dispersion (nC3) [19], and organomodified MMT powder (nC4).

TG analyses of the organomodified MMT dispersion (nC2), inorganic MMT dispersion (nC3), and organomodified MMT powder (nC4).

Thermal analysis of organomodified nMt's is more complex because the modifiers used change the physical, chemical, and, consequently, the thermal characteristics of the nMt's.

Apart from loss of free water, four regions of mass loss of organomodified nMt powders are identified in literature [19]:

- Up to 200 °C, the mass loss is associated with adsorbed interlayer water. Therefore, in the case of the inorganic MMT, a significant mass loss is observed.
- Between 200 and 500 °C the mass loss is associated with the decomposition of organic elements (modifiers and surfactants) attached to the MMT platelets usually exhibiting two peaks due to different structural arrangements. This is the area in which the main differences between the OMMT's are identified and, hence, the powder OMMT and the dispersed OMMT of this study are distinctively different, even though similar organic salts were used for the separation of the platelets.
- Between 550 and 800 °C, the mass loss is associated with the loss of structural water from the MMT, known as dihydroxylation. Again, this is the area in which mass changes are expected and observed for the inorganic MMT.

The first observation derived from Figure 6 is that the organomodification has taken place for both nC2 and nC4, which became hydrophobic. Because of this, the amount of adsorbed interlayer water is limited and; therefore, the related mass loss is limited too, as can be seen for the temperatures below 200 °C [47].

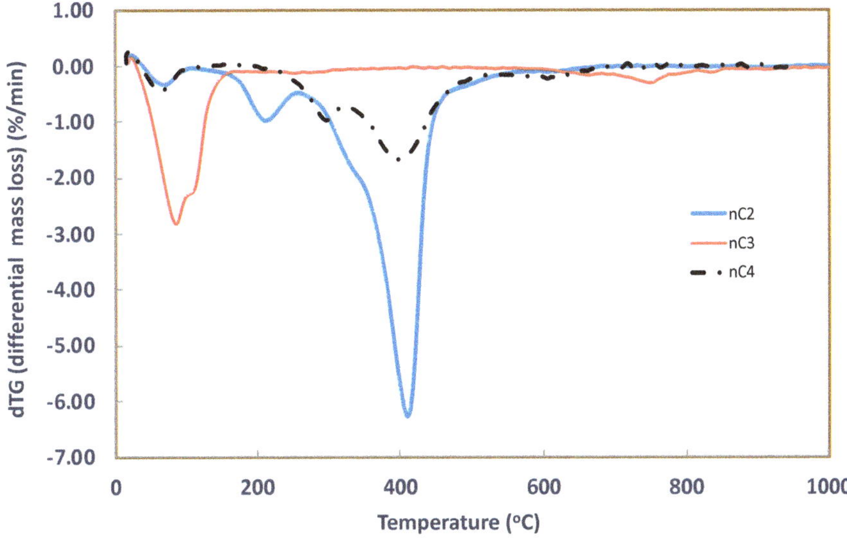

Figure 6. Comparison of Differential mass loss (dTG) of the organomodified montmorillonite (MMT) dispersion (nC2), inorganic MMT dispersion (nC3) [19], and organomodified MMT powder (nC4).

In greater detail, as far as nC4 was concerned, the TG analysis showed four endothermic peaks (Figure 6 and Table 4):

- In the temperature range of 30–130 °C, with a distinct peak at 67 °C, attributed to the loss of surface and interlayer water.
- In the temperature range of 200–330 °C, with a distinct peak at 295 °C, the decomposition of modifier bound to neighboring molecules took place.
- In the temperature range of 330–530 °C, with a distinct peak at 396 °C, the deconstruction of the modifier bound to the bentonite was completed.
- In the temperature range of 570–700 °C, with a distinct peak at 600 °C, the remaining OMMT was decomposed.

Table 4. Mass loss (%) of nC4.

	0–100 °C	100–250 °C	250–400 °C	400–500 °C	500–700 °C
nC4	1.33	0.31	13.63	8.00	2.86

Respectively, analyses for nC2 and nC3 can be found in literature [19].

3.2. Characterization of Fiber–Cement Nanohybrids

3.2.1. Flexural Strength

The standard deviation was computed for all specimens and is presented in Figure 7. For the reference paste, F.PC6LS40PVA3SP2, and for the nC2 modified pastes, it ranged from 1 to 1.4. The addition of nC3 reduced the variation between 1.0 to 1.2 MPa, whereas the standard deviation of nC4 was approximately equal to 1.1 MPa. Hence, as in the case of nMt modified pastes without *PVA* [32], the standard deviation was improved with nC3.

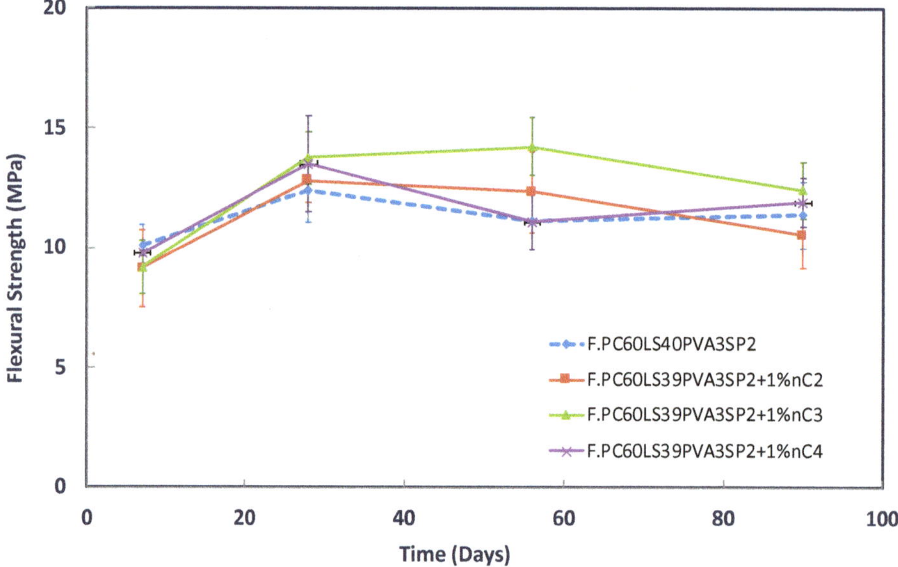

Figure 7. Flexural strength of 1% nC2, nC3 [28], and nC4 fiber–cement nanohybrids based on F.PC60LS40PVA3SP2.

As shown in Figure 7, given more curing time, the nMt modified pastes outperformed the reference paste, in contrast to [21] who suggested that the nMt particles (just like nanosilica particles) are more effective in the first seven days and their effectiveness reduces at middle ages (28 and 90 days). nC3 showed the highest flexural strength of all three. In fact, at day 28 the nC3 reinforced nanohybrids showed a 7.2% increase compared to the reference paste, at day 56 a 28% increase, and at day 90 a 9% increase in flexural strength compared to the reference formulation. nC2 showed a lower strength at day 56, which could be attributed to bad compaction. If this was the case, then all three types of nMt would not manage to maintain the strength gain after day 56, showing a relative reduction in flexural strength at day 90. It is of particular interest to note that the powder nMt did not offer any major strength improvements, neither did nC2. It should also be noted that PVA fibers are highly hydrophobic; therefore, the hydrophobic nature of nC2 and nC4 was possibly responsible for an inferior combination compared to nC3. For this it is postulated that nMt performs best when dispersed in water before being added to any cementitious formulation.

It is possible that with longer mixing time and less amount of fibers, the nMt modified specimens can be better compacted and, consequently, deliver even higher strengths.

Potential re-agglomeration of nanoparticles when added to cementitious matrices is expected to influence rheological properties of pastes, pore structure, and density, and induce weak zones in compression. It has been postulated in previous studies that the use of a sonicator or any other advanced technique, such as generation of nanofluidic droplets, could help (i) eliminate discrepancies (mixing limitations), and (ii) minimize possible agglomeration of the nMt particles in the pastes produced [31]. In the present study the first step for better mixing, compaction, and avoidance of re-agglomeration of nanoparticles or clustering of fibers was the addition of 2% superplasticizers, magnetic stirring of the nMt dispersion, and mechanical mixing of the fiber–cement nanohybrids. Results, show; however, that there is room for improvement in the techniques employed, particularly for applications that necessitate the use of fibers. In addition to this, recently a number of papers have been presented on digital or droplet based microfluidics systems, which are particularly important for biological applications [48], or for the generation and manipulation of discrete droplets inside micro-devices [49]. Therefore, although this technology is necessary for drug delivery, synthesis of biomolecules and other biological, medical, and pharmaceutical applications, it would be interesting to test it in cementitious nanohybrids in order to compare it with more traditional methods of nanoparticle addition in cementitious matrices, in future research. The effect on the fresh properties of pastes with the different nanoparticle addition methods would also be interesting to investigate and report upon.

3.2.2. Thermal Gravimetric Analyses

For these series of fiber cement pastes, the 1% nC2, nC3 and nC4 content were analyzed. It should be noted that, again, the amount of decomposing modifier and/or clay were deducted from the mass loss occurring at 400–500 °C temperature range and at 600–800 °C temperature range, respectively, as discussed in literature [19,46]. No consumption of $Ca(OH)_2$ seemed to have taken place (Figure 8A). In addition, it can be claimed that no carbonation occurred, since the lines representing the nanomodified pastes are below that of the reference paste (Figure 8B).

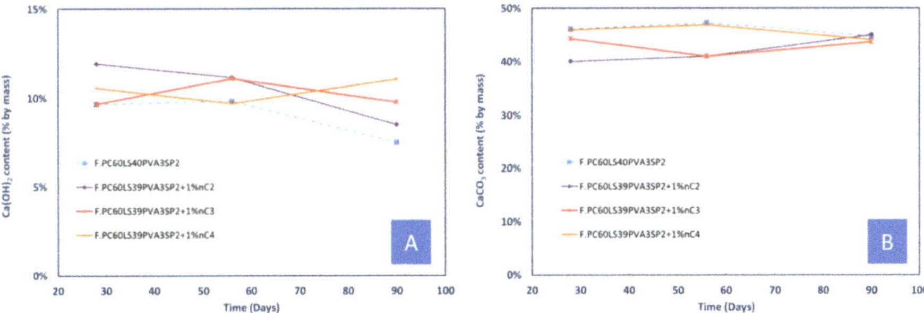

Figure 8. Effect of nMT type on (**A**) $Ca(OH)_2$ and (**B**) $CaCO_3$ content of fiber–cement nanohybrids based on F.PC60LS40PVA3SP2.

Further analysis was carried out with respect to the temperature range within which ettringite and C–S–H decompose. As shown in Figure 9, nC2 seemed to have generated greater quantities of ettringite (100–125 °C) and C–S–H, with nC3 following, given the time. In full agreement with the flexural strength results, nC3 showed the greatest production of ettringite and C–S–H at day 90. Lastly, nC4 did not show much difference compared to the reference paste, either in terms of TGA or in terms of flexural performance.

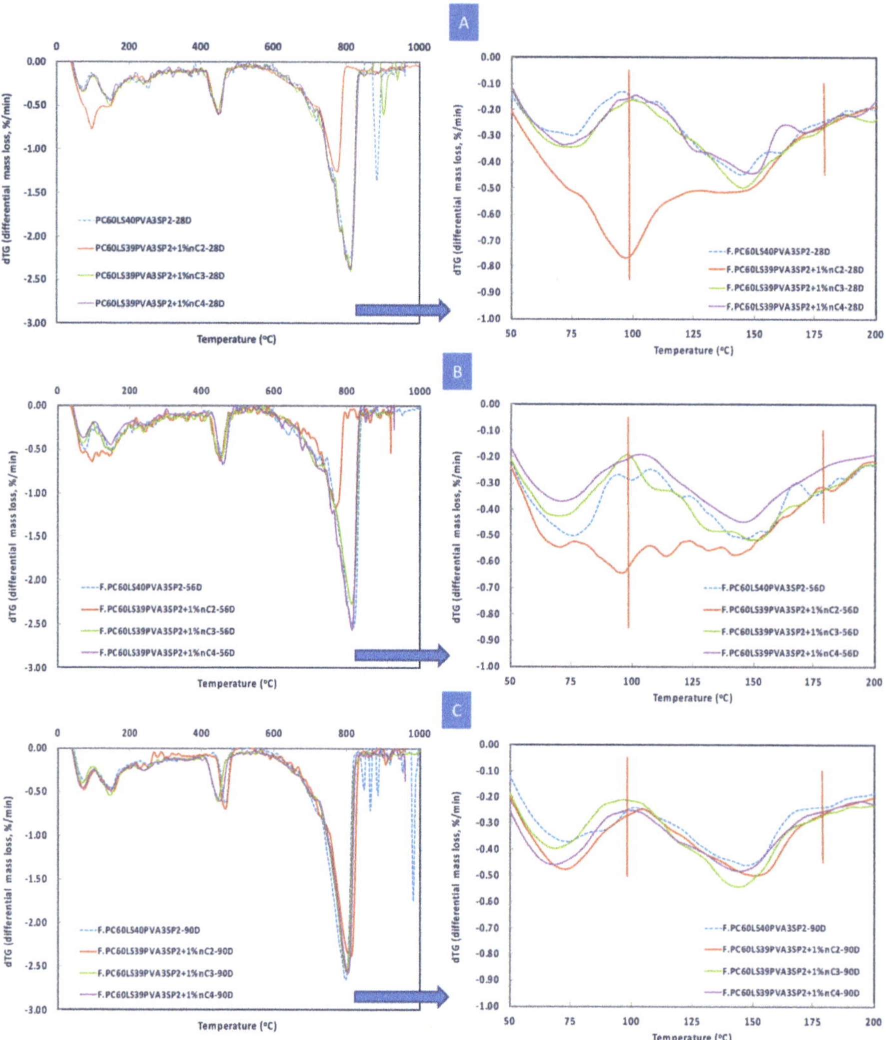

Figure 9. Differential mass loss of fiber–cement nanohybrids based on F.PC60LS40PVA3SP2 between 100–200 °C at (**A**) day 28, (**B**) day 56, and (**C**) day 90.

3.2.3. Crystallographic Analyses (XRD)

The fact that no significant $Ca(OH)_2$ consumption or carbonation took place was also confirmed by the XRD analyses executed indicatively on F.PC60LS40PVA3SP2 and the 1% nC3 and nC4 addition at day 28, as shown in Figure 10.

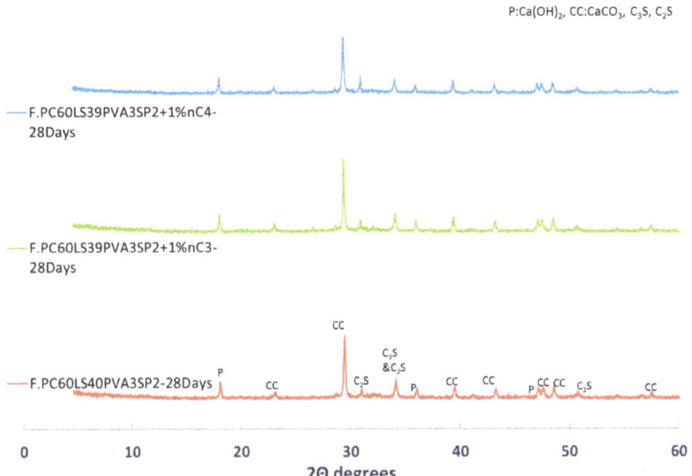

Figure 10. XRD pattern of fiber–cement nanohybrids based on F.PC60LS40PVA3SP2 at day 28—effect of nMt type.

Having completed the flexural strength tests, the thermal gravimetric and the crystallographic analyses, it can be observed that the nC2 and nC3 modified fiber pastes showed improved flexural performance; however, the difficulty of compaction of such pastes incorporating fibers and nMt can lead to increased clustering of the various particles, leading to greater strength variations. Improvements could firstly involve the nMt production method. Maybe less organic modifier should have been used for nC2 [19]. Carbon seemed to be applying a critical barrier by covering the platelets, possibly not allowing them to react as intended as stated in literature [19]. It is also possible that the excess of the organomodifier had a physical effect, as well, making the platelets re-agglomerate rather than keeping them apart. In future research, advanced ball milling techniques for the production of OMMT with anionic or non-ionic surfactants without any added water can be employed for better intercalation and thermal properties [50].

Moreover, PVA fiber content could be reduced to 1%. This would also make the nMt's in the samples be more homogeneously dispersed. Future research could include studies of fresh properties, including consistency tests in order to optimize the formulation design with respect to the fiber and superplasticizer content.

Possibly, both OMMT's (nC2 and nC4) were primarily intercalated rather than exfoliated. If they were fully exfoliated they may have performed better, as the charge would have made the platelets repulse each other and they; therefore, would have been individually available for reactions. In other words, their significant surface area would have been better utilized. Particularly in the case of the powder nMt, nC4, the platelets may have re-agglomerated within the hydrating paste. All results; therefore, suggest that nMt's should be better employed in aqueous dispersions if to be used in cementitious formulations. Adding to this, according to literature, montmorillonite K-10, on which nC4 is based, is produced via calcination [51], which is a CO_2 intensive process. The above results indicate that there should be a shift towards non-thermally treated montmorillonites, as the calcined ones do not offer such enhancement that can counteract the damage to the environment.

The inorganic nC3 dispersion was proven to contribute to flexural strength enhancement through the pozzolanic reaction. However, further research is required with respect to impermeability characteristics of pastes modified with nC3, as a tendency to adsorb water was observed in pastes containing only *PC* and *LS*. Pastes containing *PC*, *LS*, and FA, showed enhanced performance by adsorbing less water than the reference paste, but further research is required with respect to later age performance.

3.2.4. Relative Density, Water Impermeability Analyses, and Mercury Intrusion Porosimetry (MIP)

Late age, (after month three) relative density measurements were taken of the three fiber–cement nanohybrids. All measurements showed a very low standard deviation and similar values of relative density (Figure 11).

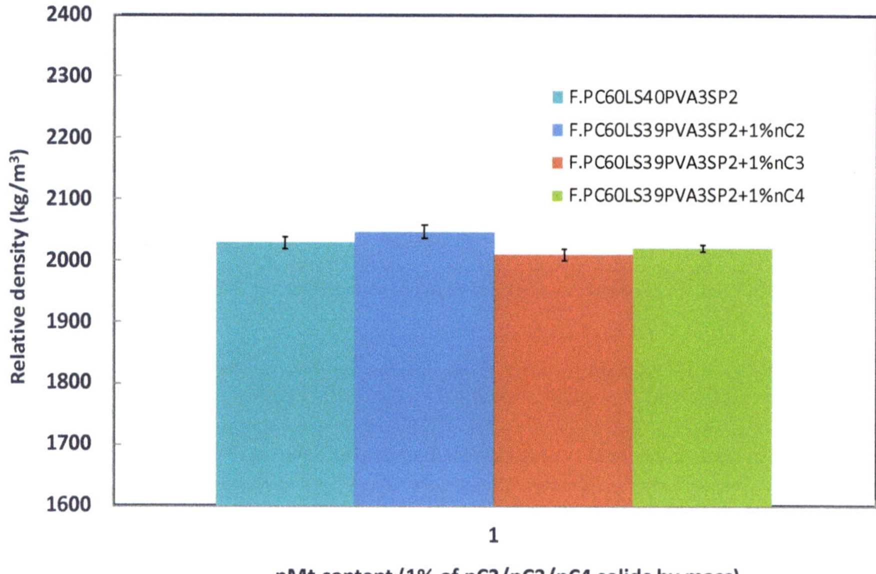

Figure 11. Effect of nMt type (at 1% dosage) on long term relative density of fiber–cement nanohybrids based on F.PC60LS40PVA3SP2.

3.2.5. Water Impermeability Tests

Water impermeability tests were carried out in accordance with BS EN 492:2012, with modifications accounting for the much smaller specimens used in this research. Nanohybrids modified by hydrophilic (nC3) or hydrophobic (nC2 and nC4) types of nMt may present water permeability issues at later ages, as shown in Figure 12. In greater detail, of the organomodified nMt's, nC2 showed the best performance at later ages, whereas the industrially-produced nC4 seemed to perform better than the reference paste only at early ages. Comparing the OMMT dispersion (nC2) and the inorganic nMt dispersion (nC3), nC3 seems to be promising, but if water impermeability is a prerequisite for certain applications, then all nMt reinforced matrices should always be checked at later ages.

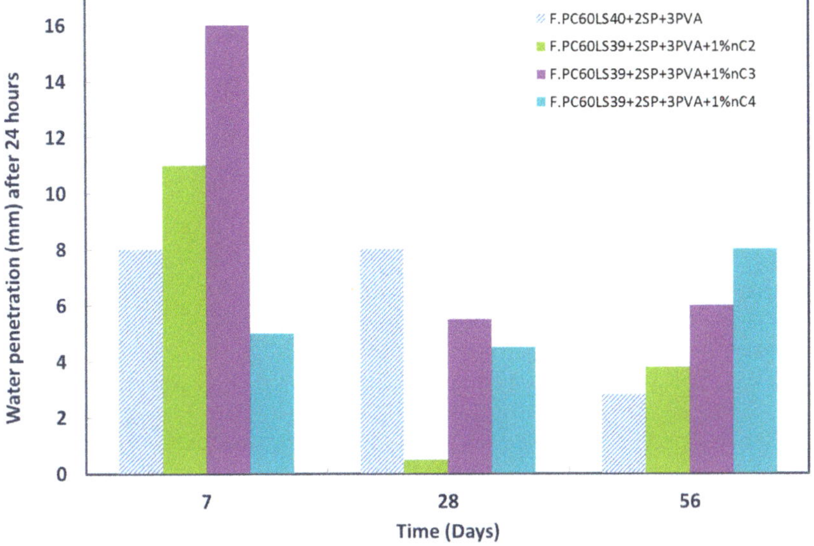

Figure 12. Effect of nMt type (at 1% dosage) on the impermeability of fiber–cement nanohybrids based on F.PC60LS40PVA3SP2.

3.2.6. Mercury Intrusion Porosimetry (MIP)

For the fiber–cement pastes based on F.PC60LS40PVA3SP2, only two formulations were tested indicatively via MIP. These were the reference paste itself (F.PC60LS40PVA3SP2) and the 1% nC3 modified paste (F.PC60LS39PVA3SP2+1%nC3), which seemed to be a successful formulation according to the analyses presented above. Both samples were tested at day 28 (Table 5).

Table 5. Mercury intrusion data summary comparison of F.PC60LS40PVA3SP2 and F.PC60LS39PVA3SP2+1%nC3.

			Mercury Intrusion Data Summary at 28 Days					
Paste	$A_{pore\text{-}Total}$ (m²/g)	$F_{pore\ Volume\text{-}Med}$ (nm)	$F_{pore\ Area\text{-}Med}$ (nm)	R_{bulk} (g/mL)	$F_{pore\text{-}Average}$ (nm)	$R_{apparent}$ (g/mL)	Porosity (%)	$V_{stem\text{-}Used}$ (%)
F.PC60LS40PVA3SP2	45.8	76.8	5.1	1.6	18.9	2.4	34.1	50.0
F.PC60LS39PVA3SP2+1%nC3	51.2	29.1	5.5	1.7	14.1	2.5	30.6	37.0

Where: $A_{pore\text{-}Total}$ = Total pore area, $F_{pore\ Volume\text{-}Med}$ = Median volume pore diameter, Fpore Area-Med = Median pore area diameter, R_{bulk} = Bulk density, $F_{pore\text{-}Average}$ = Average Pore diameter, R_{appar} = Apparnt density, $V_{stem\text{-}Used}$ = Stem Volume Used.

It can be seen that the total pore area, median volume pore diameter, and porosity were increased. This can be attributed to the presence of the fibers and mixing effects. The apparent and bulk density; however, showed small fluctuations.

Although it is acknowledged that the MIP results are not conclusive, the two formulations were tested indicatively to assist the discussion on the impermeability tests, reinforce the mechanical strength results, and provide suggestions for future research. The increase in pore diameter and porosity can be attributed to the presence of fibers, because separate studies on the *PC60LS40* matrix, without the presence of fibers and superplasticizers but with varying amounts of nC3, showed that nC3, possibly because of the better exfoliation and dispersion of the nMt platelets in the bulk of the aqueous dispersion, showed decreased porosity compared with the reference paste [29], in support of the argument that inorganic nMt can act as nucleation point and nanofiller improving the particle packing.

Further research on the pore structure of nMt and fiber-reinforced nanohybrids could involve the use of other techniques, such as X-ray tomography (CT) scan, to determine the absolute values of porosity, pore area, and other parameters. However, it should be noted that, since nC3 showed acceptable density, water permeability, and pore structure performance (along with better flexural strength and thermal properties), further research into applications that necessitate these characteristics, such as pervious concrete pavements [52], could yield interesting results.

4. Conclusions

In this paper, a successful application of the nMt's was achieved by incorporating them in fiber–cement nanohybrids. Until present, this is the first study that:

1. Extends from the characterization of nanoparticles to the characterization of fiber–cement nanohybrids.
2. Is applied to ternary fiber–cement nanohybrids directly.
3. Compares the effect of hydrophilic or hydrophobic nMt.
4. Compares the effect of dispersed hydrophobic nMt or powder hydrophobic nMt.
5. Correlates the flexural performance beyond day 28, an indeed more critical period, with the TGA results relating to the consumption of $Ca(OH)_2$ towards the production of C–S–H.

In brief, major conclusions include the following:

1. The organomodified aqueous dispersion of nMt, nC2, can be better developed as a material by studying various combinations of production methods, such as ball milling, in order for the product to necessitate lower amounts of modifier (i.e., less carbon addition).
2. The inorganic aqueous dispersion of nMt, nC3, offered strength, thermal, and microstructural improvements.
3. The industrial powder nMt, nC4, did not offer any strength, chemical, or microstructural enhancements.
4. Different methods for nMt dispersion in cementitious matrices can be assessed in future research to achieve homogeneous mixing and easy compaction, which should reflect in the pore structure and fresh properties of pastes.

Author Contributions: Conceptualization, S.P. and K.P.; methodology, S.P.; software, S.P.; analysis, S.P.; investigation, S.P.; writing—original draft preparation, S.P.; writing—review and editing, S.P.; funding acquisition, K.P.

Funding: This research was funded by the European Commission (FIBCEM project, grant No. 262954).

Acknowledgments: All partners are thanked for their input and for the supply of materials. The authors would also like to thank the Department of Chemical Engineering at the University of Bath for the use of the TG analyzer.

Conflicts of Interest: The authors declare no conflicts of interest. The funders had no role in the design of the study; in the collection, analyses or interpretation of data; in the writing of the manuscript or in the decision to publish the results.

References

1. Papatzani, S. Effect of nanosilica and montmorillonite nanoclay particles on cement hydration and microstructure. *Mater. Sci. Technol.* **2016**, *32*, 138–153. [CrossRef]
2. Sapalidis, A.A.; Katsaros, F.K.; Kanellopoulos, N.K. PVA/Montmorillonite Nanocomposites: Development and Properties. In *Nanocomposites Polymers with Analytical Methods*; Cuppoletti, J., Ed.; InTech: Horwich, UK, 2011.
3. European Commission. *Commission Recommendation of 18 October 2011 on the Definition of Nanomaterial*; 2011/696/EU 2011; OJEC: Aberdeen, UK, 2011.
4. Agubra, A.V.; Owuor, S.P.; Hosur, V.M. Influence of Nanoclay Dispersion Methods on the Mechanical Behavior of E-Glass/Epoxy Nanocomposites. *Nanomaterials* **2013**, *3*. [CrossRef]
5. Aktas, L.; Altan, M.C. Effect of nanoclay content on properties of glass–waterborne epoxy laminates at low clay loading. *Mater. Sci. Technol.* **2010**, *26*, 626–629. [CrossRef]

6. Hetzer, M.; Zhou, H.X.; Poloso, T.; De Kee, D. Influence of compatibiliser blends on mechanical and thermal properties of polymer-clay nanocomposites. *Mater. Sci. Technol.* **2011**, *27*, 53–59. [CrossRef]

7. Meng, X.Y.; Wang, Z.; Tang, T. Controlling dispersed state and exfoliation process of clay in polymer matrix. *Mater. Sci. Technol.* **2006**, *22*, 780–786. [CrossRef]

8. Utracki, L.A. *Clay-Containing Polymeric Nanocomposites—Volume 1. Shawbury, Shrewsbury, Shropshire, SY4 4NR*; Rapra Technology Limited: Telford, UK, 2004.

9. Xi, Y. Synthesis, Characterisation and Application of Organoclays. Ph.D. Thesis, Queensland University of Technology, Brisbane City, Australia, 2006.

10. Carrizosa, M.J.; Hermosin, M.C.; Koskinen, W.C.; Cornejo, J. Use of Organosmectites to Reduce Leaching Losses of Acidic Herbicides. *Soil Sci. Soc. Am. J.* **2003**, *67*, 511–517. [CrossRef]

11. Annamalai, P.K.; Martin, D.J. Can clay nanoparticles accelerate environmental biodegradation of polyolefins? *Mater. Sci. Technol.* **2014**, *30*, 593–602. [CrossRef]

12. Viseras, C.; Aguzzi, C.; Cerezo, P.; Bedmar, M.C. Biopolymer–clay nanocomposites for controlled drug delivery. *Mater. Sci. Technol.* **2008**, *24*, 1020–1026. [CrossRef]

13. Papatzani, S.; Paine, K.; Calabria-Holley, J. A comprehensive review of the models on the nanostructure of calcium silicate hydrates. *Constr. Build. Mater.* **2015**, *74*, 219–234. [CrossRef]

14. Chang, T.P.; Shih, J.Y.; Yang, K.M.; Hsiao, T.C. Material properties of Portland cement paste with nano-montmorillonite. *J. Mater. Sci.* **2007**, *42*, 7478–7487. [CrossRef]

15. Morsy, M.S.; Alsayed, S.H.; Aqel, M. Hybrid effect of carbon nanotube and nano-clay on physico-mechanical properties of cement mortar. *Constr. Build. Mater.* **2011**, *25*, 145–149. [CrossRef]

16. Kuo, W.-Y.; Huang, J.-S.; Lin, C.-H. Effects of organo-modified montmorillonite on strengths and permeability of cement mortars. *Cem. Concr. Res.* **2006**, *36*, 886–895. [CrossRef]

17. Fernandez, R.; Martirena, F.; Scrivener, K.L. The origin of the pozzolanic activity of calcined clay minerals: A comparison between kaolinite, illite and montmorillonite. *Cem. Concr. Res.* **2011**, *41*, 113–122. [CrossRef]

18. BSI. *BS EN 197-1:2011: Cement. Part 1: Composition, Specifications and Conformity Criteria for Common Cements*; BSI: London, UK, 2011.

19. Papatzani, S.; Paine, K. Inorganic and organomodified nano-montmorillonite dispersions for use as supplementary cementitious materials—A novel theory based on nanostructural studies. *Nanocomposites* **2017**, *3*, 2–19. [CrossRef]

20. He, X.; Shi, X. Chloride permeability and microstructure of Portland cement mortars incorporating nanomaterials. *Transp. Res. Rec. J. Transp. Res. Board* **2008**, *2070*, 13–21. [CrossRef]

21. Hosseini, P.; Hosseinpourpia, R.; Pajum, A.; Khodavirdi, M.M.; Izadi, H.; Vaezi, A. Effect of nano-particles and aminosilane interaction on the performances of cement-based composites: An experimental study. *Constr. Build. Mater.* **2014**, *66*, 113–124. [CrossRef]

22. Li, Z.; Gong, J.; Du, S.; Wu, J.; Li, J.; Hoffman, D.; Shi, X. Nano-montmorillonite modified foamed paste with a high volume fly ash binder. *RSC Adv.* **2017**, *7*, 9803–9812. [CrossRef]

23. Fan, Y.; Zhang, S.; Wang, Q.; Shah, S.P. Effects of nano-kaolinite clay on the freeze–thaw resistance of concrete. *Cem. Concr. Compos.* **2015**, *62*, 1–12. [CrossRef]

24. Hakamy, A.; Shaikh, F.U.A.; Low, I.M. Microstructures and mechanical properties of hemp fabric reinforced organoclay–cement nanocomposites. *Constr. Build. Mater.* **2013**, *49*, 298–307. [CrossRef]

25. Morsy, M.S.; Alsayed, S.H.; Aqel, M. Effect of nano-clay on mechanical properties and microstructure of ordinary Portland cement mortar. *Int. J. Civ. Environ. Eng. IJCEE-IJENS* **2009**, *10*, 23–27.

26. Al-Salami, A.E.; Morsy, M.S.; Taha, S.; Shoukry, H. Physico-mechanical characteristics of blended white cement pastes containing thermally activated ultrafine nano clays. *Constr. Build. Mater.* **2013**, *47*, 138–145. [CrossRef]

27. Česnienė, J.; Baltušnikas, A.; Lukošiūtė, I.; Brinkienė, K.; Kalpokaitė-Dičkuvienė, R. Influence of organoclay structural characteristics on properties and hydration of cement pastes. *Constr. Build. Mater.* **2018**, *166*, 59–71. [CrossRef]

28. Papatzani, S.; Paine, K. Dispersed Inorganic or Organomodified Montmorillonite Clay Nanoparticles for Blended Portland Cement Pastes: Effects on Microstructure and Strength. In *Nanotechnology in Construction*; Sobolev, K., Shah, S.P., Eds.; Springer International Publishing: Cham, Switzerland, 2015; pp. 131–139.

29. Papatzani, S.; Grammatikos, S.; Adl-Zarrabi, B.; Paine, K. Pore-structure and microstructural investigation of organomodified/Inorganic nano-montmorillonite cementitious nanocomposites. *Am. Inst. Phys.* **2018**, *1957*, 030004. [CrossRef]

30. Papatzani, S.; Paine, K. Optimization of Low-Carbon Footprint Quaternary and Quinary (37% Fly Ash) Cementitious Nanocomposites with Polycarboxylate or Aqueous Nanosilica Particles. *Adv. Mater. Sci. Eng.* **2019**, *2019*, 26. [CrossRef]

31. Papatzani, S.; Paine, K. Polycarboxylate/nanosilica-modified quaternary cement formulations—Enhancements and limitations. *Adv. Cem. Res.* **2018**, *30*, 256–269. [CrossRef]

32. Papatzani, S.; Paine, K. Lowering cement clinker: A thorough, performance based study on the use of nanoparticles of SiO$_2$ or montmorillonite in Portland limestone nanocomposites. *Eur. Phys. J. Plus* **2018**, *133*, 430. [CrossRef]

33. Calabria-Holley, J.; Papatzani, S.; Naden, B.; Mitchels, J.; Paine, K. Tailored montmorillonite nanoparticles and their behaviour in the alkaline cement environment. *Appl. Clay Sci.* **2017**, *143*, 67–75. [CrossRef]

34. Stefanidou, M.; Papayianni, I. Influence of nano-SiO$_2$ on the Portland cement pastes. *Compos. Part B Eng.* **2012**, *43*, 2706–2710. [CrossRef]

35. Naeem, F.; Lee, H.K.; Kim, H.K.; Nam, I.W. Flexural stress and crack sensing capabilities of MWNT/cement composites. *Compos. Struct.* **2017**, *175*, 86–100. [CrossRef]

36. Mohamed, A.M. Influence of nano materials on flexural behavior and compressive strength of concrete. *HBRC J.* **2016**, *12*, 212–225. [CrossRef]

37. Yu, P.; Wang, Z.; Lai, P.; Zhang, P.; Wang, J. Evaluation of mechanic damping properties of montmorillonite/organo-modified montmorillonite-reinforced cement paste. *Constr. Build. Mater.* **2019**, *203*, 356–365. [CrossRef]

38. Horikoshi, T.; Ogawa, A.; Saito, T.; Hoshiro, H. Properties of Polyvinylalcohol fibre as reinforcing materials for cementitious composites. *Int. Work HPFRCC Struct. Appl.* **2005**, *2*, 1–8.

39. Goodhew, P.J. *Electron Microscopy and Analysis*; Goodhew, P.J., Ed.; Wykeham Publications: London, UK; Springer: New York, NY, USA, 1975.

40. Calabria-Holley, J.; Papatzani, S. Effects of nanosilica on the calcium silicate hydrates in Portland cement-fly ash systems. *Adv. Cem. Res.* **2014**, *26*, 1–14. [CrossRef]

41. Zhang, J.; Scherer, G.W. Comparison of methods for arresting hydration of cement. *Cem. Concr. Res.* **2011**, *41*, 1024–1036. [CrossRef]

42. Struble, L.; Livesey, P.; Del Strother, P.; Bye, G. *Portland Cement*; ICE Publishing: London, UK, 2011.

43. Sinha Ray, S.; Okamoto, M. Polymer/layered silicate nanocomposites: A review from preparation to processing. *Prog. Polym. Sci.* **2003**, *28*, 1539–1641. [CrossRef]

44. Manias, E.; Touny, A.; Wu, L.; Strawhecker, K.; Lu, B.; Chung, T.C. Polypropylene/Montmorillonite Nanocomposites. Review of the Synthetic Routes and Materials Properties. *Chem. Mater.* **2001**, *13*, 3516–3523. [CrossRef]

45. Bekri-Abbes, I.; Srasra, E. Effect of mechanochemical treatment on structure and electrical properties of montmorillonite. *J. Alloys Compd.* **2016**, *671*, 34–42. [CrossRef]

46. Papatzani, S.; Badogiannis, E.G.; Paine, K. The pozzolanic properties of inorganic and organomodified nano-montmorillonite dispersions. *Constr. Build. Mater.* **2018**, *167*, 299–316. [CrossRef]

47. Nie, J.; Ke, Y.; Zheng, H.; Yi, Y.; Qin, Q.; Pan, F.; Dong, P. Preparation and Characterization of Organo Montmorillonite Modified by a Novel Gemini Surfactant. *Integr. Ferroelectr.* **2012**, *137*, 67–76. [CrossRef]

48. Bransky, A.; Korin, N.; Khoury, M.; Levenberg, S. A microfluidic droplet generator based on a piezoelectric actuator. *Lab Chip* **2009**, *9*, 516–520. [CrossRef]

49. Teh, S.-Y.; Lin, R.; Hung, L.-H.; Lee, A.P. Droplet microfluidics. *Lab Chip* **2008**, *8*, 198–220. [CrossRef] [PubMed]

50. Zhuang, G.; Zhang, Z.; Guo, J.; Liao, L.; Zhao, J. Applied Clay Science A new ball milling method to produce organo-montmorillonite from anionic and nonionic surfactants. *Appl. Clay Sci.* **2015**, *104*, 18–26. [CrossRef]

51. Cornélis, A.; Laszlo, P.; Zettler, M.W.; Das, B.; Srinivas, K.V.N.S. Montmorillonite K10. *Encycl. Reagents Org. Synth.* **2004**. [CrossRef]
52. Shakrani, S.A.; Ayob, A.; Rahim, M.A.A. A review of nanoclay applications in the pervious concrete pavement. In *AIP Conference Proceedings*; AIP Publishing: Melville, NY, USA, 2017; Volume 1885, p. 020049.

Article

A Product Development Approach in The Field of Micro-Assembly with Emphasis on Conceptual Design

Christoph Gielisch [1,*], Karl-Peter Fritz [1], Anika Noack [1] and André Zimmermann [1,2]

[1]   Hahn-Schickard Institte for Micro Assembly Technology, Allmandring 9 b, 70569 Stuttgart, Germany; Karl-Peter.Fritz@Hahn-Schickard.de (K.-P.F.); noackanika@gmail.com (A.N.); zimmermann@ifm.uni-stuttgart.de (A.Z.)

[2]   Institute for Micro Integration (IFM), University of Stuttgart, Allmandring 9 b, 70569 Stuttgart, Germany

*   Correspondence: Christoph.Gielisch@Hahn-Schickard.de; Tel.: +49-711-685-83826

Received: 30 March 2019; Accepted: 6 May 2019; Published: 10 May 2019

Abstract: Faster product lifecycles make long-term investments in machines for micro assembly riskier. Therefore, reconfigurable manufacturing systems gain more and more attention. But most companies are uncertain if a reconfigurable manufacturing system can fulfill their needs and justify the initial investment. New and improved techniques for product development have the potential to foster the utilization and decrease the investment risk for such systems. In this paper, four different methods for product development are reviewed. A set of criteria regarding micro assembly on reconfigurable manufacturing systems RMS is established. Based on those criteria and the assessment, a novel approach for a product development method is provided, which tries to combine the strengths of the beforehand presented approaches. It focuses on the conceptual design phase to overcome the customers' uncertainty in the development process. For this, an abstract representation of a micro-assembly product idea as well as a decision tree for joining processes are established and validated by real product ideas using expert interviews. The validation shows that the conceptual design phase can be used as a useful tool in the product development process in the field of micro assembly.

Keywords: product development; conceptual design; micro assembly; data structure; design for manufacturability

---

1. Introduction

Production of electronic devices has seen a decrease in product life time as well as an increase in customization over the last decades [1]. The costs for new machines in micro assembly and packaging are harder to make up and increase the investment risks, especially for smaller enterprises [2]. This trend leads to a strong need for novel production methods and systems that are able to react fast and are flexible to changing customer needs without having to build complex new systems every time a new product is introduced [3]. To achieve this, a lot of work has been done in the field of flexible and reconfigurable manufacturing systems (RMS) [4–6]. There are different reconfigurable machines available in the market for highly customized micro manufacturing needs, e.g., the OurPlant-Series of Häcker Automation. Such RMS normally consist of a core machine that provides the basic functionality like an axis system for precise positioning in X and Y-direction as well as additional hardware modules with which the user can customize the system. Examples for such hardware modules would be grippers for the pick-and-place process or camera-systems for visual detection.

Still, a lot of ideas for new products in this field get lost due to the uncertainty of whether a profitable production on a RMS is possible. RMS' tend to have higher acquisition costs. Therefore, to pay off their initial investment, one has to use them over more than one product. But with a volatile future, it is hard to predict which kind of products an enterprise will build in the next decades [7]. Decreasing investment risk can only be achieved by helping product designers and engineers to utilize their micro assembly RMS as effective as possible. For this, specialized tools are needed. With such tools, the conversion rate of product ideas to functioning products that can be built on a specific RMS is fostered, which is crucial for investment calculations.

The goal of this paper is a new product development approach that aids product designers in developing products in the field of micro assembly in a way that suits their eventual production on RMS as best as possible. Such a method can overcome the need of higher utilization by integrating production knowledge into the development process. Therefore, this product development approach must be focused on the special needs of the micro assembly. As a crucial phase in development, its emphasis is the conceptual design phase, because decisions in conceptual design have the greatest impact on the finalized product. To achieve better results in this phase, customer integration is key.

A lot of research has been done regarding product development. Section 2 of this paper investigates the most common and well-known approaches in this field. Different publications have discussed conceptual design as a part of the product development. Arnold et al. have established a morphological matrix tool for automated concept generation called MEMIC [8]. While they were able to generate new concepts for given tasks relatively quick, a case study with engineering students showed that the method was rather hard and uncomfortable to use. Different research groups targeted the improved sustainability of products during their conceptual design by using Life Cycle Assessment [9]. For this, they propose methods like a functional impact matrix that works as a visual tool [10] or a design repository [11]. Buchert and Stark propose an engineering decision support system (EDSS) for a more sustainable component choice [12]. Krus and Grantham did research on reducing the failure risk of a product by improving the conceptual design phase and introduced a risk in early design (RED) method for this [13]. Both the proposed improvements in sustainability as well as in failure risk depend on an enhanced choice of components in the conceptual design phase without integrating the customer or the production factors. Tao et al. describe that a digital twin can greatly enhance the potential of product development and manufacturing by providing consistent data, but does not introduce a data format or clear methodology for the conceptual design phase [14].

In Sections 2–5 of this paper, a new and specific product development approach in the field of micro assembly is developed. Therefore, Section 2 gives an overview of different existing product development approaches, while Section 3 defines requirements of micro assembly production and checks if the presented approaches fulfill those requirements. Section 4 combines different approaches to a new product development process that better fits micro assembly on RMS. The center of this new approach is a specialized conceptual design phase, which is introduced in Section 5. Both the new product development approach, as well as its conceptual design phase, are thoroughly discussed in Section 6. Finally, Section 7 outlines all results and findings and gives a first brief outlook over possible new studies as well as the future planned work of the authors.

2. Traditional Product Development Approaches

In this chapter, four different well-known product development approaches are presented. Systematic research on methods of formalized conceptual product development has begun in the 1940s with Value Engineering. Its base is to define the sub-functions of a new product that follows a clear scheme. Sub-functions are described with a verb-object pair, e.g., a bus's sub-function would be to transport passengers [15]. Later two methods were built upon this research: In the Soviet Union, Genrich Altshuller published his Theory of Inventive Problem Solving, while in Europe different methods can be subsumed with the Systematic Approach of Pahl and Beitz. In the last decades, Concurrent Engineering, as well as Lean Development methods, are getting a lot of attention as the basis for the development of new products.

2.1. Theory of Inventive Problem Solving (TRIZ)

The Soviet scientist Altshuller has published his Theory of Inventive Problem Solving (Russian abbreviation TRIZ) in 1984 [16]. He and his group were able to analyze over 2 million patents. They found a set of about 40 principles and a few hundred physical, chemical, and geometrical effects with which they were able to describe most patents. Furthermore, they noticed that most patents try to find pareto efficient improvements to already existing solutions, ergo, solutions where one is able to improve something without having to deteriorate something else.

TRIZ is a complex system with different elements. Of main interest for product development is the Algorithm for Inventive Problem-Solving (Russian abbreviation ARIZ). Altshuller tried to find systematic ways to formalize the innovation process. His technique is suitable for incremental, less for disruptive innovations. In ARIZ, a developer first has to use a divide and conquer approach to break down a problem in an existing product in small pieces. These pieces are called mini-problems. Mini-problems should describe a system conflict in the product due to physical contradictions. The goal of this process is to reduce the physical properties that affect a mini-problem to a minimum amount without losing necessary dependencies. To solve the physical contradictions Altshuller defined own methods. Based on his patent analysis he developed a matrix of physical properties. Each cell of this matrix describes a physical contradiction of two properties. A set of inventive principles is assigned to each cell to provide a developer with possibilities to resolve the physical contradiction of mini-problems.

2.2. Systematic Approach of Pahl and Beitz (SAPB)

In the 1970s, the German researchers Pahl and Beitz established their systematic approach for the development of new products [17]. Prior to the start of the approach, an idea for a new product has to be found. Based on this idea a task needs to be formulated, e.g., the development of a new product. This task is transformed into a set of requirements and constraints regarding development. The SAPB is then structured on four main phases that are presented in Figure 1.

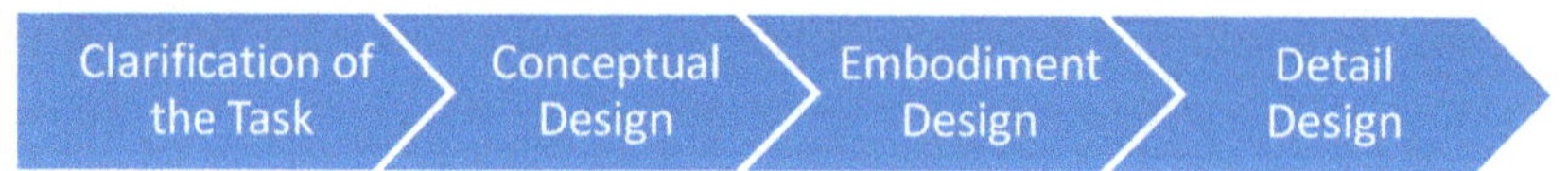

Figure 1. Phases of the Systematic Approach of Pahl and Beitz.

1. Clarification of the task

The general technological feasibility, as well as a possible market, should be proven. The task needs to be precisely clarified via specifications in a proposal.

2. Conceptual Design

With the beforehand identified specifications, a first conceptual design is worked out. This conceptual design incorporates functional structures to solve identified problems. It should show possible solution principles, but no concrete implementation. The solution principles must be evaluated against technical and economic criteria.

3. Embodiment Design

Based on the solution, principles first concrete layouts are developed. Those layouts must be carefully evaluated against technical and economic criteria. With multiple iterations, the layout can be refined and optimized to a complete design.

4.   Detail Design

In the last step, the new product design is finalized by working out its specific details and creating the according to complete data, e.g., CAD files. All records must be checked, all steps documented.

Before and after each step there is a strict deliverable. This deliverable can lead to a leap back into a prior phase to redo some of the already performed steps, due to former unseen problems or new possibilities.

2.3. Concurrent Engineering (CE)

Concurrent Engineering (sometimes also called Simultaneous Engineering) is a product development approach that aims to shorten the development time of a new product by intertwining different development sequences, as shown in Figure 2. Since the 1980s and 1990s, it has been widely established in different industries [18,19].

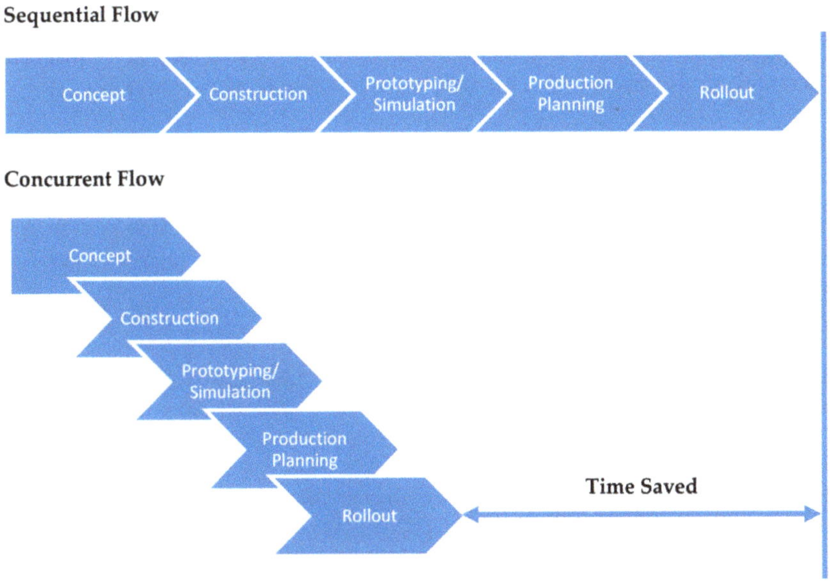

Sequential Flow

Concurrent Flow

Figure 2. Time savings due to Concurrent Engineering (based on Reference [20]).

Its focus is broader than the SAPB and does not only look at the different steps needed to find a product design. It also takes the production planning and manufacturing of a product into account. It does that by watering down the border between different development steps. For example, the production planning of a new product gets started before the final product design has been established. Even if the production planning gets more complicated, the time, as well as the communication benefit of overlapping development phases, overcompensates the complexity management effort. Also, mistakes that lead to costly redesigns can be noticed earlier, if the later development phases start sooner. This helps in reducing development costs. Concurrent Engineering often depends on interdisciplinary teams with a product related core-team and different subject-related members, which change in profession and number based on the actual project status.

2.4. Lean Development

Lean Development tries to transfer the general principles of lean production, mainly to minimize waste, to the process of developing new products. Therefore, a definition of waste has to be found.

In the development process, waste can be described as any effort which has been put into a new product without generating any appropriate value for the customer [21]. Several methods exist to help in minimizing this kind of waste. One is the Minimum Viable Product (MVP) [22]. Its fundamental idea is that no one can predict the customer value of a new product precisely. So, the best method to prevent needless development effort is to get customer feedback as fast as possible. When developing a new product, one should try to focus on the very core value the product has to offer to the customer. In a next step, one must find the fastest and most inexpensive way to build up a prototype that satisfies the expected customer need. This prototype is the MVP. It does not need to incorporate any production specific characteristics. If it is suitable to get feedback with, it can even be a complete fake. If the customer feedback is positive, the MVP can gradually be advanced to more sophisticated prototypes that better fit production needs. Unlike the SAPB, CE, and TRIZ, LD is able to completely change the original idea behind a new product. It does not define a concrete task a product has to fulfill, but only focuses on satisfying the customer need. On the other hand, it is harder to predict how much time a project will take using this approach.

3. Applicability of the Presented Approaches for Micro Assembly on RMS

All of the beforehand given product development methods and approaches are designed without a specific field of application in mind. While this is handy for a broader usage, it comes with the disadvantage of not being optimized for a particular use case. This work focuses on product development for micro assembly on RMS. To design a novel approach in this field, characteristics that influence the product development of micro-assembled products, especially micro-assembled products on RMS, have to be found. Based on those characteristics, an evaluation matrix must be arranged. With this matrix, the presented approaches can be assessed to find out their strengths and weaknesses regarding micro assembly on RMS.

3.1. Challenges for Product Development on Micro Assembly RMS

Many operations in producing micro assembly products have very small tolerances and cannot be done by hand so micro assembly is a highly automated manufacturing process with already existing RMS. Examples for such micro assembly products are optical modules for an endoscopic application that need to be actively aligned during production to ensure proper functionality or special rotary sensors with a very high accuracy. The most micro-assembled products are built for business-to-business rather than consumer sales. The primary goal of the product development process is to foster the utilization of RMS to decrease the investment risk. Operations should be done by using as much existing standard devices and as few newly built special devices as possible. These constraints lead to a set of four different characteristics that are important for the product development in the field of micro assembly.

- Clear Structure

Many product ideas in the field of micro assembly are developed during a commission work and are not directly based on marketing or consumer studies by the developer. This fact demands a clear, billable structure for working together with a contracting partner as well as focusing the work on technical problems. For example, when developing and producing a new kind of sensor, a contractor has a set of requirements regarding function, connection and constraints. Any solution fulfilling those is acceptable for the contractor, but it must be delivered in a beforehand set time and budget following a reasonable development cycle containing clear status updates.

- High Integration of Production especially for Micro Assembly RMS

Due to the need of high automation, micro assembly lacks the possibility to use manual production steps for operations where no machine or machine module exists. So, if the design of a new product does not suit the already existing standard modules of a micro assembly RMS, specialized modules must be developed and built, which is costly and time-consuming. So, to optimize the use of a micro assembly RMS, production must be integrated in the whole development process. For instance, an RMS owner could possess a manipulation module with which it is possible to move a part in every direction and tilt it by some degree but impossible to rotate the part. In this case the product design must not need the rotation during the production process to be able to use this existing manipulation module.

- Design for Manufacturability

In micro assembly, fulfilling customer needs often means documenting their technical requirements and identifying clever solutions to solve them. While this sounds simple at first, it inherits some problems regarding product development. If the technical requirements are very strictly given, there may be only a few possibilities to find a solution for them and those might not easily be implemented on an RMS. This can lead to the situation where the more work the client has already put into pre-designing a potential product, the more expensive its production can get. A possible client should be engaged by the product development approach to get in touch with the development team as soon as possible and with a very rough idea for the product itself to optimize the RMS utilizing possibilities.

- Time Consumption

Shorter product lifecycles are an important trend in micro assembly and neighboring industry fields. If a customer needs a new sensor for a technical device that has an annual release cycle, e.g., a new smartphone, the development and production of this sensor must be fast enough to catch up with this annual cycle. The product development process must be fast enough to satisfy the general innovation speed. So, the more time consuming an approach is, the less suitable it is in the field of micro assembly.

3.2. Pairwise Comparison

To find out their strengths and weaknesses, all described product development approaches need to be assessed. The focus of this work is the readiness for the use in the field of the micro assembly with RMS. In 3.1 four criteria for this specific task are derived which can be used for comparing the different approaches. Each product development approach has been paired with each other approach. This leads to a total number of six pairs. For every pair of approaches in each of the four criteria, it has been qualitatively decided, if one or another approach is better or if both fulfill the criterion equally good. The winner of a comparison obtains two points, the other zero. If it is a draw, both approaches get one point. The maximum amounts of points an approach could get by this method is ten. The final score of an approach in a criterion is, therefore, determined by its collected amount of points within the pairwise comparison divided by ten.

3.3. Evaluation of Different Product Development Approaches Regarding Micro Assembly on RMS

The four product development approaches described in Section 2 are assessed by using the established pairwise comparison. The results of the pairwise comparison of all product development approaches are shown in Table 1. Of all approaches, SAPB has the clearest structure when it comes to developing a new product. CE does this job in a rather broad way, which helps to integrate production and cutting time costs, but is not as detailed as SAPB in designing and developing the product itself. TRIZ has good techniques to help solve very specialized problems within the development of a product. But these techniques are only loosely connected, and one can get lost in the details. LD mainly works with the MVP. While this is good for fitting a product to a customer need, it is not perfectly suited for contracting work that dominates the field of micro assembly on RMS. LD is clearly and deliberately less structured than the other methods.

When it comes to integration of production into the product development process, CE is doing best with the intertwining of product and process design, as well as the interdisciplinary teams. This is very handy when developing products for a micro assembly RMS. SAPB only designs the product, but has no clear and defined links to production, even if those links could be possible during the later design phases. TRIZ is a similar case. Only if the mini-problems are formulated in a way that takes production in account it may be considered, otherwise not. LD production is an inferior problem that needs to be solved after the customer need is satisfied. Nevertheless, the strong iterative and heavily-prototyped procedure can be very helpful in designing production processes, too.

Table 1. Result of the pairwise comparison of the different product development approaches regarding the derived criteria for micro assembly on reconfigurable manufacturing systems (RMS).

	TRIZ	SAPB	CE	LD
Clear Structure	50%	100%	50%	0%
High Integration of Production	33%	33%	100%	33%
Design for Manufacturability	17%	17%	67%	100%
Time Consumption	33%	67%	100%	0%

Designing a product so that it fits the RMS is of uppermost importance in the field of micro assembly. This can only be done by encouraging the potential customer to be involved in the development process. SAPB defines the clarification of the task before the conceptual design. Often this happens by creating specification documents. In a lot of cases, these documents are written before an exact process can be defined and solely focus on the product itself. If they are part of a binding contract such documents are hard to change in the later development phases, even if it is clear that they do not fit production needs and harm the development without any benefit for the underlying product. TRIZ has a similar problem. CE is able to reduce it by integrating production experts in the interdisciplinary core team without completely solving the problem. LD integrates the customer in the development of the product and does not depend on clear specifications. The strong iterative process has no strict goal besides satisfying the customer need. This can greatly help to adapt the product design to production requirements in the later design phases.

The main premise of CE is to shorten the needed development time by overlapping the different phases of the development process. LD and TRIZ both need a lot of time to execute the different given development methods. SAPB is more streamlined but still lacks the optimization CE has. So, the best technique for a fast, less time-consummating product development approach is CE.

Overall, there is no approach that fits all the needs that are found in Section 2 for a micro assembly-based product development for RMS. Each approach has different strengths and weaknesses. But a lot of the presented methods are combinable. In the next step, a novel product development approach is proposed based on a recombination of the existing presented and assessed techniques.

4. New Combined Product Development Approach in the Field of Micro Assembly

To establish a new product development approach in the field of micro assembly, the different presented methods are recombined using their strengths and avoiding their weaknesses regarding micro assembly on RMS derived in Section 3.1. Figure 3 gives an overview of the found novel combined product development approach. It is structured in three sections indicated by the vertical bars on the left:

- In the first section some highlighted resources are shown that are used by the development approach;
- In the process section, the four different development phases are arranged chronologically from left to right;
- The data format section defines data types that can or should be used in the corresponding development phases;
- In the team section the two horizontal streams show the composition of the employees working on the new product in the corresponding development phases.

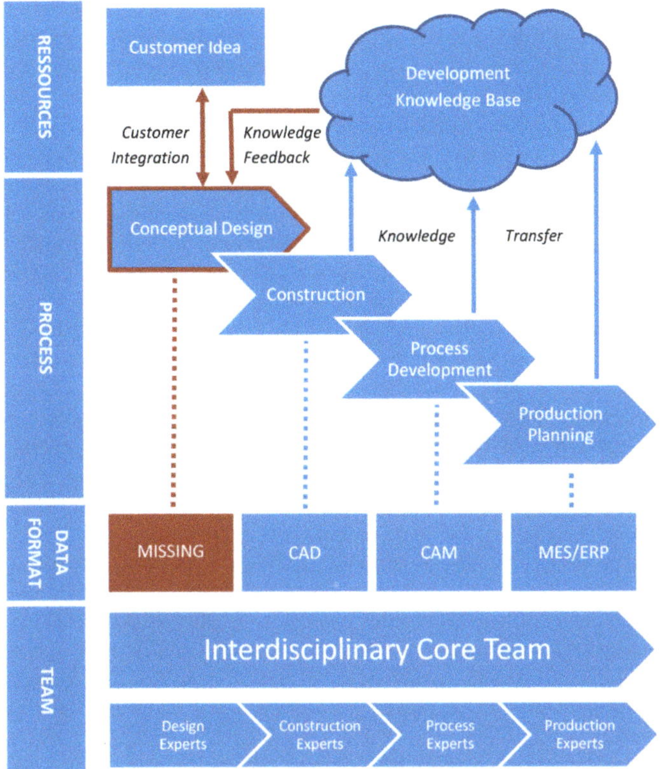

Figure 3. A novel Combined Product Development approach.

Following the challenges for product development in micro assembly on RMS described in Section 3, four goals must be targeted when designing a new product development approach in this field: Clear structure, integration of production, design for manufacturability, and optimized time consumption. As an initial beginning, the clear structure criterion was selected to establish a first basis on which further improvements can be done to satisfy the other goals. Therefore, SAPB has been used as a starting frame for the development process, as it has shown the best results in the clear structure criterion in the pairwise comparison. Based on this choice, the SAPB has been altered to match the results of the remaining three criteria.

For the second goal, the integration of production, CE has ranked first in the evaluation, so the combined approach is using the identified strengths of it. SAPB does not strictly include methods for processing or production. The combined approach extends the thematic range of the SAPB and, following the CE, also introduces an interdisciplinary core team to guarantee a smoother transition between the different phases. Two phases are added after the original ones of the SAPB to bridge the gap towards manufacturing, a process development and a production planning phase. The process development phase uses CAM (computer aided manufacturing) data files, while the production planning can be done via MES (manufacturing execution systems) and ERP (enterprise resource planning). The embodiment and the detail phases of the SAPB are unified into a new construction phase. The main reason for this is that both phases use the same kind of data format and are done by the same kind of experts. Still, all the activities of Pahl and Beitz have determined the embodiment and detailing remain the same and are done in the aggregated new construction phase. In this construction phase, TRIZ techniques can be used to obtain faster and better results. As a data format computer-aided design (CAD) files are used. Following the CE, the combined approach requires an interdisciplinary core team that consists of employees out of every field of the four phases, so design, construction, process, and production. The core team is supported by additional experts based on the actual phase the product development is in.

Design for manufacturability is the third and eventually most important goal. The combined approach must establish a system that improves the assembly of the new products before they have been constructed. For this, the combined approach integrates the task clarification phase into the conceptual design phase to prevent harmful decisions regarding assembly and production while doing the design for a new product. To achieve this, three things need to be done. Firstly, the customer is integrated into the conceptual design phase. Potential customers clarify their exact tasks while they are doing the conceptual design together with micro assembly engineers. The LD idea of iteratively adding features and checking if the so newly created MVP is an improvement can be very helpful to achieve this. The conceptual design phase must establish a formalized method for this customer integration. Secondly, a specialized data format has to be found to enable customers to model their product ideas. Contrary to the other three phases such a data format does not exist for the specialized use case of micro assembly on RMS. The last point is transferring construction, process and production knowledge to a development knowledge base, which is able to generate fast feedback if a production of a planned product idea on a micro assembly RMS is feasible. All three points must work together to provide a conceptual design phase for micro assembly on RMS.

The final goal is optimized time consumption. Reaching this goal is done by using CE techniques. Each phase is overlapping with the subsequent one, so work for the following phase is done before the actual phase has been finished. Due to this phase overlap the time consumption, the strength of the CE can be realized in the combined approach. But it is necessary to work with interdisciplinary teams when doing so to fully reach this time potential and to avoid inefficiencies based on the higher complexity and stronger need of coordination compared to a pure SAPB approach.

5. Conceptual Design for Micro Assembly Product Development

The beforehand presented novel combined product development approach for products in the field of micro assembly depends on strong methods during the conceptual design phase. The main goal is to achieve the best possible design for manufacturability on an RMS. Therefore, the owners of a new product idea in the field of micro assembly, in many cases the contractor in a business-to-business relationship, must be convinced that they can benefit from using tools and methods to conceptually model their idea in a way that suits production on RMS.

Similar to using an MVP in LD, a user can iteratively alter his product model and is getting feedback whether production on a micro assembly RMS is feasible or not. Rather than optimizing the MVP based on a customer need, the abstracted model is optimized towards manufacturability on RMS. But it is using the same idea of getting feedback as fast as possible and during a development phase in which it is still possible to change the whole product idea without generating a lot of costs. This cost reduction should enhance the method approval of potential users. Still, they may not be experts in this field, so the methods must not be too complicated to apply. Based on these thoughts three main goals are set for the conceptual design system:

1. Establish a system that allows the user as much freedom as possible for the design of a micro-assembled product without losing the ability to systematically analyze the given data to automatically provide information regarding the feasibility of a production on a micro-assembly RMS.
2. Allow the system to work reasonably well with a given set of micro assembly RMS.
3. Enable users to work with the system without extensive training.

A lot of physical problems behave differently in the micro world compared to the macro world. Also, some degrees of freedom that one typically finds in assembly might not apply to micro assembly. Hence, RMS for micro assembly are highly specialized machines. The conceptual design system has to consider this. To implement such tools, a formalized system, as well as a specific data structure to model product ideas, are needed. There already exist commonly known data structures for the later phases of the combined approach. In the construction phase, CAD data files like STEP are used. Process development can be done by using CAM data structures, production planning benefits from data management in Manufacturing Execution Systems (MES) and Enterprise Resource Planning (ERP). There is no comparable data format for the conceptual work on micro-assembled products.

To establish a system for the conceptual design phase, first an abstract product representation for micro assembly products must be found. Based on this representation, data types and their properties can be defined. Those properties must be able to successfully translate a modeled product idea to a micro-assembly RMS. For this particular step decision trees are needed. Lastly, the method must be validated by real examples.

5.1. Abstract Product Representation for the Conceptual Design

The abstract product representation aims to be as simple as possible without losing the ability to describe micro-assembled products sufficiently. Three different data types were used for the representation of the conceptual design: Components, relations, and connection points. A micro-assembled product normally consists of different elements like a substrate, chips, capacitors, wires, and so forth. These elements are the components. In an assembled product, the components are linked. This can be a geometrical but also an electrical or optical link. This is the second data type, the relation. Often a relation is not valid for a whole component but a special place anywhere on, in, or near the component. This special place, called connection point, can be something like a joint, a thread, or a face and is the last data type. Figure 4 shows a typical setup for the abstract product representation including different components, relations, and connection points.

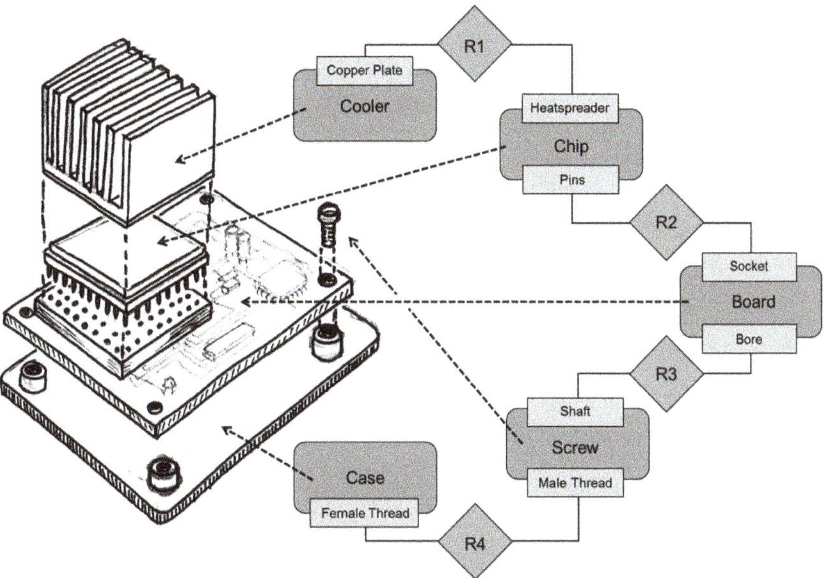

Figure 4. Typical setup of the Abstract Product Representation including components (rounded box), relations (diamond box), and connection points (squared box).

5.2. Data Type Properties

As a first orientation, the DIN 32,563 (Production equipment for microsystems–System for classification of components for microsystems) can be used [23]. Based upon the DIN and own micro-assembly knowledge, eight property fields for the components are selected: Identifier, geometrical base form, weight, temperature consistency, material properties, feeding possibilities, picking possibilities, and connection points. A detailed view on all properties in the data fields is given in Table 2.

Table 2. Properties of the component data type.

Data Field	Possible Properties
Identifier	Name of Component
Geometrical Base Form	Box, Cylinder, Tube, Cone, Sphere
Weight	Weight in Gramm
Temperature Consistency	Minimum Temperature, Maximum Temperature
Material Properties	Magnetism, Shock Resistance, Clean Room Class, Optical Property
Feeding Possibilities	Tray, Bulk, Tube, Belt
Picking Possibilities	Vacuum Gripper, Parallel Gripper, Radial Gripper, Three-Point Gripper, Magnetic Gripper
Connection Points	List of Identifiers

Similarly, the property fields for the relation and the connection point were chosen. For the relation, three property fields are established: Connecting point partners, spatial properties, and physical requirements. The connection points have four fields: Identifier, type, location, and material properties. All fields for both data types can be found in Tables 3 and 4.

Table 3. Properties of the relation data type.

Data Field	Possible Properties
Connection Point Partners	Identifier Point 1, Identifier Point 2
Spatial Properties	Position (Above, Beneath, Ahead, Behind, Left, Right, In, Through), Loose, Offset, Tolerance
Physical Requirements	Electric Conductivity, Optics, Stiffness, Stability

Table 4. Properties of the Connection Point Data Type.

Data Field	Possible Properties
Identifier	Name of Connection Point
Type	Free Surface, Free Volume, Clamping, Jack, Plug, SMD
Location	Above, Beneath, Ahead, Behind, Left, Right, In, Through
Material Properties	Metallization, Roughness, Optics

5.3. Decision Trees

The crucial step for the proposed conceptual design approach for micro-assembly is the early link between the product design and the assembly process that is needed for the fabrication of the product. This means that in a very early stage of the product development process feedback on the manufacturability of the product is provided. For this the link between the properties of the product and the subsequent assembly process is needed.

In order to standardize this link, decision trees for different phases of the micro-assembly process were developed in Reference [24] that allow a comparatively detailed choice of a specific process step based on the given properties of the product, represented by the above-mentioned data types. When applying the decision tree methodology, the decisions that have to be taken are reduced to a sequence of simple questions that can be easily answered, leading to a suggestion of the appropriate assembly process.

Exemplarily the decision tree for the joining processes is shown in Figure 5. The decision tree leads to a specific choice between 4 possible joining processes: UV-cured adhesive bonding, adhesive bonding, soldering, and a mere positioning, whereas the choice of the adhesive (whether conductive or non-conductive) can be seen as an additional feature that does not affect the decision on the process and, therefore, needs not to be considered in the decision tree. For each decision tree, different properties of the abstract product representation are used. As shown in the figure, this decision for the joining process is taken based on the following properties:

- Component properties: Weight, temperature consistency;
- connection point properties: Metallization;
- relation properties: Stability, electric conductivity, tolerance.

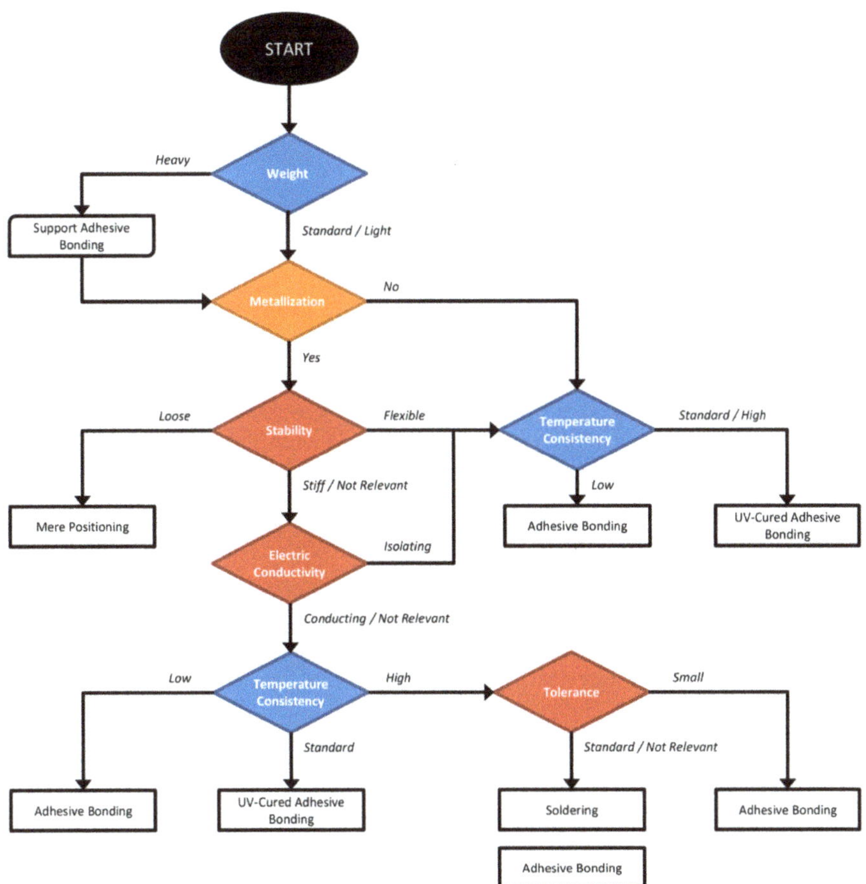

Figure 5. Decision tree for the Joining Processes. Blue are component properties, orange connection point properties, and red relation properties (based on Reference [24]).

An analog approach is shown in Reference [24] for the decision on the sequence of the assembly processes. The methodology was established based on the analysis of a cross-section of several different micro-assembly products and validated with four specific products:

- Printed circuit board (PCB) based optical module for a rotary encoder;
- surface-mounted device (SMD) assembly of a Flex-PCB;
- molded Interconnected Devices (MID) based light module;
- MID based sensor holder for a hall sensor element.

All the aforementioned examples turned out to confirm the proposed methodology using the decision trees. Further validation with more examples during the ongoing development of the whole conceptual approach is planned.

5.4. Qualitative Validation of the Conceptual Design

For each data type of the abstract product representation (component, relation, and connection point) a paper survey was arranged. The paper surveys obtain all properties of the different data types as given in Tables 2–4. With the collected information out of the surveys, the decision trees as presented in Section 5.3 can be used to find concrete processing ways. These processing ways can then be linked to machine configurations of micro-assembly RMS. Thereby, the conceptual design phase can be simulated and as a result also qualitatively validated by using the surveys and the decision trees.

This simulation has been used to assess the conceptual design phase approach. Three teams were built, each consisting of product developers, machine experts for a specific micro-assembly RMS, and software developers. For the validation they used different, real existing micro-assembly product ideas and filled in the corresponding paper surveys. During this process the following findings were made by the experts:

1. Most of the characteristics of the product ideas could successfully be modeled by using the paper surveys.
2. If all properties are given, specific processing ways can be found for the selected micro-assembly product ideas by using the decision trees.
3. There is currently no possibility to satisfactorily describe a component that is formed or reshaped during the assembly process, e.g., the molding of a cavity with glue.
4. Analyzing the found processing ways, RMS experts were able to derive machine configurations that suit the product ideas. Still, not all processes could be implemented with standard modules. Some processes require specialized modules for the RMS that need to be designed and constructed separately.
5. Because of the sheer amount of demanded properties, the system is not easy to use, especially if a product idea needs a lot of components and relations. Often the precise value for a property is unknown or deliberately not defined at this moment. Sometimes properties need to be filled in without being necessary for the later steps because previously made decisions have already made them unnecessary.
6. It would be possible to implement the method as an automated software solution.
7. The method does not generate a clear processing order, only a collection of required processing steps.

6. Discussion

Based on the review of four distinct well-known product development approaches and four criteria for challenges on production in the field of micro-assembly, a new combined product development approach was introduced. This approach has a focus on the conceptual design phase. Therefore, it introduces an abstract product representation as well as specialized decision trees. Both the new combined product development approach, as well as the conceptual design phase, are discussed in this chapter.

6.1. Discussion of the Product Development Approach

The pairwise comparison of the presented product development approaches shows that no single approach fulfills all requirements for a micro-assembly production on RMS. Still, only four approaches were examined. When selecting the described approaches, the goal was to choose distinct methods. A lot of product development approaches are very similar. They are highly optimized in the way they work, but their foundation tends to be alike. For example, Roth [25], Koller [26], and Rodenacker [27] have published their own approaches for product development. However, all of these approaches use a base structure that is similar to the one the SAPB has. Therefore, they were not considered during the comparison. Their estimated additional value towards the SAPB was too small for further consideration. An equal decision was made with LD. There are a lot of different techniques that are very similar to the described approach. The presented method is just a selection that had to be made. Nevertheless, the recent work in the field of product development methods should continuously be checked, and more and new approaches will be considered and involved.

To assess the product development approaches, four criteria for micro-assembly production on RMS were defined. Those criteria may not be applicable to all potential companies in the field of micro-assembly. Also, the pairwise comparison is a valuation that may not correctly portray all potential users of the product development approach. For example, an enterprise needs not to be focused on products for business-to-business sales or fast development cycles. In such cases, the established combined product development approach has to be adapted and altered to the special demands of the company.

The new combined product development approach has a strong focus on conceptual design because design decisions have the greatest impact in this phase. Also, the needed customer integration can be obtained best in this development phase. The conceptual design phase is an inherent part of the original SAPB and widely used in a lot of different publications [8–12]. Still, as shown in Figure 3, several research gaps could be identified. There is no general data format for the conceptual design phase that fits the needs of the micro-assembly. Often function-based modeling methods [10] or component databases with decision systems [11,12] are used. Both systems are either too complex or too restrictive. Therefore, customer integration must be done by using the three identified goals in Section 5. To enhance this customer integration by better feedback as well as guidance a development knowledge base is proposed, which would greatly benefit from digital twin techniques [14].

6.2. Discussion of the Conceptual Design for Micro-Assembly on RMS

Three goals for the new conceptual design phase were set. The first goal is to allow the users as much freedom in designing their product ideas as possible, but still be able to automatically prove the feasibility of the modeled ideas. Overall, this goal has been fairly reached. Most but not all characteristics could be modeled by the validators. Especially, forming or reshaping of components is not possible. The used paper surveys are only a proof of concept. So, an automated output of the method is not generated at the moment. However, the validation showed that this is feasible as long as all properties of the data types are given.

The second goal is that the system should work reasonably well with micro-assembly RMS. The validation came to the conclusion that not all steps of a micro-assembly can be done by only using standard modules of the RMS. The participating machine experts for micro-assembly on RMS estimate that roughly 80% of all operations required for a micro-assembly can be carried out by using standard modules, the remaining 20% need to be done by special modules.

The last goal was to provide a system that is relatively easy to use. Validation shows that this has not been reached yet. Paper surveys are a very static tool for data acquisition. They are not able to react to a given input and they are also not able to visualize possible inputs to help the user. Software implementations can help reducing these problems by aiding the user with graphical interfaces as well as automatically hiding and showing input forms based on the already given data. Yet, for the automated output, the system still relies on a full set of properties that may not be certain at the moment the conceptual design phase is done.

All in all, the conceptual design phase could be validated as a working tool to model product ideas in the field of micro-assembly and proof their feasibility regarding production on RMS. But the described approach still has limitations and does not fulfill all the goals that were set. The method needs to be further refined and enhanced. One main starting point is the clear orientation towards describing the final product model during the conceptual design phase. While this seems to be the right way when creating a method for product development it comes with some downsides. In the conceptual design phase, there exists no construction or detailed design of a product, but instead, mostly an idea what a finished product should do, which operational and physical principles can be used to achieve this and some first thoughts on how this may be implemented. Often neither materials nor geometrical proportions of the used components are certain. This leads to the situation that some of the properties cannot be defined and also should not be defined. But without those properties, the system is not able to propose a processing way. Also, to get a processing order, complex algorithms have to be implemented that need a lot of additional data to function correctly.

7. Conclusions and Outlook

The four distinct product development approaches SAPB, TRIZ, CE, and LD are presented and evaluated by a set of four criteria that have been specifically designed for the needs of micro-assembly production on RMS: Clear structure, high integration of production, design for manufacturability, and time consumption. Based on the evaluation a new combined approach is introduced, which successfully combines strengths from the four existing methods. The core of this combined approach is a novel conceptual design phase. It consists of an abstract product representation with three data types. Each data type has a set of properties that is particularly built for micro-assembly products. Using those properties together with a set of decision trees, one can derive processes the RMS has to fulfill. Validation shows that this conceptual design phase can be used to model product ideas and to generate matching machine configurations for RMS, but still has room for improvements.

7.1. Outlook

Something that should be done in future works is the better quantitative evaluation of requirements companies in the field of micro-assembly have regarding production on RMS. The four identified criteria might not be applicable to all micro-assembly manufactories. This could be done by more expert interviews to qualitatively gather additional criteria followed by an empirical survey of selected enterprises.

For the conceptual design phase, several improvements are possible. Mainly the easiness of use must be enhanced. Therefore, a shortcut in modeling the product idea could be established. Instead of basing the modeling on the idea of a complete product, one could describe the different processes needed to build the product. For each process step, the user only needs to define a basic process type like joining or shaping and the involved input and output components. Later the user can but does not have to provide additional information regarding the process or the material and geometry of the components. This additional information can be inquired based on the corresponding decision tree of the selected process type. Not forcing the user to enter all the information the system needs to find a singular solution makes it easier for the user to obtain fast results, even if they are not one but a set of possible machine configurations. To improve user experience, a separate evaluation function should rank all the possible machine configurations and suggest the best one to the user.

While this system comes with the disadvantage of requesting the user to have some knowledge of processing a lot of advantages could be realized by it. The problem of not being able to correctly model shaping or reforming of components can be solved by this approach. Also, nonexistent processes can be simply replaced by a wildcard-like black box process that is later turned into a black box hardware module on the RMS. All in all, this process-based enhancement of the conceptual design phase could give the user more freedom in modeling an idea, reducing the amount of data that has to be given, shortening the time a first result can be presented and improving the overall easiness of use. However, the conceptual design phase still needs to be implemented as a software solution to reach its full potential.

7.2. Future Work

The presented method to enhance the product development in the field of micro-assembly with emphasis on conceptual design is only a first start. Further research and development are needed, so a conceptual design phase with a process-based model is planned for future work. For this a set of different fundamental process types must be found. A matching feature-based database for micro-assembly RMS as well as a configurator software must be implemented and thoroughly tested. The configurator software must be able to integrate customers in the design process following the established rules in Section 5 of this paper. Therefore, the introduced decision trees can be used to guide the customers. Tests of such a new method using real hardware should be done as well.

Author Contributions: Conceptualization, C.G., A.N. and K.-P.F.; methodology, C.G. and A.N.; validation, C.G., A.N., and K.-P.F.; formal analysis, C.G.; investigation, C.G., A.N., and K.-P.F.; data curation, C.G. and A.N.; writing—original draft preparation, C.G..; writing—review and editing, C.G., K.-P.F., and A.Z.; visualization, C.G.; supervision, K.-P.F. and A.Z.; project administration, C.G. and K.-P.F.; funding acquisition, K.-P.F.

Funding: This research was funded by the Federal Ministry of Education and Research of Germany in the framework of 'Methodische Werkzeuge zur Erhöhung der Wandlungsfähigkeit von Mikromontage-Anlagen bei Entwicklung, Konfiguration und Monitoring (MIKROKOMO), grant number 02P15A120.

Conflicts of Interest: The authors declare no conflict of interest.

References

1. Abele, E.; Elzenheimer, J.; Liebeck, T.; Meyer, T. Globalization and Decentralization of Manufacturing. In *Reconfigurable Manufacturing Systems and Transformable Factories*; Dashchenko, A.I., Ed.; Springer: Berlin/Heidelberg, Germany, 2006; pp. 3–13, ISBN 978-3-540-29397-2.

2. Abele, E.; Liebeck, T.; Wörn, A. Measuring Flexibility in Investment Decisions for Manufacturing Systems. *CIRP Ann.* **2006**, *55*, 433–436. [CrossRef]

3. Jovane, F.; Westkämper, E.; Williams, D. *The ManuFuture Road: Towards Competitive and Sustainable High-Adding-Value Manufacturing*; Springer Science & Business Media: New York, NY, USA, 2008; ISBN 978-3-540-77012-1.

4. Koren, Y.; Wang, W.; Gu, X. Value creation through design for scalability of reconfigurable manufacturing systems. *Int. J. Prod. Res.* **2017**, *55*, 1227–1242. [CrossRef]

5. Hees, A.; Reinhart, G. Approach for Production Planning in Reconfigurable Manufacturing Systems. *Procedia CIRP* **2015**, *33*, 70–75. [CrossRef]

6. Koren, Y.; Shpitalni, M. Design of reconfigurable manufacturing systems. *J. Manuf. Syst.* **2010**, *29*, 130–141. [CrossRef]

7. Doheny, M.; Nagali, V.; Weig, F. Agile operations for volatile times. *McKinsey Q.* **2012**, *3*, 126–131.

8. Arnold, C.R.B.; Stone, R.B.; McAdams, D.A. MEMIC: An Interactive Morphological Matrix Tool for Automated Concept Generation. In Proceedings of the 2008 IEE Annual Conference and Expo/Industrial Engineering Research Conference (IERC '08), Vancouver, BC, Canada, January 2008; pp. 1196–1201. Available online: https://prosedesign.tamu.edu/publications/memic-_an_interactive_morph.pdf (accessed on 6 May 2019).

9. Kishawy, H.A.; Hegab, H.; Saad, E. Design for Sustainable Manufacturing: Approach, Implementation, and Assessment. *Sustainability* **2018**, *10*, 3604. [CrossRef]

10. Devanathan, S.; Ramanujan, D.; Bernstein, W.Z.; Zhao, F.; Ramani, K. Integration of Sustainability into Early Design Through the Function Impact Matrix. *J. Mech. Des.* **2010**, *132*, 081004. [CrossRef]

11. Bohm, M.R.; Haapala, K.R.; Poppa, K.; Stone, R.B.; Tumer, I.Y. Integrating Life Cycle Assessment into the Conceptual Phase of Design Using a Design Repository. *J. Mech. Des.* **2010**, *132*, 091005. [CrossRef]

12. Buchert, T.; Stark, R. Decision Support Tool to Derive Sustainable Product Configurations as a Basis for Conceptual Design. Available online: https://www.designsociety.org/publication/40653/DECISION+SUPPORT+TOOL+TO+DERIVE+SUSTAINABLE+PRODUCT+CONFIGURATIONS+AS+A+BASIS+FOR+CONCEPTUAL+DESIGN (accessed on 26 April 2019).

13. Krus, D.; Grantham, K. *A Step Toward Risk Mitigation During Conceptual Product Design: Component Selection for Risk Reduction*; American Society of Mechanical Engineers: New York, NY, USA, 2010; pp. 561–573.

14. Tao, F.; Cheng, J.; Qi, Q.; Zhang, M.; Zhang, H.; Sui, F. Digital twin-driven product design, manufacturing and service with big data. *Int. J. Adv. Manuf. Technol.* **2018**, *94*, 3563–3576. [CrossRef]

15. Miles, L.D. *Techniques of Value Analysis and Engineering*, 3rd ed.; Lawrence, D., Ed.; Miles Foundation: Plymouth, MA, USA, 2015.

16. Altshuller, G.S. *Creativity as an Exact Science: The Theory of the Solution of Inventive Problems*; Stud. Cyber; Gordon and Breach: Amsterdam, The Netherlands, 1984; ISBN 978-0-677-21230-2.

17. Pahl, G.; Beitz, W. *Engineering Design: A Systematic Approach*; Springer Science & Business Media: New York, NY, USA, 2013; ISBN 978-1-4471-3581-4.

18. Stjepandic, J.; Wognum, N.; Verhagen, W. *Concurrent Engineering in the 21st Century*; Springer: Berlin/Heidelberg, Germany, 2015; ISBN 978-3-319-13775-9.

19. Sohlenius, G. Concurrent Engineering. *CIRP Ann.* **1992**, *41*, 645–655. [CrossRef]

20. Warschat, J.; Bullinger, H.-J. *Forschungs- und Entwicklungsmanagement: Simulataneous Engineering, Projektmanagement, Produktplanung, Rapid Product Development*, 1st ed.; Vieweg+Teubner: Wiesbaden, Germany, 1997.

21. Liker, J.K.; Morgan, J.M. The Toyota Way in Services: The Case of Lean Product Development. *Acad. Manag. Perspect.* **2006**, *20*, 5–20. [CrossRef]

22. Münch, J.; Fagerholm, F.; Johnson, P.; Pirttilahti, J.; Torkkel, J.; Jäarvinen, J. Creating Minimum Viable Products in Industry-Academia Collaborations. In *Proceedings of the Lean Enterprise Software and Systems, Galway, Ireland, 1–4 December 2013*; Fitzgerald, B., Conboy, K., Power, K., Valerdi, R., Morgan, L., Stol, K.-J., Eds.; Springer: Berlin/Heidelberg, Germany, 2013; pp. 137–151.

23. *DIN 32563:2002-03, Fertigungsmittel für Mikrosysteme_- Klassifizierungssystem für Mikrobauteile*; Beuth Verlag GmbH: Berlin, Germany, 2002.

24. Noack, A. Wandlungsfähige Mikromontage—Auf dem Weg zur TransferFab 4.0. Master's Thesis, University of Stuttgart, Stuttgart, Germany, 2017.

25. Roth, K. *Konstruieren mit Konstruktionskatalogen: Band 1: Konstruktionslehre*; Springer: New York, NY, USA, 2011; ISBN 978-3-642-17466-7.

26. Koller, R. *Konstruktionslehre für den Maschinenbau: Grundlagen des Methodischen Konstruierens*; Springer: New York, NY, USA, 2013; ISBN 978-3-662-12183-2.

27. Rodenacker, W.G. Methodisches Konstruieren. In *Methodisches Konstruieren: Grundlagen, Methodik, Praktische Beispiele*; Rodenacker, W.G., Ed.; Konstruktionsbücher; Springer: Berlin/Heidelberg, Germany, 1984; pp. 37–227, ISBN 978-3-642-96852-5.

 *applied sciences*

Article

Low-Temperature Plasma Nitriding of Mini-/Micro-Tools and Parts by Table-Top System

Tatsuhiko Aizawa [1,*], Hiroshi Morita [2] and Kenji Wasa [3]

[1]   Surface Engineering Design Laboratory, Shibaura Institute of Technology, Tokyo 144-0045, Japan
[2]   Nano Coat & Film Laboratory, Limited Liability Company, Tokyo 144-0045, Japan;
     hiroshi-m@mtc.biglobe.ne.jp
[3]   MicroTeX Labs, Limited Liability Company, Tokyo 144-0051, Japan; wasa@mui.biglobe.ne.jp
*   Correspondence: taizawa@sic.shibaura-it.ac.jp; Tel.: +81-3-6424-8615

Received: 24 March 2019; Accepted: 19 April 2019; Published: 23 April 2019

Featured Application: Surface modification of mini- and micro-nozzles for dispensing systems.

Abstract: Miniature products and components must be surface treated to improve their wear resistance and corrosion toughness. Among various processes, low-temperature plasma nitriding was employed to harden the outer and inner surfaces of micro-nozzles and to strengthen the micro-springs. A table-top nitriding system was developed even for simultaneous treatment of nozzles and springs. A single AISI316 micro-nozzle was nitrided at 673 K for 7.2 ks to have a surface hardness of 2000 HV0.02 and nitrogen solute content up to 10 mass%. In particular, the inner and outer surfaces of a micro-nozzle outlet were uniformly nitrided. In addition, the surface contact angle increased from 40° for bare stainless steels to 104° only by low-temperature plasma nitriding. A stack of micro-nozzles was simultaneously nitrided for mass production. Micro-springs were also nitrided to improve their stiffness for medical application.

Keywords: plasma nitriding; micro-nozzle; micro-spring; nitrogen supersaturation; hardening; hydrophobicity; stiffness control

---

1. Introduction

MEMS (Mechanical Electric Micro-System), miniature mechanical systems, and mini-and micro-tools have a significant risk of wear and corrosion without suitable surface treatment to each application for each component [1]. In particular, mechanical tools and parts in the order of mm and sub-mm ranges, must have sufficient wear resistance and corrosion-toughness for operations even in severe conditions [2]. Dry coatings by PVD (Physical Vapor Deposition) and CVD (Chemical Vapor Deposition) are the first policy to protect them from wearing and corrosion [3]. Their deposition layer is often limited by several to 10 μm, and inner surfaces as well as holes are difficult to coat [4]. Among some candidate alternatives, low-temperature plasma nitriding has been highlighted to provide the thick nitrided layer up to 0.1 mm with higher hardness than 1200 HV and less nitride precipitates [5–8]. This plasma nitriding at a lower temperature than 700 K was characterized by the nitrogen supersaturation; after [5], this processing was expected to be applied to various surface treatments such as carburizing and nitrocarburizing as the S-phase engineering. In addition, this nitrogen supersaturation process accompanied by two-phase nano-structuring to harden and strengthen the stainless steel parts and members as pointed in [8,9]. As demonstrated in [9–11], the corrosion toughness was also improved in these nitrogen supersaturated stainless steels. Furthermore, inner surfaces and small holes in the dies and punches are efficiently nitrided and hardened with more ease than coatings [12,13]. These intrinsic features of low-temperature plasma nitriding were accommodated to miniature tools, and even parts, by scaling down the chamber size [14,15]; e.g., the hollow cathode device assisted the high-density plasma nitriding process in the smaller chamber systems.

In the present paper, a table-top plasma nitriding system is developed for low-temperature plasma processing of mini- and micro-tools. Micro-nozzles as well as micro-springs are plasma nitrided at 673 K for 7.2 ks to describe their nitriding and hardening behavior. In particular, a AISI316 micro-nozzle specimen is utilized to analyze the nitrogen super-saturation on the outer and inner surfaces as well as its outlet. In addition, its surface contact angle is measured to prove that the nitrided surfaces become hydrophobic. Furthermore, a stack of micro-nozzles is also homogeneously nitrided to demonstrate the capacity of the present system in application to industries. Two micro-springs are also nitrided to demonstrate that they have higher stiffness by 10 % than before nitriding. The present table-top system works to locally increase the stiffness and hardness of various tools and parts by selective plasma nitriding.

2. Experimental Procedure

2.1. Down-Sized Plasma Nitriding System

The high-density RF (Radio-Frequency)/DC (Direct Current) plasma nitriding system is downsized by 1/20 as suitable equipment for surface treatment of miniature mechanical elements in the dimensional size range from mm to sub-mm. Figure 1a depicts a table-top plasma nitriding system with a chamber size of $\phi 350$ mm × 150 mm. This chamber is automatically operated to open and move upward for the experiment setup. Every experimental operation is performed on the touch-panel in the controller. All commands as well as data acquisition are driven by a process computer through wireless communication.

There are two setting-up modes in this system, for nitriding a single component and for simultaneously nitriding a stack of elements up to 24 pieces. In the following, the former setup is used to describe the nitriding behavior of a single micro-nozzle. The latter is also employed to demonstrate the feasibility of simultaneous nitriding in mass production.

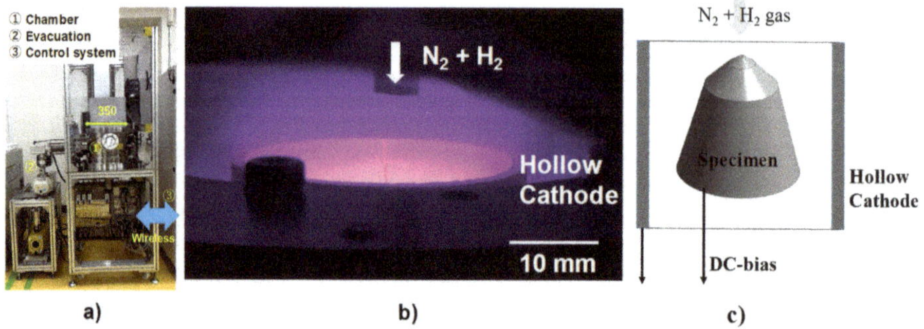

Figure 1. Down-sized plasma nitriding for surface treatment of the miniature mechanical elements. (**a**) Experimental apparatus, (**b**) hollow cathode to intensify the ion density in $N_2 + H_2$ plasmas, and (**c**) illustration on the hollow cathode device.

The hollow cathode device was utilized to increase the N_2^+ ion and NH-radical densities in the experimental setup to work in both modes. Figure 1b depicts an experimental setup for nitriding of a single micro-nozzle. As also illustrated in Figure 1c, $N_2 + H_2$ mixture gas was introduced to this DC-biased hollow. Since the DC-bias was also applied to this micro-nozzle, the ignited RF (Radio Frequency)-plasma was also confined inside the micro-nozzle by the hollow cathode effect.

2.2. Nitriding Conditions

The plasma nitriding conditions are summarized in Table 1. The DC-bias is parametrically varied to investigate the sputtering effect on the nitriding process. After [9], the nitrogen and hydrogen flow rate ratio was controlled to be constant by 160 mL/min for nitrogen and 30 mL/min for hydrogen, respectively. In situ plasma diagnosis in [9] revealed that the highest NH radical density against N_2^+ ion density was yielded around this flow rate ratio.

Table 1. High-density plasma nitriding conditions.

Item	Parameters
RF-Voltage	250 V
DC-bias	−300 V, −400 V, −500 V
Pressure	70 Pa
Temperature	673 K
Duration	7.2 ks

2.3. Specimens

Austenitic stainless steel type AISI316 plates and micro-nozzles were prepared for this low-temperature plasma nitriding. Figure 2a depicts a typical micro nozzle for a dispensing system with the outlet diameter of 1 mm at the top. AISI316 plate with a diameter of 25 mm and thickness of 5 mm, was employed to describe the nitriding and hardening behavior by using this table-top plasma nitriding system. Two types of micro-springs were also prepared for nitriding to improve their stiffness, as shown in Figure 2b.

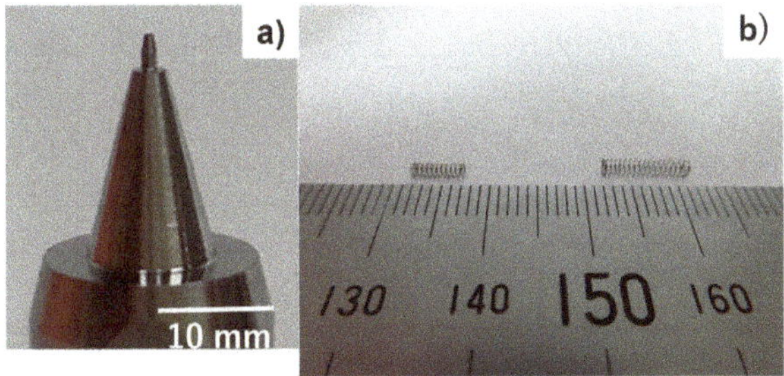

Figure 2. Miniature, mechanical parts for plasma nitriding. (**a**) Micro-nozzle with the outlet diameter of 1 mm, and (**b**) micro-springs for medical applications.

2.4. Observation and Measurement

SEM (Scanning Electron Microscopy; JSDM-IT300LV, JEOL Ltd., Akishima, Tokyo, Japan) with EDX (Electron Dispersive X-Ray Spectroscopy; Pegasus, EDAX, Inc., Minato-ku, Tokyo, Japan) were utilized to describe the nitrided specimen and to analyze the nitrogen solute content distribution on the surfaces. Micro-Vickers hardness testing (HM-210C; Mitsutoyo Co., Ltd., Kawasaki, Japan) was also employed to measure the surface hardness for various applied weights.

3. Results

3.1. Nitriding Behavior of AISI316 Substrates

AISI316 plate was nitrided at 673 K for 7.2 ks by parametrically varying the DC-bias as listed in Table 1. The nitrogen ion density as well as the NH radical density increases with increasing the DC-bias. More nitrogen atoms penetrate from the substrate surface and diffuse into its depth [6,9]. Figure 3a depicts the variation of measured nitrogen solute content by EDX with increasing the DC-bias.

The nitrogen solute content at the surface increases monotonically with DC-bias. In the following nitriding experiments, this DC-bias is fixed to be constant by −500 V. The surface roughing took place when this DC bias became lower than −500 V.

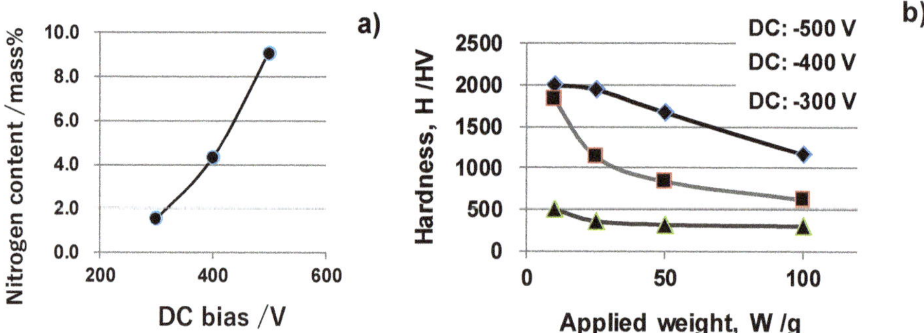

Figure 3. Effect of DC-bias on the nitrogen content and hardness. (**a**) Variation of the nitrogen solute content, and, (**b**) variation of hardness profile.

In the micro-Vickers hardness testing, the matrix hardness affects the measured hardness when the indentation depth exceeds (the nitrided layer thickness)/6 [12,16]. In fact, the measured hardness of thin nitrided layer became lower than the surface hardness by increasing the applied weight in testing [12,13]. Figure 3b depicts the variation of hardness with the increasing applied weight in hardness testing. In case of the nitriding by DC-bias of −400 V, the hardness significantly reduces toward the matrix hardness of 250 HV with an increase in the loading weights. Higher DC-bias is also needed to attain a nitrided layer sufficiently thick enough to keep high hardness in depth.

3.2. Nitriding of a Single Micro-Nozzle

AISI316 micro-nozzle specimen in Figure 2a was also plasma nitrided at 673 K for 7.2 ks. Figure 4a shows the nitrided micro-nozzle. The spot for EDX was controlled to move from the edge of the outlet toward the step of the nozzle body along the line-A in Figure 4a. Figure 4b depicts the nitrogen solute distribution along this line-A, from the top of the outlet down to the nozzle body. Both the nozzle outlet and the nozzle body surfaces are uniformly nitrided to have high nitrogen contents, except for the vicinity of a step between two regions. On the surface of this step, the electron beam in EDX significantly scattered to lower the measured intensity. This average intensity in Figure 4b corresponds to the nitrogen content by 8.6 mass%. This nitrogen content is equivalent to the measured content of 9 mass% for the nitrided AISI316 plate in Figure 3a.

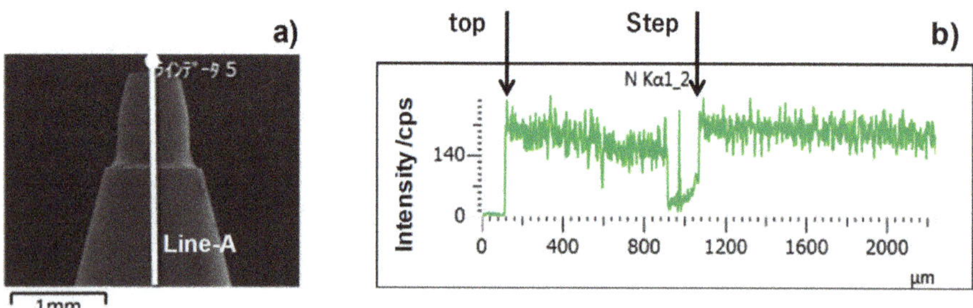

Figure 4. Plasma nitrided AISI316 micro-nozzle at 673 K for 7.2 ks. (**a**) Outlook of nitrided micro-nozzle, and, (**b**) nitrogen mapping measured from the top of the outlet down to the nozzle body.

As had been discussed in [12,13], the inner surface of the outlet channel in the micro-nozzle could be hardened when using the nozzle channel as a hollow cathode. A single micro-channel with the outlet diameter of 0.2 mm was also used as the hollow cathode in the present plasma nitriding to evaluate the hardness across the nozzle thickness.

Figure 5b depicts the hardness distribution across the thickness of nitrided micro-nozzle. The average hardness becomes more than 1000 HV0.1 at the vicinity of the channel toward the outlet. The surface hardness of the nitrided AISI316 plate by 100 g or 1 N, becomes 1000 to 1100 HV in Figure 3b. This suggests that a single micro-nozzle is nitrided and hardened to be 1000 HV not only on its outer surface but also on its inner surface even at 673 K for 7.2 ks.

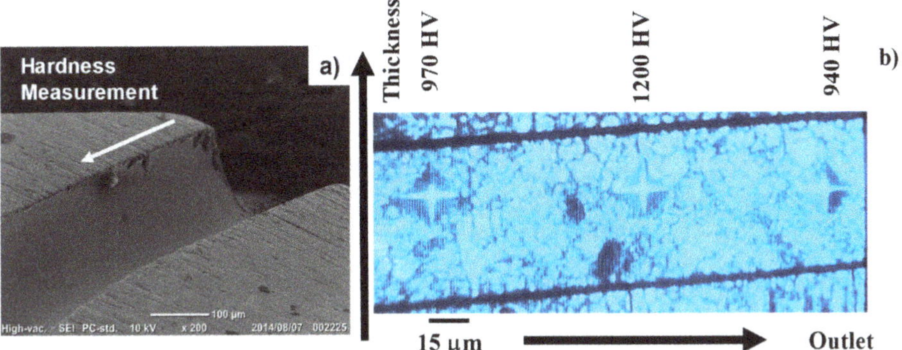

Figure 5. Inner nitriding of the nozzle hole. (**a**) A cross-section of plasma nitrided micro-nozzle with the outlet diameter of 0.2 mm, and, (**b**) micro-hardness distribution near the inner surface of nitrided micro-nozzle at 673 K for 7.2 ks.

3.3. Wettability Control by Plasma Nitriding

Bare stainless steel is hydrophilic as its contact angle against the pure water is only 40°. This is because the metallic surface including the stainless steels has higher surface energy. The contact angle measurement was performed to modify this surface property by nitriding the micro-nozzle for dispensing the droplet from the outlet hole with low adhesion. Figure 6 depicts the pure water, swelling on the nitrided AISI316 plate at 673 K for 7.2 ks. The original hydrophilic surface changes itself to a hydrophobic surface with the contact angle (θ) of 104°. This suggests that this micro-nozzle outlet could be hydrophobic enough to dispense the droplet with low adhesion to the outlet [12,15].

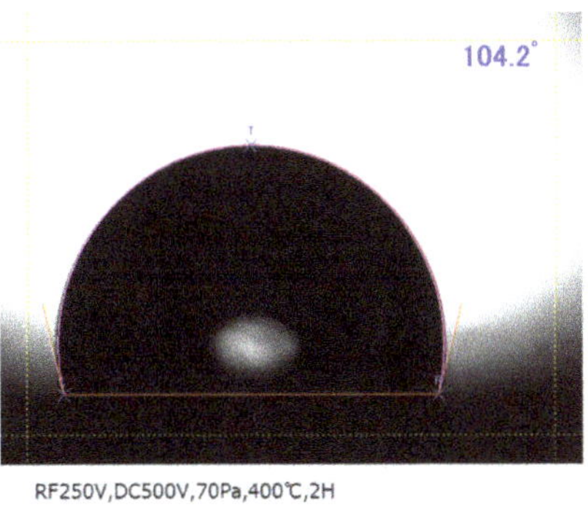

Figure 6. Measurement of the contact angle for a pure water droplet on the nitrided AISI316 surface.at 673 K for 7.2 ks.

3.4. Simultaneous Nitriding of Micro-Nozzles

This table-top plasma nitriding system can simultaneously nitride multiple micro-nozzles in a stack. Using the rotating mechanism in Figure 7a, twenty-four micro-nozzles were nitrided at 673 K for 7.2 ks at the same plasma sheath conditions. EDX was utilized to measure the nitrogen content for each micro-nozzle in a single stack.

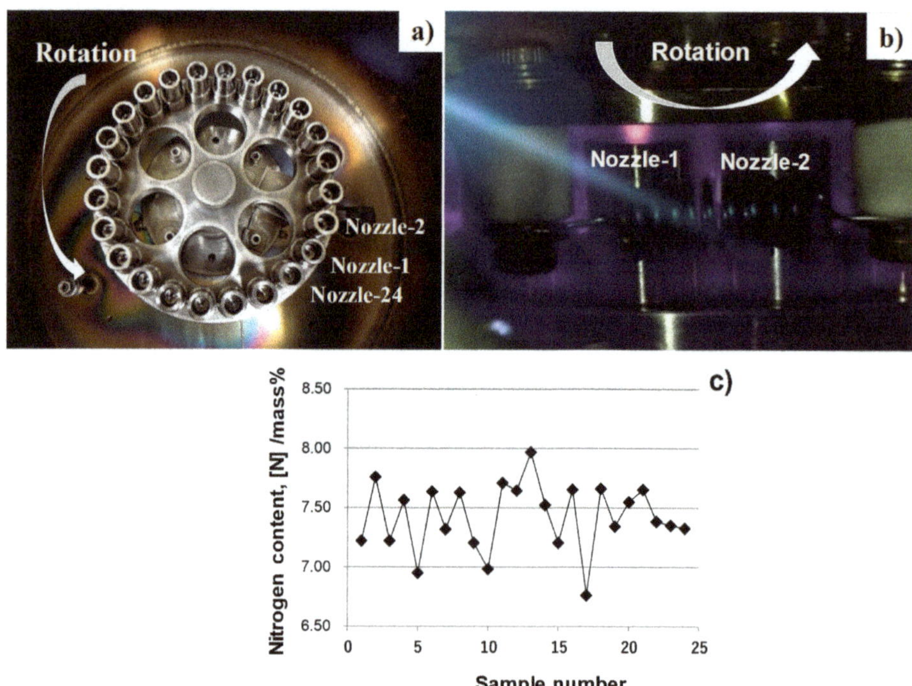

Figure 7. Simultaneous nitriding behaver of 24 micro-nozzles. (a) Experimental setup of 24 micro-nozzles into the DC-biased plate with heaters, (b) An experimental set-up for simultaneous plasma-nitriding of 24 micro-nozzles, and, (c) Statistical distribution of the surface nitrogen content among 24 nitrided micro-nozzles at 673 K for 7.2 ks.

Three positions were selected for this EDX analysis on each micro-nozzle surface; e.g., the top of the outlet, the step between the outlet and body, and, the center of the body in Figure 2a. Figure 7a shows the initial setup of 24 micro-nozzles into the DC-biased plate with electric hears. As depicted in Figure 7b, the stack of micro-nozzles is rotated in the plasma sheath. Figure 7c shows the measured nitrogen content for each micro-nozzle by EDX. A deviation of average nitrogen contents is low among twenty-four micro-nozzles in the same stack. After statistical analysis, the mean nitrogen content is estimated to be 7.4 mass% and the standard deviation, 0.3 mass%. Nitriding took place homogeneously and is the equivalent to the nitriding of a single micro-nozzle in Figure 3.

In addition to this nitrogen mapping, the chromium content was also measured; the average content is 16.4 mass% and its standard deviation, 0.4 mass%. This proves that chromium content in these nitrided micro-nozzles remains the same as before nitriding and that no CrN is precipitated by this plasma nitriding.

3.5. Simultaneous Nitriding of Micro-Springs

The micro-spring constant (k) is determined by its geometry and number of turns in its length. Since its diameter, thickness, and length are strictly specified in each medical application, k is difficult to control once it is fabricated as a spring. Two micro-springs were prepared to demonstrate the possibility to increase their spring constants by nitriding their inner surfaces; e.g., a shorter microspring-1 and a longer micro-spring-2 as listed in Table 2.

Table 2. Geometry and dimensions of two micro-springs.

Item	Microspring-1	Microspring-2
Wire diameter (d)	0.2	0.15
Coil diameter (D)	1.7	1.5
Length (L)	5.7	9.5
Number of turns (N)	10	16

One end of micro-spring was fixed to the metallic jig, to which DC-bias was applied. Its inner cylindrical space was utilized as a hollow cathode to nitride its inner surfaces at 673 K. The duration time was also constant by 7.2 ks. It is difficult to measure the hardness and microstructure of thin wire in these micro-springs. Their spring constant is employed to evaluate the strengthening by plasma nitriding of the spring wires. The uniaxial tensile testing with small-scaled load cell was employed to measure the applied load with the minimum resolution of 10^{-3} N or 0.1 gram-weight. Figure 8 depicts the load-displacement relations between an original spring and two nitrided micro-springs in type-1. The spring constant is 9.3 mN/mm for the original microspring-1, while k = 10 mN/mm for the nitrided one. The spring constant is enhanced by 8% through the plasma nitriding. In case of the microspring-2, k = 7.3 mN/mm is increased up to 8.0 mN/mm by nitriding. This increase of spring constant by 8 to 10 % reveals that the inner surface nitriding has a significant influence on the strengthening of micro-springs.

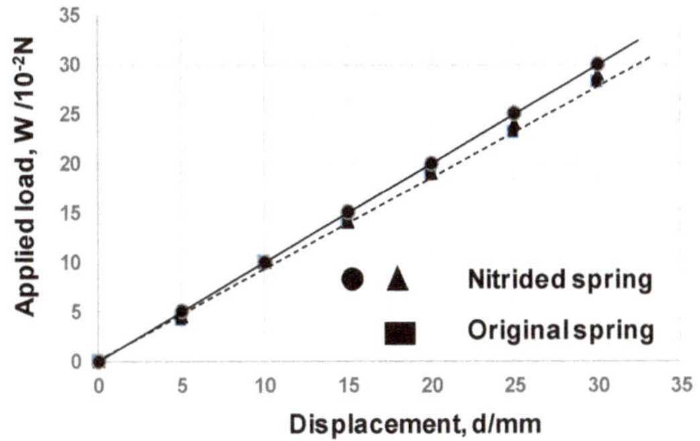

Figure 8. The load–displacement relationship for the original spring and two nitrided springs.

4. Discussion

In the deposition process by PVD (Physical Vapor Deposition) and CVD (Chemical Vapor Deposition) coating onto the inner surfaces, the adhesion probability of coating particles is high enough to coat at the vicinity of micro-nozzle inlet but significantly decreases with the distance toward the outlet [17]. Hence, it is very difficult to deposit the inner surfaces of a micro-nozzle hole smaller than 1 mm in diameter and to coat the curved surfaces in the micro-spring with uniform thickness. In the present plasma nitriding, the inner surfaces of the micro-nozzle as well as micro-springs are subjected to the plasma sheath by the hollow cathode effect. Small RF-plasma is confined to the inside of the micro-nozzle by the application of DC-bias. As shown in Figure 5, its inner surface is nitrided and hardened up to 1000 HV. This reveals that the tool life of micro-nozzles could be prolonged by mini- and micro-sized plasma control for nitriding of their inner surfaces [12,13,18].

The stiffness of the coil-springs (k) is defined by the following equation,

$$k = G \times d^4/(8 \times N \times D^3), \tag{1}$$

where G is the effective shear modulus of the stainless steel wire including the residual stresses in the fabrication of springs from wires and so forth. Since the micro-springs were self-standing on the DC-biased metallic fixture during nitriding, the geometry of coil spring and number of turns are insensitive to the perturbed modification by nitriding; e.g., $\Delta D = \Delta N = 0$ in nitriding. The ratio of perturbed stiffness (Δk) to k is then given by a variation of Equation (1),

$$\Delta k/k = \Delta G/G + 4\,(\Delta d/d), \tag{2}$$

where ΔG is a perturbed effective shear modulus and Δd, a perturbed wire diameter.

Since Δd is negligibly small in measurement, the main contribution to the increase of ($\Delta k/k$) in Equation (2) is ($\Delta G/G$) during nitriding. Consider that the torsional shear strain is induced on the cross-section of the stainless steel wire in the coil spring during its elongation or compression. The inner surface of the micro-coil is hardened by nitriding to reduce this torsional strain even when applying the same load to the micro-springs. This reveals that the equivalent shear modulus of wires is significantly increased up to 8 to 10 % by the inner nitriding to enhance the stiffness of micro-springs. In medical applications, a top-edge of tweezers and forceps as well as these micro-springs might well have higher stiffness to catch and hold the targeted objects precisely in operation. Selective nitriding of these tools by the present system provides a means to control their stiffness in part.

5. Conclusions

A table-top high-density plasma nitriding system is developed to super-saturate a single AISI316 micro-nozzle, as well as a stack of AISI316 micro-nozzles, with nitrogen. DC-bias is optimized to be -500 V to fabricate the nitrided micro-nozzles with a higher surface nitrogen content than 8 mass% and higher hardness than 1200 HV0.1 by nitriding at 673 K for 7.2 ks. In particular, the hardness at the depth of 40 μm reaches to 1000 HV0.1. Due to this nitrogen super-saturation into the depth, the original hydrophilic nozzle surface can be modified to be hydrophobic with $\theta > 100°$. This high content nitrogen super-saturation also takes place homogeneously even for a stack of 24 micro-nozzles. The present table-top system is also useful to increase the stiffness of micro-springs by 10%. Selective hardening of inner surfaces results in the significant increase of equivalent shear modulus in wires of springs. This proves that the present system works to control the local stiffness in various medical tools and parts by selective nitriding.

Author Contributions: Design, conceptualization and methodology of miniature nitriding system by T.A.; experiments by K.W. and H.M.; writing-original draft preparation by T.A.; writing—review and editing by this T.A. and H.M.

Funding: This study was financially supported in part by MEXT-project on the supporting industries in 2018.

Acknowledgments: The authors would like to express their gratitude to S. Hashimoto (TECDIA, Co. Ltd., Japan), A. Farghali and S. Kurozumi (SIT) for their help in experiments.

Conflicts of Interest: The authors declare no conflict of interest.

References

1. Ku, I.S.Y.; Reddyhoff, T.; Holmes, A.S.; Spikes, H.A. Wear of silicon surfaces in MEMS. *Wear* **2011**, *271*, 1050–1058. [CrossRef]
2. Kamenik, R.; Pilc, J.; Varga, D.; Martincek, J.; Sadilek, M. Identification of tool wear intensity during miniature machining of austenitic steels and titanium. *Proc. Eng.* **2017**, *192*, 410–415. [CrossRef]

3. Erdemir, C.D. Tribology of diamond-like carbon films: Recent progress and future prospects. *J. Phys.* **2006**, *39*, 311–327. [CrossRef]

4. Tamagaki, H.; Haga, J.; Ito, H.; Umeda, A. A MF-AC plasma enhanced CVD technology for high rate deposition of DLC. In Proceedings of the 57th Society of Vacuum Coaters (SVC) Annual Technical Conference, Chicago, IL, USA, 3–8 May 2014; pp. 215–220.

5. Dong, H. S-phase surface engineering of Fe-Cr, Co-Cr and Ni-Cr alloys. *Int. Mater. Rev.* **2011**, *55*, 65–98. [CrossRef]

6. Farghali, A.; Aizawa, T. Phase transformation induced by high nitrogen content solid solution in the martensitic stainless steels. *Mater. Trans.* **2017**, *58*, 697–700. [CrossRef]

7. Lu, S.; Zhao, X.; Wang, S.; Li, J.; Wei, W.; Hu, J. Performance enhancement by plasma nitriding at low gas pressure for 304 austenitic stainless steel. *Vacuum* **2017**, *145*, 334–339. [CrossRef]

8. Farghali, A.; Aizawa, T. Nitrogen super-saturation process in the AISI420 martensitic stainless steels by low temperature plasma nitriding. *ISIJ Int.* **2018**, *58*, 401–407. [CrossRef]

9. Aizawa, T. Low Temperature Plasma Nitriding of Austenitic Stainless Steels. In *Stainless Steels and Alloys*; IntechOpen: London, UK, 2019; Chapter 3; pp. 31–50.

10. Ferreira, L.M.; Brunatto, S.F.; Cardoso, R.P. Martensitic stainless steels low-temperature nitriding dependence on substrate composition. *Mater. Res.* **2015**, *18*, 622–627. [CrossRef]

11. Aizawa, T.; Yoshihara, S.-I. Microtexturing into AISI420 dies for fine piercing of micropatterns into metallic sheets. *J. JSTP* **2019**, *60*, 53–57. [CrossRef]

12. Aizawa, T.; Wasa, K. Plasma nitriding of inner surfaces in the mini- and micro-nozzles for joining. In Proceedings of the 11th International Conference on Multi-Material Micro Manufacture and 10th International Workshop on Microfactories (4M/IWMF2016), Kgs. Lyngby, Denmark, 13–15 September 2016; pp. 145–148.

13. Aizawa, T.; Wasa, K. Low temperature plasma nitriding of inner surfaces in the stainless steel mini-/micro-pipes and nozzles for industries. *Micromachines* **2017**, *8*, 157. [CrossRef]

14. Aizawa, T.; Sugita, Y. Distributed plasma nitriding systems for surface treatment of miniature functional products. In Proceedings of the 9th 4M/ICOMM, Milan, Italy, 31 March–2 April 2015; pp. 449–453.

15. Aizawa, T.; Morita, H.; Wasa, K. Low temperature plasma nitriding of mini-/micro-tools and parts. In Proceedings of the 2nd WCMNM, Portoroz, Slovenia, 18–20 September 2018; pp. 207–210.

16. Smith, R.L.; Sandland, G.E. An accurate method of determining the hardness of metals with particular reference to those of a high degree of hardness. *Proc. Inst. Mech. Eng.* **1992**, *102*, 623–641. [CrossRef]

17. Donovan, R.P.; Yamamoto, T.; Periasamy, R. Particle deposition, adhesion and removal. *MRS* **1993**, *315*, 3–15. [CrossRef]

18. Schoenbach, K.H. High-pressure hollow cathode discharges. *Plasma Sources Sci. Technol.* **1997**, *6*, 468–477. [CrossRef]

 *applied sciences*

Article

Tunable Silver Nanoparticle Arrays by Hot Embossing and Sputter Deposition for Surface-Enhanced Raman Scattering

Chu-Yu Huang * and Ming-Shiuan Tsai

Department of Mechanical Engineering, National Chung Hsing University, Taichung 402, Taiwan; bsmithqwer1@gmail.com
* Correspondence: tomhuang@nchu.edu.tw

Received: 19 March 2019; Accepted: 15 April 2019; Published: 19 April 2019

Featured Application: The SERS active substrates developed in this study have great potential in areas such as analytical chemistry, biochemistry, and environmental science.

Abstract: Surface-enhanced Raman scattering (SERS) spectroscopy has attracted a lot of attention over the past 30 years. Due to its extreme sensitivity and label-free detection capability, it has shown great potential in areas such as analytical chemistry, biochemistry, and environmental science. However, the major challenge is to manufacture large-scale highly SERS active substrates with high controllability, good reproducibility, and low cost. In this study, we report a novel method to fabricate uniform silver nanoparticle arrays with tunable particle sizes and interparticle gaps. Using hot embossing and sputtering techniques, we were able to batch produce the silver nanoparticle arrays SERS active substrate with consistent quality and low cost. We showed that the proposed SERS active substrate has good uniformity and high reproducibility. Experimental results show that the SERS enhancement factor is affected by silver nanoparticles size and interparticle gaps. Furthermore, the enhancement factor of the SERS signal obtained from Rhodamine 6G (R6G) probe molecules was as high as 1.12×10^7. Therefore, the developed method is very promising for use in many SERS applications.

Keywords: SERS; Surface-enhanced Raman scattering; nanosphere array; nanocone array; hot embossing; nanoimprinting

---

1. Introduction

Surface-enhanced Raman spectroscopy (SERS) technology has attracted widespread attention since its discovery in 1974 [1]. Comparing to the normal Raman scattering process, SERS is capable of enhancing Raman scattering of analytes by up to a million times or more [2]. It has the potential to provide a very fast and sensitive method of detecting chemicals and biomolecules, which is useful for applications that require fast and highly sensitive detection [3,4]. For example, Caro [5] et al. used a SERS probe for intracellular imaging, SERS signals were strong enough and could be detected even from inside cells. SERS can also be employed to study weak interaction between protein and alizarin [6].

Many SERS substrates use colloidal clusters of noble metal nanoparticles or noble metals with a rough surface to enhance SERS signals. For example, An [7] et al. used silver nanoparticles, tri-iron tetroxide, and carbon cores to form multilayered microsphere particles SERS substrate to detect pentachlorophenol (PCP), diethylhexyl phthalate (DEHP), and trinitrotoluene (TNT). In 2014, Au-Ag-S substrate developed by Cao et al. [8] was used for surface-enhanced Raman detection and photocatalytic degradation of DEHP and DEHA. Liu et al. [9] developed an alloy of gold and silver nanoparticles urchin shape (hollow Au-Ag alloy nanourchins, HAAA-NUs) as a SERS active substrate to detect 10^{-15} mol/L of DEHP. Most of these studies focus on achieving large enhancement factors, but fail to address the uniformity and reproducibility of these substrates. Due to the random distribution of nanoparticles on the substrate, the interparticle gap size is difficult to control. Therefore, it is difficult to uniformly control the generation of hot spots, resulting in large signal intensity variations, which is detrimental to quantitative analysis. Some researchers have tried to use linker molecules to control the distance between colloidal nanoparticles. For example, Anderson [10] et al. made silver nanoparticle array tethered to a silver film using an appropriate tethering linker such as a dithiol or diamine. However, these linker molecules usually require a dedicated environment (some pH levels or temperatures) to serve the desired purpose, which limits their applications. In addition, linker molecules can block the analyte and prevent it from attaching to the plasmonic surface, which results in a low SERS signal of the analyte. Furthermore, it is possible for the linker molecules to introduce background noise, making the desired SERS signal difficult to measure.

The lack of uniform, reproducible and low-cost SERS active substrates suitable for quantitative analysis remains a major obstacle to the widespread use of SERS for routine analysis [11]. Many research efforts have focused on the development of bottom-up and top-down approaches for the manufacture of controllable and reproducible metal nanostructures for SERS applications [12]. These approaches include: anodic aluminum oxide (AAO) templates assist techniques, lithographic techniques, and oblique angle deposition (OAD) technique. For the metal nanoparticle arrays created using AAO templates, Mu et al. [13] used electroless deposition and adjusted the pH and temperature of the gold plating bath to control the plating rate to achieve the desired particle size and interparticle gap. Lee [14] used densely packed nanowires fabricated from AAO templates. Nevertheless, these approaches require sacrificing the AAO template each time to produce an ordered gold nanoparticle array, which results in increased time and cost expense.

For the lithographic techniques, Sánchez-Iglesias et al. [15] uses block copolymer micellar nanolithography to create seeds for chemical growth of uniform Ag nanoparticle arrays. However, controlling seeds growth at the same rate is not easy, which makes it difficult to obtain a large-sized uniform Ag nanoparticle array. Another disadvantage of lithography is that if standard lithography is used, in order to create a few nanometer gap features, extreme ultraviolet (EUV) will be required, which is very expensive and disadvantageous for low-cost applications.

OAD technology is another method of making SERS active substrates. A number of studies have been conducted to fabricate nanostructured SERS substrates by OAD technology [16,17]. It is a simple method of fabricating SERS substrates, and these fabricated substrates exhibit good sensitivity and uniformity. However, to fabricate SERS substrates by the OAD method, one will need a custom designed electron-beam/sputtering evaporation system. In addition, the vapor incident angle to the substrate normal is around 86°, which means that only a small portion of the material is deposited on the substrate, which results in increased manufacturing costs. Lastly, the anisotropic character of the optical properties, as well as the SERS responses of the substrates [4], lead to the need of special angle coupling of the excitation laser which makes it not convenient to use.

In this study, we used a hot embossing technique to create a uniform and closely packed nanocone arrays on a substrate. Furthermore, by utilizing the self-aggregating nature of nanoscale metal materials, we were able to sputter uniformly distributed silver nanoparticles at the tip of the nanocone array structure. Using this method, we can create a uniform silver nanosphere array with very small gaps between the nanoparticles to create hot spots evenly. We used this structure as our SERS active substrate for rapid detection of low concentrations of analytes. The analytes tested in this study included Rhodamine 6G (R6G) and DEHP. R6G is a highly fluorescent rhodamine family dye, and it is commonly used in SERS experiments as a probe molecule. DEHP is a PVC plasticizer commonly used in food-related containers, such as plastic packaging (PVC cling film), plastic bags, plastic bowls, and plastic cups. According to the Taiwan Ministry of Health, DEHP is defined as an environmental hormone and is defined as a Class 2B carcinogen. The ability to quickly detect low concentrations of DEHP helps prevent people from eating food contaminated with DEHP, which will benefit people's health.

The developed method can be used to fabricate disposable, low cost and highly sensitive SERS substrates. The results show a strong and uniform SERS enhancement effect. It not only opens up possibilities for using SERS in routine food safety analytics but also helps SERS to be widespread in many other applications as well.

2. Methodology

The method used for the fabrication of silver nanosphere arrays decorated on a polycarbonate substrate with self-organized, hexagonal close-packed nanocone is illustrated schematically in Figure 1. First, a nanocone-shaped groove array structure of aluminum substrate was fabricated by using anodic aluminum self-assembly technique [18–20]. To form a cone-shaped cavity, the anodization and pore widening processes were alternately repeated several times, which creates a top widened and bottom narrow holes, the diameter of these holes are reduced gradually from the top to the bottom of these holes. The fabrication process of the nanocone cavity array of the aluminum substrate including electrolytic polishing, anodic treatment, removal of anodic aluminum oxide. Briefly, a 99.999% purity aluminum substrate was polished in a solution mixed with perchloric acid and anhydrous alcohol in a volume ratio of 1:3.5 and applied with a voltage of 25 V for 2 min. The polished aluminum substrate was anodic oxidation treated in 0.3 M oxalic acid solution with 50 volts for two hours. The anodic aluminum oxide layer was removed by placing the substrate in a 5 wt% phosphoric acid solution at temperature 35 °C for 1 h. The anodization and anodic aluminum oxide removal processes were alternately repeated 5 times to obtain the desired nanocone cavity array structure.

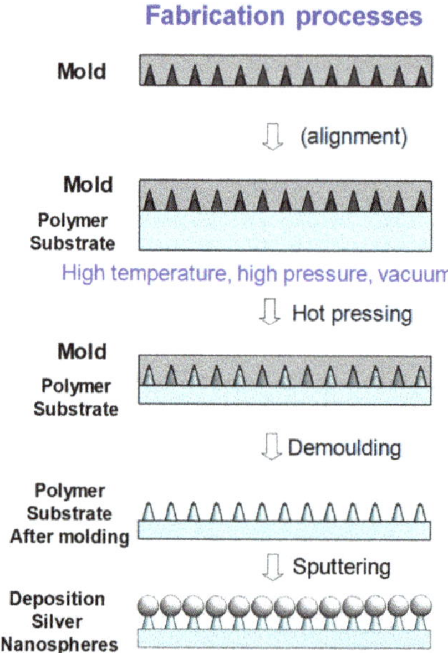

Figure 1. Schematic illustration of the fabrication processes. The nanostructured mold was pressed on a Polycarbonate (pc) plastic substrate using a hot-press molding machine. The nanostructured-nickel mold was heated and pressed on the pc surface for several minutes. After cool down to room temperature, the pc substrate with replicated nanostructure was released from the master mold.

Then a Nickel replica of the nanocone-shaped groove array was obtained by electroforming the nanostructured AAO layer. The nanocone-shaped groove array structured Nickel mold was used as a master mold. Figure 1. illustrated the hot pressing process. The nanocone-shape groove structured Nickel mold was imprinted on a Polycarbonate(pc) plastic substrate using a hot-press molding machine for heating and pressing the nanostructured nickel mold on pc surface for several minutes. After cool down to room temperature, the pc substrate was released from the master mold. Using this method, we were able to produce many nanocone arrays (moth eye like structure as illustrated in Figure 1) structured pc substrates with low-cost and same quality.

The silver nanoparticles were formed on top of nanocones using sputtering. A dc magnetron sputtering system (Cressington 108 Sputter Coater) was used. Due to the self-aggregating nature of nanoscale metallic materials, the sputtered silver nanoparticles tend to self-aggregate at the tip of the nanocones, and finally form silver nanospheres at the tip of each nanocone (as illustrated in Figure 1). The diameter and interspacing of the nanospheres can be varied by adjusting the sputtering parameters. In order to obtain a uniform and a closely packed array of nanospheres, the appropriate sputtering parameter is needed. In this study, we varied the sputtering duration while the sputtering current and gas pressure were kept at constant (20 mA and 0.02 mbar respectively). The nanostructured polycarbonate substrates were sputtered for different durations to obtain nanospheres with various diameters, and at the same time, the interparticle gaps of the nanospheres were changed accordingly.

3. Experiments

A scanning electron microscope (SEM, JEOL JSM-7800F) was used to examine the surface morphology of the nanocone array and the silver nanosphere array fabricated on the nanocone array. EDS analysis was also performed using the same SEM system. UV–vis absorption spectrum measurements were performed using a HITACHI U3900 spectrophotometer at room temperature in the range of 400–800 nm. The SERS active substrates were placed perpendicular to the light beam to obtain the absorption spectra.

For the SERS studies, the developed surface enhanced Raman scattering sensors were tested using analytes including R6G and DEHP. The developed nanostructured SERS substrates were first immersed in different concentrations of R6G solution and DEHP methanol solution for 12 h, then rinsed with deionized water and dried in air. The SERS spectrometer used was a microscope Raman spectrometer (Tokyo Instruments. Nanofinder 30) with 632.8 nm HeNe laser excitation, and the power of the laser source was set to 0.1 mW. The focused laser spot on the surface of the sample has a diameter of approximately 2 μm and a penetration depth of about 4 μm. The integration time per measurement is 10 s.

4. Results and Discussion

4.1. Characterization of the Developed Substrates

SEM image showed in Figure 2 are the hot imprinted nanocone array on the PC substrate. The spacing between the nanocones is about 130 nm, the width of the nanocone at the base is about 90 nm, and the height of the nanocone is about 140 nm. Silver nanospheres were then deposited by sputtering and the size of these nanospheres was controlled by sputtering duration. Figure 3a–f show SEM images of these silver nanospheres with different sputtering durations of 50, 100, 150, 200, 250 and 300 s, respectively. As can be seen from these figures, the silver nanospheres are uniformly decorated on top of the nanocone array substrate. It can be seen that the longer the duration, the larger the diameter of these nanospheres. However, If the sputtering duration is too long, the nanospheres will start to merge. Figure 3d shows that the diameter of the nanospheres is about 120 nm when the sputtering duration is set to 200 s. This results in a gap between the nanospheres in the range of a few nanometers and is therefore suitable for generating many intense hot spots, which is very helpful for enhancing the SERS effect. Figure 3e,f shows that when the sputtering duration is set to 250 and 300 s, the silver nanospheres start to merge together.

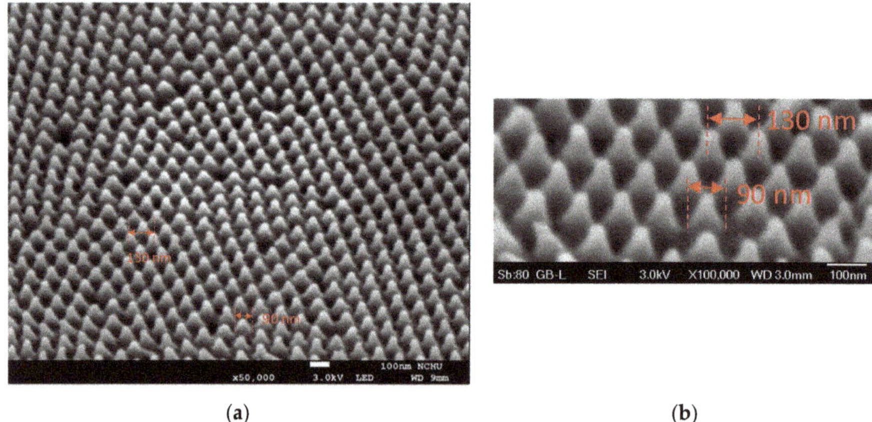

(a)   (b)

Figure 2. (a,b) shows the 45 degree view of the hot imprinted nanocone array. (b) shows the nanocone image at 100,000× magnification for easier to observe the morphology of nanocone.

We show in Figure 4 the relationship between the sputtering time and the size of the obtained nanospheres and the gap between the nanospheres. It can be observed that before the nanospheres merge, there is a linear relationship between the size of the nanospheres and the gap them. Therefore, we can adjust the sputtering time accordingly to change the gap size of the nanospheres to optimize the SERS enhancement. After carefully adjusting the sputtering parameters, we were able to make a uniform array of silver nanospheres with a few nanometers interspacing. These uniform and closely packed silver nanospheres can produce a large number of uniformly distributed plasmonic hotspots, which results in high surface enhancement factors and good uniformity. The advantages of this method are that we can prepare a uniform array of nanocones very quickly and at low cost, and the size of the silver nanoparticles can be easily adjusted to change the gap size between them.

EDS was used to analyze the silver nanospheres decorated nanocone array substrate. The EDS spectrum (Figure 5b) shows that the SERS active substrate consisted of 77.5% Ag and 11.48% C. The EDS mapping of Ag of SERS substrate is shown in Figure 5c. From the figure, we can see that the elemental mapping of Ag demonstrates a uniform distribution of the Ag nanosphere on the SERS active substrate.

The UV–vis absorption spectrum was used to analyze the surface plasmon resonance (SPR) characteristics of the proposed SERS active substrates. A series of absorption spectra for the SERS substrate with different sputtering durations (different nanospheres diameters) are shown in Figure 6. As the sputtering duration increase from 100 to 200 s, the diameters of the nanosphere increase from 70 to 125 nm, the absorption peak shifts to the red from 450 to 680 nm. We can observe that the SERS substrate with 200 s sputtering duration shows a broad peak at a wavelength of ~680 nm, and it is closer to the laser excitation wavelength of 632.8 nm. We believe that this absorption peak is caused by the excitation of surface plasmons. Ideally, it would be optimal to use a laser with a wavelength close to 680 nm. The reason we used the 632.8 nm laser is that our Raman system only has a 632.8 nm wavelength laser. If a 680 nm laser is used as the excitation source, it can be expected that the SERS substrate can have a better performance. Comparing these SERS substrates with different sputtering durations, the SERS substrate with a 200 s sputtering duration also showed the highest SERS signal intensity in subsequent experiments (see Figure 7).

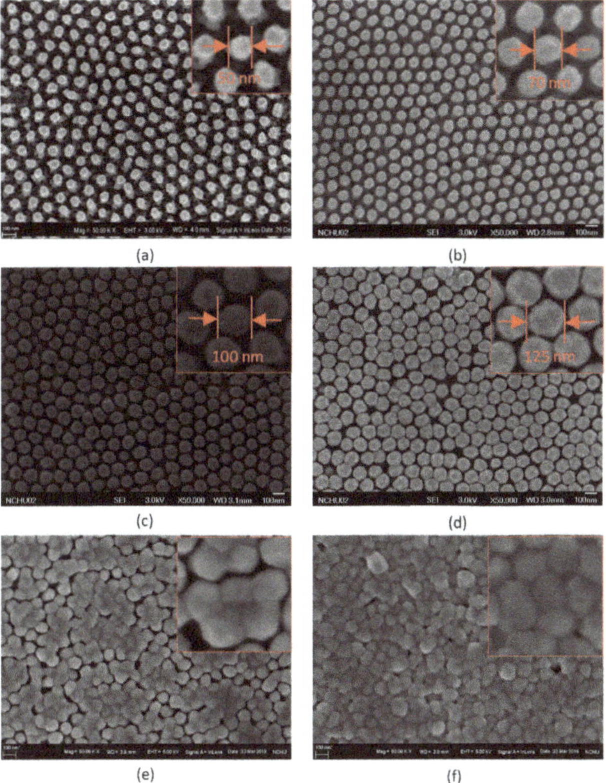

Figure 3. The SEM images show silver nanospheres uniformly distributed on the nanocone array substrate. (**a–f**) show different sputtering duration of 50, 100, 150, 200, 250 and 300 s, respectively. Top right corner of each image shows the silver nanospheres at higher magnification for easier to observe their morphology. It can be seen that the longer the duration, the larger the diameter of these nanospheres. If the sputtering duration is too long, the nanospheres will start to merge. By carefully adjusting the sputtering duration, we can fine-tune the diameter of the nanosphere so that we can control the gaps between the nanospheres in the range of a few nanometers.

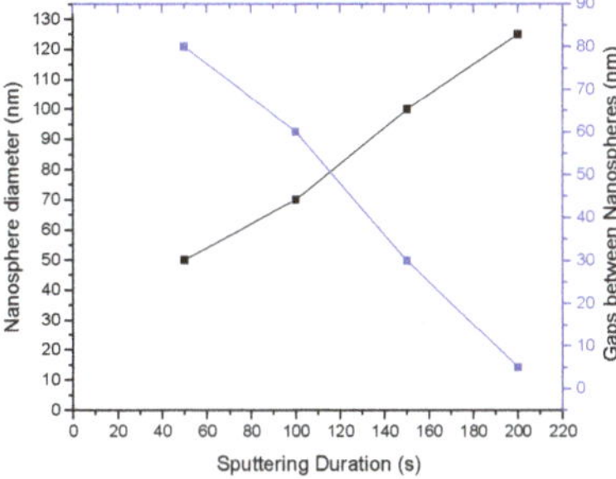

Figure 4. The relationship between the sputtering duration and the size of the obtained nanospheres and the gaps between the nanospheres.

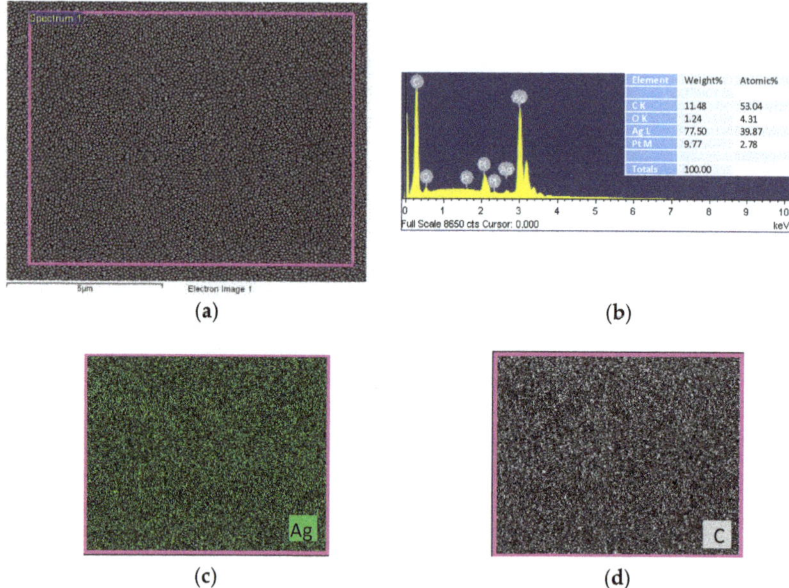

(a) (b)

(c) (d)

Figure 5. (**a**) Shows the SEM image of the SERS active substrate used for EDS analysis (**b**) Shows the EDS spectrum, it shows that the SERS active substrate consisted of 77.5% Ag and 11.48% C. (**c**) shows the EDS mapping of Ag of the SERS active substrate. (**d**) Shows the EDS mapping of C of the SERS active substrate.

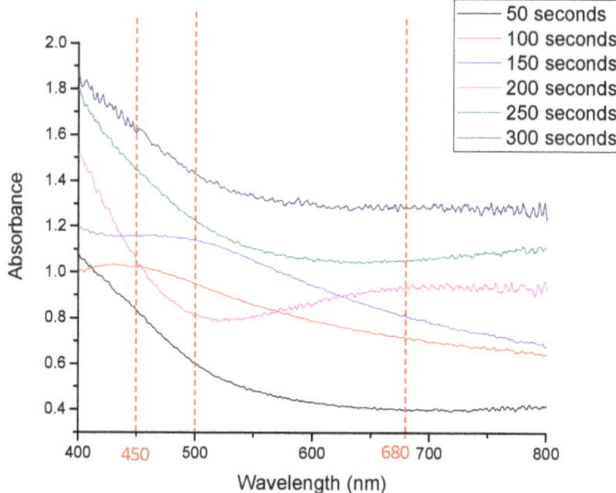

Figure 6. UV-vis absorption spectra of SERS active substrate with different sputtering durations of 50, 100, 150, 200, 250 and 300 s, respectively.

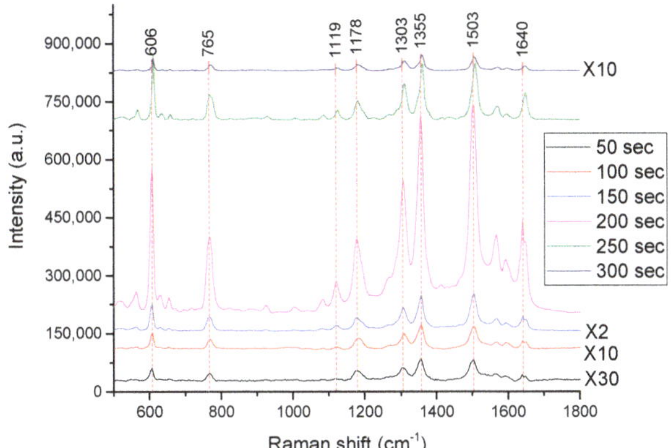

Figure 7. The SERS spectra of the R6G molecules on the developed SERS active substrate with sputtering durations of 50, 100, 150, 200, 250 and 300 s, respectively.

4.2. Performance of the SERS Substrates

In order to see how the different substrate morphologies affect the SERS signal, we measured SERS spectra of R6G solution using substrates with different sputtering times. Figure 7 shows the results of measured SERS spectra of a 10^{-6} M concentration of R6G solution using substrates with different sputtering times. We can see that before the nanospheres merge, as the sputtering duration becomes longer, the gap between the nanospheres becomes smaller and the SERS signal intensity increases rapidly. After the nanospheres merge, the SERS intensity begins to decline. By plotting the peak intensities of 1503 cm^{-1} in the R6G SERS spectra against the corresponding inter-nanosphere gaps, we found that before the nanospheres merge, the intensity of the SERS signal is a logarithmic function of the inter-nanosphere gap, as shown in Figure 8. This result is consistent with the findings of Mu et al. [13]. It also indicates that when the inter-nanosphere gaps are as small as 10 nm or less, we can obtain strong hot spots, so that a good SERS enhancement effect can be obtained.

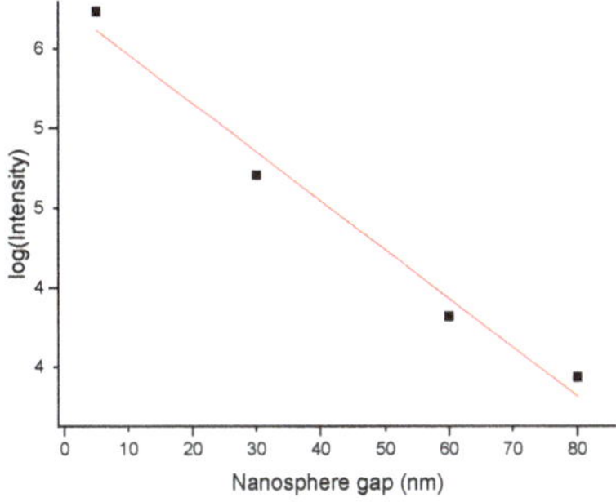

Figure 8. Intensities of the 1503 cm^{-1} SERS signal of R6G recorded as a logarithmic function of the nanosphere gap size.

Figure 9 shows the SERS spectra of blank SERS substrate, bulk R6G, and R6G solutions with concentrations of 1×10^{-6} M, 1×10^{-7} M, 1×10^{-8} M, and 1×10^{-9} M, respectively, using our SERS substrate with sputtering duration of 200 s. The Raman band assignment of the main peaks for R6G were given in Figure 9. Characteristic peaks of R6G at 607, 765, 1176, 1304, 1355, 1503 and 1640 cm^{-1} were clearly observed from these spectra. These spectra demonstrate the good enhancement factor of the SERS substrate we have developed. As can be seen from the figure, the signal from the SERS system itself is very small compared to the SERS signal of the analyte and does not mask the measured analyte signal. One might question that polycarbonate is a Raman-active compound that can interfere with molecular signals for SERS detection. Here we show that proper decoration of the polycarbonate surface with silver metal can suppress the background Raman signal of the polycarbonate substrate and also prevent melting and degradation of the polycarbonate substrate when exposed to the focused laser beam. This result is consistent with the findings of Geissler et al. [21].

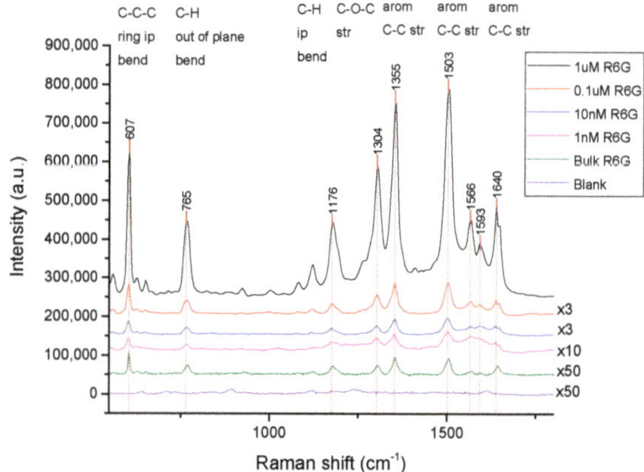

Figure 9. The SERS spectra of blank SERS substrate, bulk R6G, and various concentrations of R6G solution (1×10^{-6} M, 1×10^{-7} M, 1×10^{-8} M, and 1×10^{-9} M, respectively) with our SERS substrates. The Raman band assignment of the main peaks of R6G is given in the figure.

To estimate the enhancement factor (EF) values of R6G, the following commonly used formula [2] was used:

$$EF = (I_{SERS}/I_{bulk})/(N_{bulk}/N_{SERS}) \tag{1}$$

where I_{SERS} and I_{bulk} are the vibration intensity of the SERS and the normal Raman spectra of R6G molecules (the vibration band at 1503 cm^{-1} was selected in this study), respectively. N_{SERS} and N_{bulk} are the number of molecules irradiated by laser spots under SERS and normal Raman of bulk sample condition, respectively. The average surface density of rhodamine 6G (R6G) molecules in densely packed monolayers is reported to be approximately one R6G molecule per 4 nm^2 [22]. Then, the surface coverage of the R6G monolayer on the SERS substrate is 4.15×10^{-11} mol/cm^2 (molecular density = $1/(4 \times 10^{-14})/(6 \times 10^{23})$ mol/cm). For the laser spot of 2 µm in diameter, the N_{SERS} has a value of 1.3×10^{-18} mol ($N_{SERS} = 4.15 \times 10^{-11}$ mol/cm$^2 \times \pi \times 1$ µm^2). For bulk samples, the sample volume is the product of the laser spot area (about 2 µm diameter) and the penetration depth of the focused laser beam (~2 µm). Assuming that the density of the bulk R6G is 0.79 g·cm^{-3}, N_{bulk} can be calculated as 1.04×10^{-14} mol ($N_{bulk} = 0.79$ g·cm$^{-3} \times \pi \times 1$ µm$^2 \times 4$ µm/479.01 g·mol$^{-1} = 2.08 \times 10^{-14}$ mol). For the vibration band at 1503 cm^{-1}, the I_{SERS} and I_{bulk} are 554082 and 791, respectively. Therefore, the EF was calculated to be 1.12×10^7, which is quite high compared to other studies [23,24]. This result is assuming the SERS substrate is densely packed with R6G molecules after immersed in 10^{-6} M R6G solution for 12 h. However, the R6G is likely to cover the substrate at a lower density. Therefore, the SERS enhancement factor calculated here is the lower limits.

In order to calculate the reproducibility of the SERS active substrate, SERS spectra of R6G were obtained at 10 random locations of the SERS substrates with sputtering duration of 200 s. Figure 10a shows the measured spectra from 10 random positions. The SERS intensities of the Raman characteristic peak of the 1503 cm^{-1} band from these locations are further displayed in Figure 10b. The relative standard deviation (RSD) of intensities of the 1503 cm^{-1} Raman characteristic peak was calculated to be 7.28%, which is relatively low compared to other studies [25–27], indicating good uniformity and reproducibility of the fabricated SERS substrates.

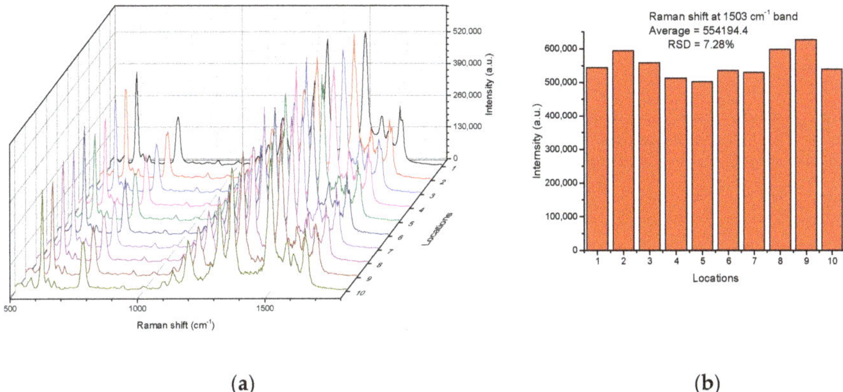

(a)   **(b)**

Figure 10. (**a**) The SERS spectra of R6G measured from 10 random positions of the SERS substrate. (**b**) The SERS intensities of the Raman characteristic peak of the 1503 cm^{-1} band from these locations.

The developed SERS substrates were also tested with low concentration DEHP solution. Figure 11 shows measured SERS spectra for DEHP concentrations ranging from 1000 ppm to 2 ppm on SERS substrates. The Raman band assignment of the main peaks for DEHP are given in Figure 11. Raman peaks of DEHP at 1055 and 1433 cm^{-1} are aromatic ring breathing and C=C stretch, respectively. The characteristic peaks of DEHP at 1055, 1433 cm^{-1} are clearly observed from the spectrum. The SERS substrates were found to be able to detect a low concentration (2 ppm) DEHP methanol solution.

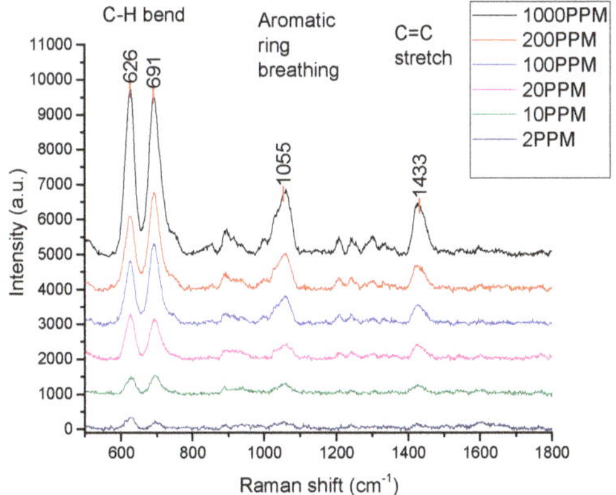

Figure 11. SERS spectra of different concentrations of DEHP (from 1000 ppm to 2 ppm) on the nanostructured SERS substrate. The Raman band assignment of the main peaks of DEHP is given in the figure.

5. Conclusions

In conclusion, we have presented a simple and inexpensive method for synthesizing a uniform nanostructured SERS active substrate with tunable nanosphere diameter and interparticle gaps. The SERS active substrate was fabricated through hot embossing and sputtering deposition. The nanosphere diameter and interparticle gaps were controlled by adjusting the sputtering duration to obtain a uniform and a densely packed silver nanosphere array substrate. According to the SERS experiments, the nanostructured polycarbonate substrate combined with uniformly distributed and closely packed silver nanospheres showed a strong SERS enhancement factor up to 1.12×10^7. The developed SERS substrate has been further applied to the sensitive detection of DEHP molecules and demonstrated its practical application potential in food contamination detection. Therefore, the developed SERS active substrate is highly promising for use in rapid chemical and biomolecular detection applications.

Author Contributions: Conceptualization, C.-Y.H.; methodology, C.-Y.H.; investigation, M.-S.T.; data curation, M.-S.T.; writing—original draft preparation, C.-Y.H.; writing—review and editing, C.-Y.H.; supervision, C.-Y.H.; project administration, C.-Y.H.; funding acquisition, C.-Y.H.

Funding: This research was funded by Ministry of Science and Technology, Taiwan, grant number MOST 107-2221-E-005-045.

Acknowledgments: The authors would like to thank the Ministry of Science and Technology of the Republic of China, Taiwan, for financially supporting this research under Contract No. Grant MOST 107-2221-E-005-045. We would also like to thank Vida Biotechnology Co., Ltd. for their help in the fabrication of PC substrates.

Conflicts of Interest: The authors declare no conflict of interest.

References

1. Fleischmann, M.; Hendra, P.J.; McQuillan, A.J. Raman spectra of pyridine adsorbed at a silver electrode. *Chem. Phys. Lett.* **1974**, *26*, 163–166. [CrossRef]

2. Le Ru, E.C.; Blackie, E.; Meyer, M.; Etchegoin, P.G. Surface enhanced raman scattering enhancement factors: A comprehensive study. *J. Phys. Chem.* **2007**, *111*, 13794–13803. [CrossRef]

3. Bantz, K.C.; Meyer, A.F.; Wittenberg, N.J.; Im, H.; Kurtuluş, Ö.; Lee, S.H.; Lindquist, N.C.; Oh, S.H.; Haynes, C.L. Recent progress in SERS biosensing. *Phys. Chem. Chem. Phys.* **2011**, *13*, 11551–11567. [CrossRef] [PubMed]

4. Alvarez-Puebla, R.A.; Liz-Marzán, L.M. SERS-Based diagnosis and biodetection. *Small* **2010**, *6*, 604–610. [CrossRef] [PubMed]

5. Caro, C.; Quaresma, P.; Pereira, E.; Franco, J.; Pernia Leal, M.; García-Martín, M.L.; Royo, J.L.; Oliva-Montero, J.M.; Merkling, P.J.; Zaderenko, A.P.; et al. Synthesis and characterization of elongated-shaped silver nanoparticles as a biocompatible anisotropic SERS probe for intracellular imaging: Theoretical modeling and experimental verification. *Nanomaterials* **2019**, *9*, 256. [CrossRef] [PubMed]

6. Cañamares, M.V.; Sevilla, P.; Sanchez-Cortes, S.; Garcia-Ramos, A.V. Surface-enhanced Raman scattering study of the interaction of red dye alizarin with ovalbumin. *Biopolym. Orig. Res. Biomol.* **2006**, *82*, 405–409. [CrossRef]

7. An, Q.; Zhang, P.; Li, J.M.; Ma, W.F.; Guo, J.; Hu, J.; Wang, C.C. Silver-coated magnetite–carbon core–shell microspheres as substrate-enhanced SERS probes for detection of trace persistent organic pollutants. *Nanoscale* **2012**, *4*, 5210–5216. [CrossRef] [PubMed]

8. Cao, Q.; Che, R. Tailoring Au–Ag–S Composite Microstructures in One-Pot for Both SERS Detection and Photocatalytic Degradation of Plasticizers DEHA and DEHP. *ACS Appl. Mater. Interfaces* **2014**, *6*, 7020–7027. [CrossRef] [PubMed]

9. Liu, Z.; Yang, Z.; Peng, B.; Cao, C.; Zhang, C.; You, H.; Xiong, Q.; Li, Z.; Fang, J. Highly Sensitive, Uniform, and Reproducible Surface-Enhanced Raman Spectroscopy from Hollow Au-Ag Alloy Nanourchins. *Adv. Mater.* **2014**, *26*, 2431–2439. [CrossRef]

10. Anderson, D.J.; Moskovits, M. A SERS-Active System Based on Silver Nanoparticles Tethered to a Deposited Silver Film. *J. Phys. Chem. B* **2006**, *110*, 13722–13727. [CrossRef]

11. Cialla, D.; März, A.; Böhme, R.; Theil, F.; Weber, K.; Schmitt, M.; Popp, J. Surface-enhanced Raman spectroscopy (SERS): Progress and trends. *Anal. Bioanal. Chem.* **2012**, *403*, 27–54. [CrossRef] [PubMed]

12. Love, S.A.; Marquis, B.J.; Haynes, C.L. Recent advances in nanomaterial plasmonics: Fundamental studies. *Appl. Spectrosc.* **2008**, *62*, 346A–362A. [CrossRef]

13. Mu, C.; Zhang, J.P.; Xu, D. Au nanoparticle arrays with tunable particle gaps by template-assisted electroless deposition for high performance surface-enhanced Raman scattering. *Nanotechnology* **2010**, *21*, 015604. [CrossRef] [PubMed]

14. Lee, S.J.; Guan, Z.; Xu, H.; Moskovits, M. Surface-Enhanced Raman Spectroscopy and Nanogeometry: The Plasmonic Origin of SERS. *J. Phys. Chem.* **2007**, *111*, 17985–17988. [CrossRef]

15. Sánchez-Iglesias, A.; Aldeanueva-Potel, P.; Ni, W.; Pérez-Juste, J.; Pastoriza-Santos, I.; Alvarez-Puebla, R.A.; Mbenkum, B.N.; Liz-Marzán, L.M. Chemical seeded growth of Ag nanoparticle arrays and their application as reproducible SERS substrates. *Nano Today* **2010**, *5*, 21–27. [CrossRef]

16. Nuntawong, N.; Eiamchai, P.; Wong-ek, B.; Horprathum, M.; Limwichean, K.; Patthanasettakul, V.; Chindaudom, P. Shelf time effect on SERS effectiveness of silver nanorod prepared by OAD technique. *Vacuum* **2013**, *88*, 23–27. [CrossRef]

17. Singh, J.P.; Chu, H.; Abell, J.; Tripp, R.A.; Zhao, Y. Flexible and mechanical strain resistant large area SERS active substrates. *Nanoscale* **2012**, *4*, 3410–3414. [CrossRef] [PubMed]

18. Runge, J.M. *The Metallurgy of Anodizing Aluminum*; Springer International Publishing: Cham, Switzerland, 2018.

19. Brzózka, A.; Szeliga, D.; Kurowska-Tabor, E.; Sulka, G.D. Synthesis of copper nanocone array electrodes and its electrocatalytic properties toward hydrogen peroxide reduction. *Mater. Lett.* **2016**, *174*, 66–70. [CrossRef]

20. Huang, C.Y.; Tsai, M.S. Fabrication of 3D nano-hemispherical cavity array plasmonic substrate for SERS applications. *Int. J. Optomechatron.* **2018**, *12*, 40–52. [CrossRef]

21. Geissler, M.; Li, K.; Cui, B.; Clime, L.; Veres, T. Plastic Substrates for Surface-Enhanced Raman Scattering. *J. Phys. Chem. C* **2009**, *113*, 17296–17300. [CrossRef]

22. Freeman, R.G.; Grabar, K.C.; Allison, K.J.; Bright, R.M.; Davis, J.A.; Guthrie, A.P.; Hommer, M.B.; Jackson, M.A.; Smith, P.C.; Walter, D.G.; et al. Self-Assembled Metal Colloid Monolayers: An Approach to SERS Substrates. *Science* **1995**, *267*, 1629–1632. [CrossRef] [PubMed]

23. Araújo, A.; Caro, C.; Mendes, M.J.; Nunes, D.; Fortunato, E.; Franco, R.; Águas, H.; Martins, R. Highly efficient nanoplasmonic SERS on cardboard packaging substrates. *Nanotechnology* **2014**, *25*, 415202. [CrossRef] [PubMed]

24. Araújo, A.; Pimentel, A.; Oliveira, M.J.; Mendes, M.J.; Franco, R.; Fortunato, E.; Águas, H.; Martins, R. Direct growth of plasmonic nanorod forests on paper substrates for low-cost flexible 3D SERS platforms. *Flex. Print. Electron.* **2017**, *2*, 014001. [CrossRef]

25. Caro, C.; Gámez, F.; Zaderenko, A.P. Preparation of Surface-Enhanced Raman Scattering Substrates Based on Immobilized Silver-Capped Nanoparticles. *J. Spectrosc.* **2018**, *2018*, 1–9. [CrossRef]

26. Wu, W.; Liu, L.; Dai, Z.; Liu, J.; Yang, S.; Zhou, L.; Xiao, X.; Jiang, C.; Roy, V.A. Low-Cost, Disposable, Flexible and Highly Reproducible Screen Printed SERS Substrates for the Detection of Various Chemicals. *Sci. Rep.* **2015**, *5*, 10208. [CrossRef] [PubMed]

27. Sree Satya Bharati, M.; Byram, C.; Soma, V.R. Femtosecond Laser Fabricated Ag@Au and Cu@Au Alloy Nanoparticles for Surface Enhanced Raman Spectroscopy Based Trace Explosives Detection. *Front. Phys.* **2018**, *6*, 28. [CrossRef]

Article

Effect of Process Parameters on the Generated Surface Roughness of Down-Facing Surfaces in Selective Laser Melting

Amal Charles [1],*, Ahmed Elkaseer [1,2], Lore Thijs [3], Veit Hagenmeyer [1] and Steffen Scholz [1,4]

[1] Institute for Automation and Applied Informatics, Karlsruhe Institute of Technology, 76344 Eggenstein-Leopoldshafen, Germany; ahmed.elkaseer@kit.edu (A.E.); veit.hagenmeyer@kit.edu (V.H.); steffen.scholz@kit.edu (S.S.)

[2] Faculty of Engineering, Port Said University, Port Fuad 42526, Egypt

[3] Direct Metal Printing Engineering, 3D Systems, 3001 Leuven, Belgium; lore.thijs@3dsystems.com

[4] Karlsruhe Nano Micro Facility, Hermann-von-Helmholtz-Platz 1, 76344 Eggenstein-Leopoldshafen, Germany

* Correspondence: amal.charles@kit.edu; Tel.: +49-721-608-25851

Received: 26 February 2019; Accepted: 20 March 2019; Published: 26 March 2019

Featured Application: This paper presents the first investigations towards closed loop feedback control of the selective laser melting (SLM) process. Insight gained from this work can be applied to facilitate in-process optimization of the SLM process for maximizing part quality and minimizing surface roughness.

Abstract: Additive manufacturing provides a number of benefits in terms of infinite freedom to design complex parts and reduced lead-times while globally reducing the size of supply chains as it brings all production processes under one roof. However, additive manufacturing (AM) lags far behind conventional manufacturing in terms of surface quality. This proves a hindrance for many companies considering investment in AM. The aim of this work is to investigate the effect of varying process parameters on the resultant roughness of the down-facing surfaces in selective laser melting (SLM). A systematic experimental study was carried out and the effects of the interaction of the different parameters and their effect on the surface roughness (Sa) were analyzed. It was found that the interaction and interdependency between parameters were of greatest significance to the obtainable surface roughness, though their effects vary greatly depending on the applied levels. This behavior was mainly attributed to the difference in energy absorbed by the powder. Predictive process models for optimization of process parameters for minimizing the obtained Sa in 45° and 35° down-facing surface, individually, were achieved with average error percentages of 5% and 6.3%, respectively, however further investigation is still warranted.

Keywords: additive manufacturing; selective laser melting; surface roughness; design of experiments; Ti6Al4V

---

1. Introduction

Since the advent of the first additive manufacturing (AM) technique, the Stereolithography process by 3D Systems in 1987, additive manufacturing has been under continuous and rapid development to meet growing industrial demands. Formerly, it was mainly used as a prototyping technique for pre-production, testing and analysis. However, different additive manufacturing techniques, which have recently emerged, play a significant role in modern industries [1,2].

This is principally because additive manufacturing has shown high potential to produce intricate 3D geometries with short lead-times at a relatively low cost. This helps strengthen supply chains and boosts their profitability and flexibly [3]. Presently, AM has been successfully exploited in electronic, aerospace and biomedical industries where highly specialized and customizable components are required [4–7]. The selective laser melting (SLM) technique makes especially great strides in the field of metal AM. This is especially achievable with the commercial emergence of titanium and nickel based super alloys that have exceptional material properties [7,8].

The SLM process is a powder bed additive manufacturing technique which uses a laser as a power source. The interaction between the laser and the metal powder causes the powder to selectively melt according to the desired slices. Once one layer is scanned, the platform is moved down by the height of this layer and another layer of powder is applied on top of the formerly built layer. This process of melting and bonding layers together continues successively until the desired part is built.

The non-melted powder remains in the build chamber and provides support to the part being built. This non-melted powder can be subsequently removed after building is completed and sieved; it can therefore be reused in successive builds [9].

Though noticeable progress has been made, there have been some challenging issues that still need addressing to allow AM and SLM to be used among mainstream manufacturing processes. Limited precision of the fabricated products and the repeatability of the processes are considered especially high technological barriers to the maturation of additive manufacturing techniques [4]. In particular, AM parts are often built with high surface roughness, which necessitates some post processing steps to refine the resultant roughness and makes it suitable for a wide range of engineering applications. These post-processing steps are rather expensive and time consuming.

The SLM process exhibits the so-called staircase effect in both up-facing and down-facing surfaces as depicted in Figure 1.

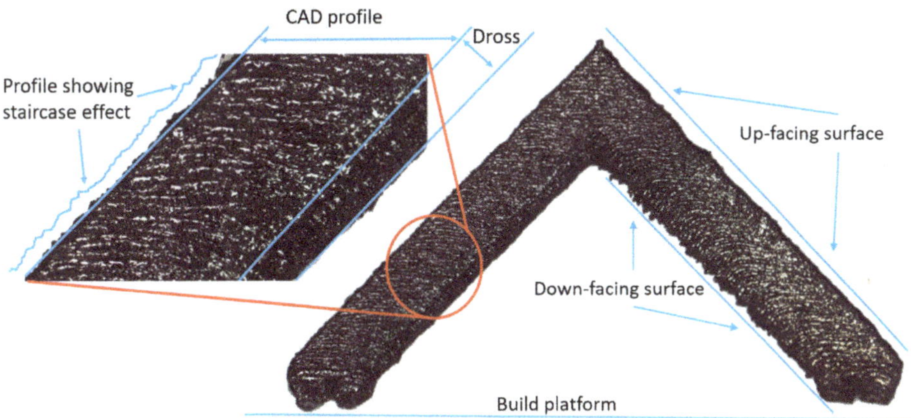

Figure 1. The staircase effect in up-facing and dross formation in down-facing surfaces in additive manufacturing (AM) parts.

This staircase effect contributes to the increased roughness of these surfaces. Down-facing surfaces, especially ones that are at an angle less than 45°, with respect to the build platform, show very high roughness. This is mainly attributed to the formation of dross and spatter due to the high laser absorptivity of powders compared to the solid metal in the bulk of the part as can be seen on Figure 2. The surface topology of parts produced by SLM are highly dependent on their orientation. This is why, in order to produce parts with good surface quality, down-facing surfaces with angles less than 45° are usually avoided by reorienting the part. Otherwise, there is a need for the building of support structures. However, this in turn results in the increase of process steps, in particular, removal of the support could exhibit defects, such as burr formation, leading to even higher roughness [10].

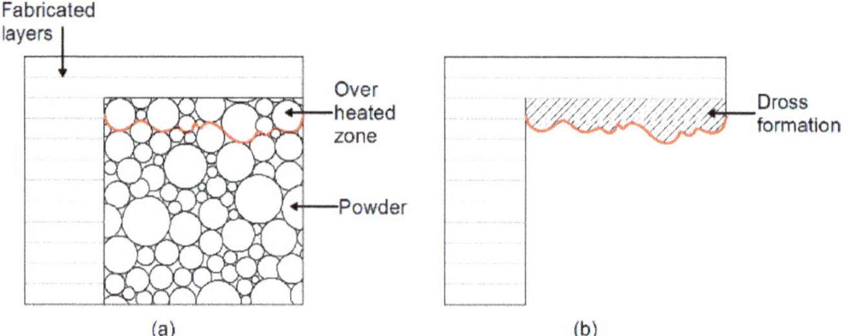

Figure 2. (**a**) A depiction of overheated zone above loose powder and (**b**) the resulting dross formation in the final part.

There have been some research attempts which try to correlate process parameters with quality marks of the parts, Sufiiarov et al. concluded that parts built with a 30 μm layer thickness demonstrated higher strength and lower elongations than parts built with a 50 μm layer thickness [11]. Wang et al. concluded that the mechanical properties of Inconel 718 parts did not vary along the build height of the parts [12].

Shipley et al. conducted a process optimization with the primary goal of maximizing part density, they identified the cooling rate to be important for the microstructural evolution and proposed methods to maintain a high energy input, which results in high cooling rates, thereby tailoring the microstructure and reducing residual stresses [13]. Evaluation of dimensional accuracy is a popular topic in AM research [14] as these technologies hold a lot of potential for application in research as well as industry. In this context, residual stresses are an important problem faced by the SLM process. These stresses can cause warping and deformation in parts thereby affecting the final dimensional accuracy and surface quality of printed parts. Therefore, many researchers have focused on the reduction of residual stresses and resulting substantial progress has been made in this field [15–17]. However, little research has been devoted to characterize the surface roughness of down-facing surfaces [18,19] and the optimization of parameters specifically for down-facing surfaces is, so far, an unexplored topic. In this context, the motivation for this work is to investigate and correlate the effects of different build parameters on the surface roughness of down-facing surfaces when the build parameters are only varied within the plane of the down-facing surface and its immediately adjacent volume as seen in Figure 3.

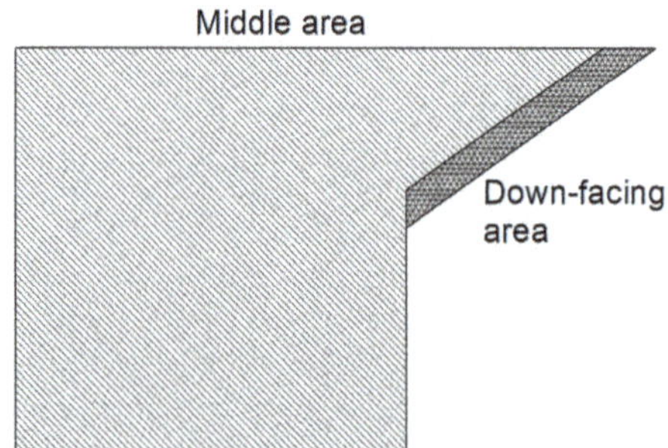

Figure 3. Illustration of areas printed with down-facing parameters.

2. Materials and Methods

2.1. Parameters

The parameters selected for this research work were laser power, scan speed and scan spacing. These were the parameters chosen as they are the ones considered most significant by a large body of research conducted in the metal AM field and have hence been the focus of many parameter optimization research works with regards to the SLM technique and powder bed AM techniques [20,21]. These parameters can easily be varied within the software of most printers and hence these are the parameters being considered in this work as well. It is to be noted that the parameters were varied only for the down-facing surfaces of the build as seen in Figure 3. The remainder of the part was built using the standard build parameters, as recommended by 3D Systems, for 60 μm layer thickness. The different levels of the parameter settings after rounding off can be seen in Table 1.

Table 1. Selected parameters and their levels.

Value	Laser Power	Scan Speed	Scan Spacing
−1	50	200	50
−0.59	90	465	60
0	150	850	75
0.59	210	1235	90
1	250	1500	100

2.2. Design of Experiments

Central composite method was used to design the experimental trials. This model was used as the limits for the factor settings. The design of experiments used to fabricate the test pieces can be seen in Table 2.

Table 2. Design of experiments.

Trial	Laser Power (W)	Scan Speed (mm/s)	Scan Spacing (μm)
1	90	465	60
2	90	465	90
3	90	1235	60
4	90	1235	90
5	210	465	60
6	210	465	90
7	210	1235	60
8	210	1235	90
9	50	850	75
10	250	850	75
11	150	200	75
12	150	1500	75
13	150	850	50
14	150	850	100
15	150	850	75
16	150	850	75
17	150	850	75
18	150	850	75
19	150	850	75
20	150	850	75
21	150	850	75
22	150	850	75
23	150	850	75
24	150	850	75

2.3. Test Piece

The test piece was designed to enable measure of the roughness of the down-facing surfaces. Consequently, the test pieces were designed to have a down-facing surface area of 10 mm × 20 mm and overhang inclinations of 45° and 35° as seen in Figure 4.

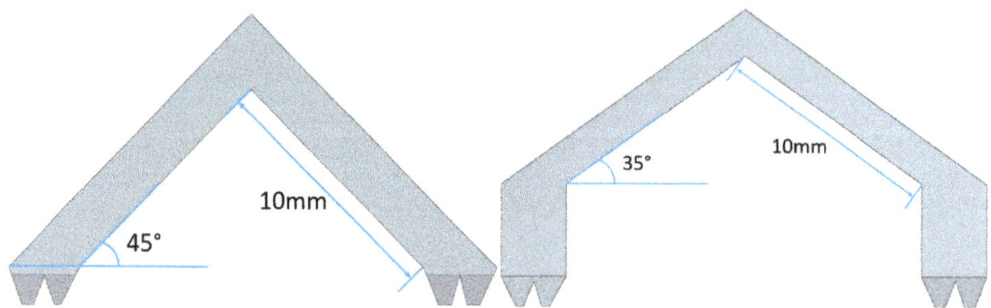

Figure 4. Depiction of CAD models with 45° and 35° overhangs.

2.4. Additive Manufacturing

The test pieces were designed using CAD Software Solid Edge (ST9, Siemens PLM software) and were directly imported into 3DXpert™ software (v13.0, 3D Systems) for the slicing, positioning and pre-processing of the build files. A 3D Systems ProX® DMP 320 machine (3D Systems, Leuven, Belgium) was used to perform the printing. The parts were heat treated before removal from the build platform in order to prevent warpage. The parts depicted in this paper were given a hexagonal cell scanning strategy and built with a 60 μm layer thickness. All parts were built using LaserForm Ti gr23 (A) powder.

2.5. Characterisation of Manufactured Parts

A Sensofar S neox 3D surface profiler (Sensofar, Barcelona, Spain) was used for the measurement of the surface roughness. The samples were subject to ultrasonic cleaning with water in order to detach any loose powder on the surface prior to measurement. A focus variation technique was used to scan the topography of a square of 4 mm × 4 mm dimension at the center of the down-facing surface for all samples. This scanned topography was then used to generate the areal roughness parameters.

3. Results and Statistical Analysis

A careful visual examination was conducted for all samples to characterize the visual appearance of the down-facing surfaces. Test pieces were visually examined to detect the presence of bright spots that could indicate spots of large spatter or dross presence, as shown in Figure 5.

Increased surface irregularities were observed in the 35° down-facing surface seen in Figure 5 (right) in the form of bright spots. This inferred the presence of spots of large spatter and dross formations. As can also be seen in the above image, this phenomenon was far less prominent in the 45° down-facing surface (left). This indicated a higher surface roughness in the 35° samples than the 45° samples. This indication was further confirmed by the experimental measurement results.

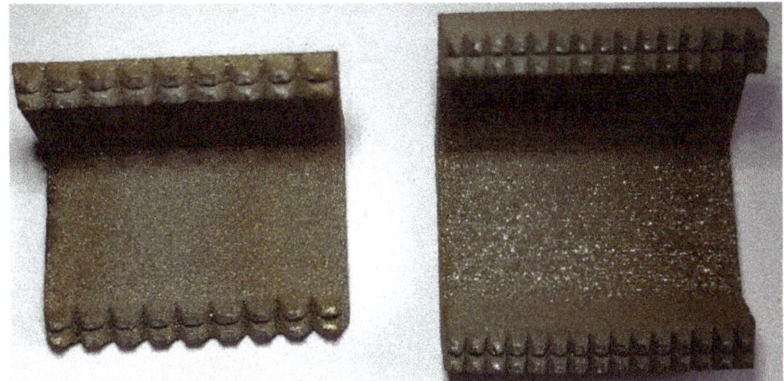

Figure 5. Visual observation of the down-facing surfaces shows significant irregularities of the surface in the 35° test piece (**right**) when compared to the 45° test piece (**left**).

3.1. Surface Roughness

The arithmetic mean height (*Sa*) parameter, which is an areal surface roughness parameter, expresses, as an absolute value, the difference in height of each point measured compared to the arithmetical mean of the surface. The *Sa* parameter has become one of the more popular methods for characterizing surface roughness of AM manufacturing parts because of the tendency of AM parts to have highly irregular and rough surfaces. Therefore, this parameter was chosen as the first quality mark for further process analysis.

The results presented herein in Figures 6 and 7 are for test pieces with an overhang angle of 45° and 35°, respectively.

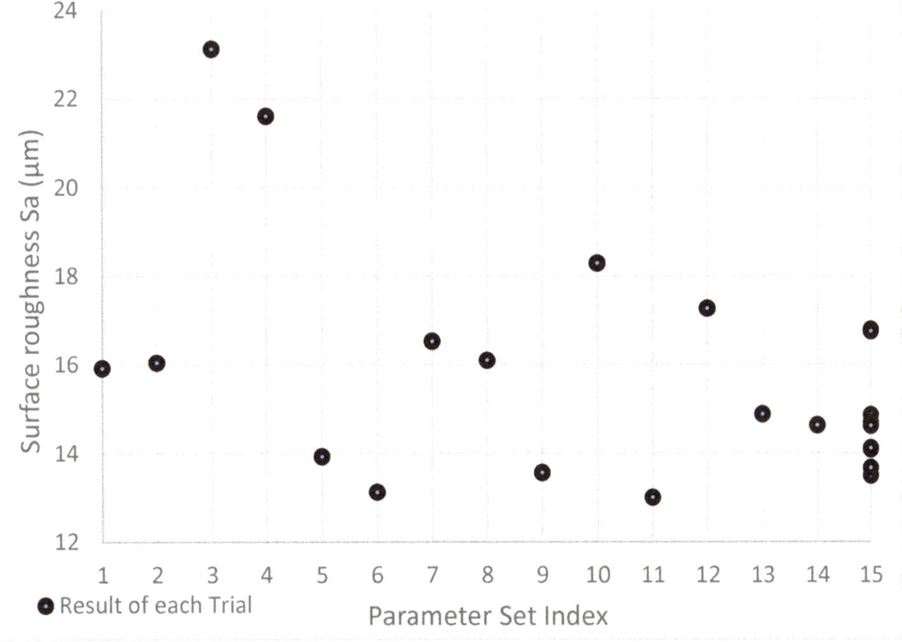

Figure 6. Replicate plot for 45° down-facing surfaces: Surface roughness (*Sa*) for each trial number.

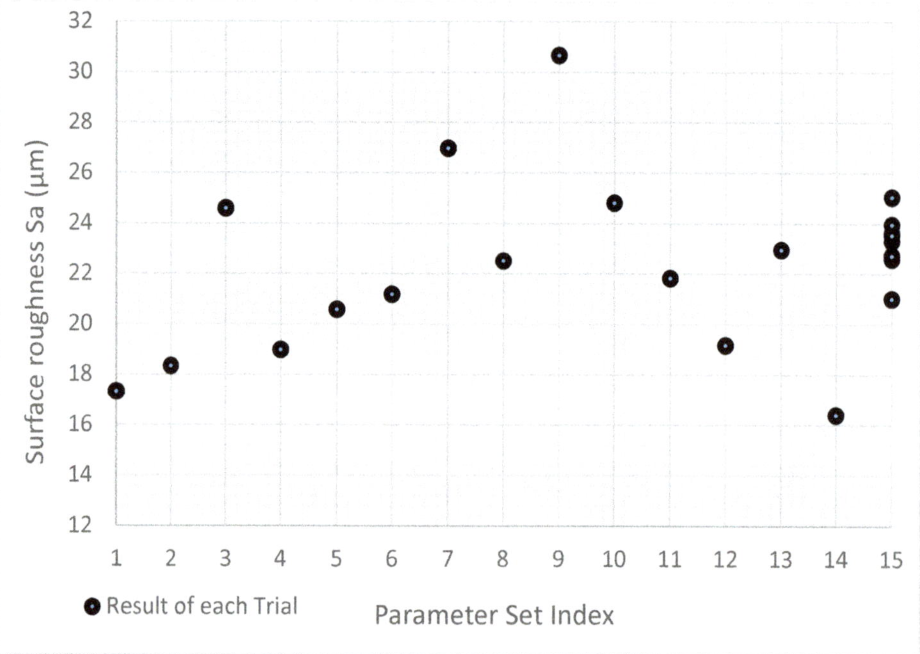

Figure 7. Replicate plot for 35° down-facing surfaces: Surface roughness (*Sa*) for each trial number.

The replicate plot shown in Figures 6 and 7 depicts the process as a stable process, thereby maximizing the possibility for developing a process model using these experimental results. By plotting the interaction effects of the various parameters on the obtained *Sa* as seen in Figures 8 and 9, clear trends can be seen thereby allowing us to draw conclusions.

3.2. Effect of Interaction of Parameters on Sa

3.2.1. Laser Power and Scan Speed

- 45° down-facing surface

Looking at the two graphs of laser power and scan speed in Figure 8, it is clear that for any given laser power, the *Sa* value increases with an increasing scan speed. The rate of increase in *Sa* varies at different laser powers. At lower laser powers, there is a rapid increase in *Sa* with increasing scan speed, while at higher laser powers the increase is only gradual.

The second graph depicts the decrease in *Sa* at different scan speeds when increasing the laser power up to a certain point, after which the *Sa* begins to gradually increase again. The laser power point at which the *Sa* begins to rise again varies amongst the different scan speeds.

- 35° down-facing surface

The 35° down-facing surfaces depict a different effect from the 45° surfaces. In this case, as seen in Figure 9 at the different laser powers, the *Sa* increases up to a scan speed of 1000 mm/s after which it begins to reduce once again.

3.2.2. Laser Power and Scan Spacing

- 45° down-facing surface

The interaction between laser power and scan spacing also shows clear trends. The *Sa* at the different laser powers gradually decreases when increasing scan spacing, however it tends to increase once it has passed the scan spacing of 80 μm.

The second graph shows that at all scan spacing levels, the *Sa* decreases up to a certain laser power (between 150 and 200 W) after which it begins to increase once again. While the laser power of 150 W showed the lowest *Sa* values.

- 35° down-facing surface

The effect of these parameters on the 35° down-facing surface also shows some difference when compared with the 45° down-facing surface, as seen in Figure 9. When increasing the scan spacing at different laser powers, the *Sa* values increase up to a scan spacing between 60 and 70 μm, after which they decrease. The laser power of 150 W showed the lowest *Sa* values.

3.2.3. Scan Speed and Scan Spacing

- 45° down-facing surface

It is quite clear from the first graph that at all scan spacings, the *Sa* value increases with an increase in scan speed. The second graph shows that at all the different scan speeds the *Sa* value will decrease up to a certain scan spacing value (80 μm) after which the *Sa* value increases once again. This is consistent with the observations made between laser power and scan spacing as well.

- 35° down-facing surface

In Figure 9, it can be seen that the effect of each scan spacing on the *Sa* shows an increase up to a certain scan speed, after which it begins to decrease once again. This point differs for each scan spacing, as can be seen in the graphs.

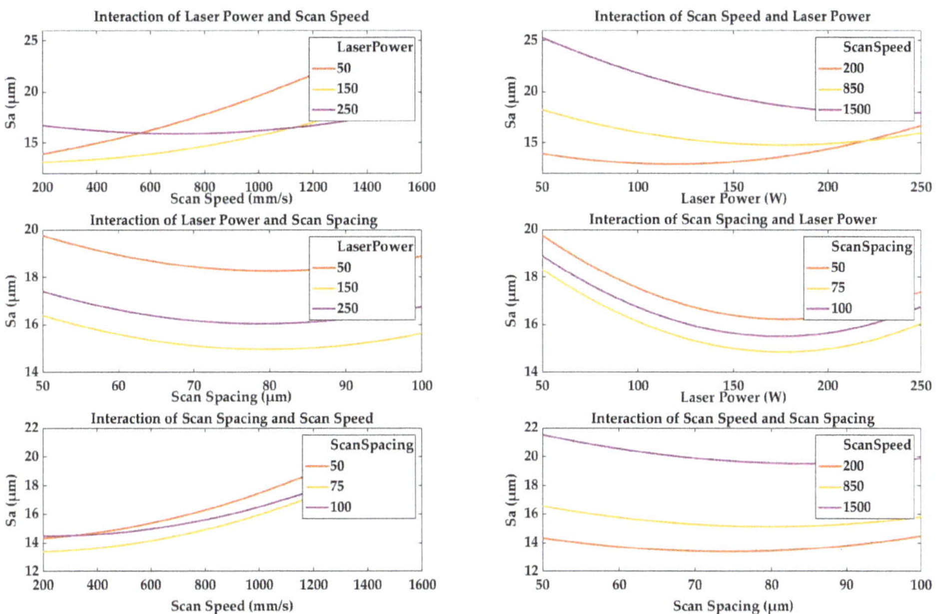

Figure 8. Interaction effects of parameters with the measured *Sa* for 45° down-facing surfaces.

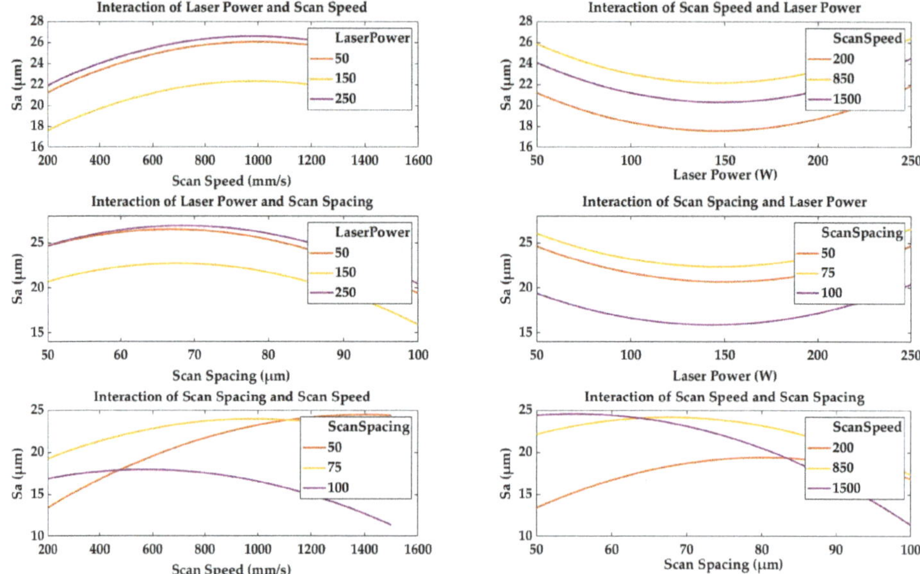

Figure 9. Interaction effects of parameters with the measured *Sa* for 35° down-facing surface.

The data processing and statistical modelling were completed using MATLAB (R2018b, MathWorks). A linear regression model with interaction effects was used to describe the relationship between parameters (laser power, scan speed and scan spacing) and *Sa*. The linear model was fit to the measured raw data, and this model was then used to generate the interaction plots as seen in Figures 8 and 9. These plots show the estimated effect on the response from changing each variable value, averaging out the effect of other parameters. This plot also shows the estimated effect when the other variable values are fixed at certain values as seen in Figures 8 and 9.

This same model also makes it possible to generate the prediction slice plots as seen in Figures 10 and 11. The prediction slice plots in Figures 10 and 11 show the main effects for all parameter values. The green line in each panel shows the change in the response variable as a function of the parameter values when all other predictor values are kept constant. The dashed red curves in each panel depict the 95% confidence bounds for the predicted response variable.

The plots shown in Figures 8 and 9 illustrate the effect of each predictor on the *Sa* model, with the objective being to minimize the obtainable surface roughness for the down-facing surface by understanding the effect of the down-facing parameters. It is now possible to use this predictive model in order to obtain parameters when minimizing the *Sa*. Obtaining a smoother surface by simply controlling the process parameters extends the current capabilities of printing inclined and down-facing surfaces without requiring the printing of support structures, thereby contributing to a reduction in post-processing times. To obtain an as-printed *Sa* of 12 µm would illustrate an improvement over current capabilities when printing with default parameters. Figure 9 suggests a set of parameters in order to achieve a *Sa* of 12.73 µm, the corresponding normalized values for each process parameter were: Laser power = −0.35 (115 W), scan speed = −1 (465 mm/s) and scan spacing = −0.02 (approx. 75 µm). For the 35° down-facing surface, a prediction of 20 µm would illustrate an improvement over current capabilities and with use of the developed model it was possible to predict process parameters to achieve this, as seen in Figure 11. The normalized value of parameters suggested were: Laser power = 0 (150 W), scan speed = 0.18164 (968 mm/s) and scan spacing = 0.56836 (89 µm). The predicted *Sa* and measured *Sa* values were determined to have an average error of 5% for 45° down-facing surfaces and 6.3% for 35° down-facing surfaces, which are considered to be acceptable at this stage. However, further experiments and measurements are being conducted in order to further feed and develop the model.

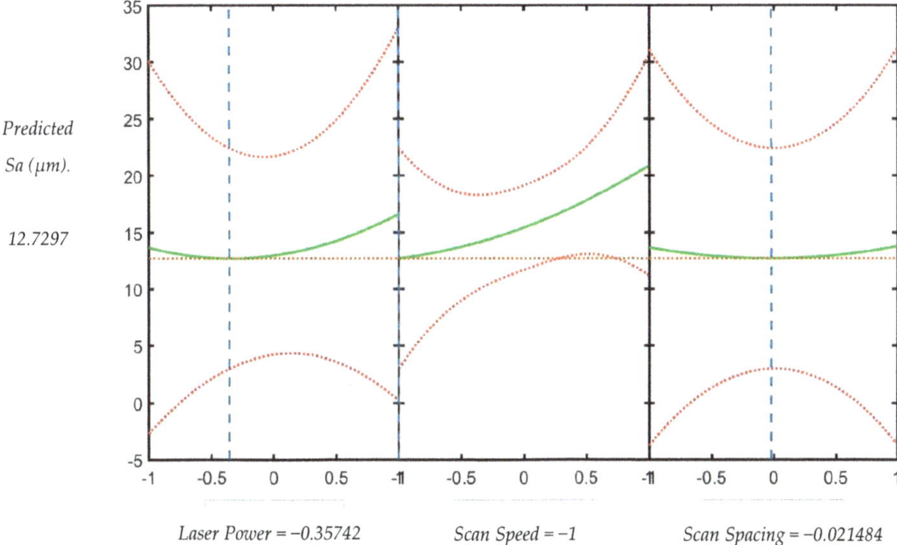

Figure 10. Effect of each parameter (normalized values) on the surface roughness model for down-facing surfaces.

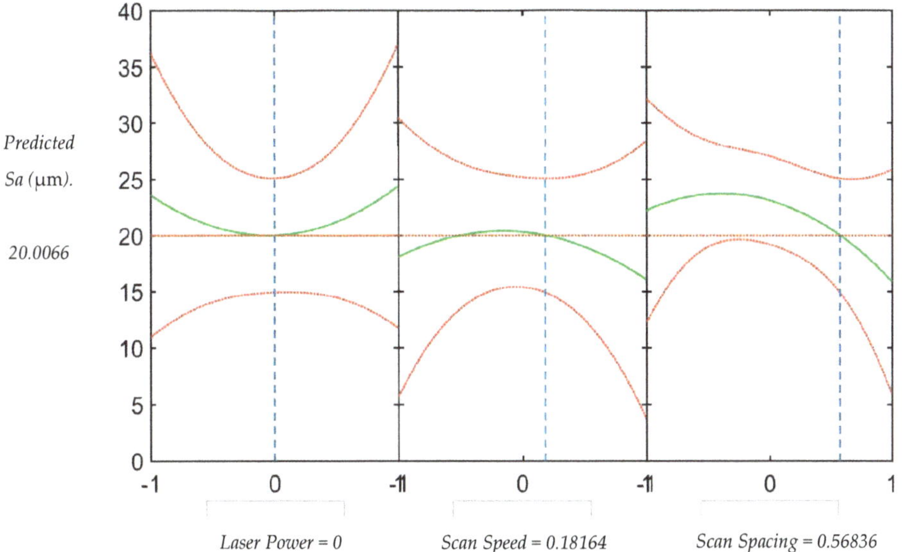

Figure 11. Effect of each parameter (normalized values) on the surface roughness model for a 35° down-facing surface.

4. Discussion

A close observation of Figure 12 confirms that one of the main reasons for the measured high *Sa* values for down-facing surfaces was unmelted or partially melted powder. The presence of partially melted powder on the down-facing surface was almost unavoidable. This was caused during the printing process when the meltpool/dross that was formed came into contact with loose unsupported powder, thereby causing the partial melting of this lose powder and thereby the attachment/embedding of the loose powder within the surface of the desired printed part.

This aspect, coupled with the dross formation, which is formed by the full melting of loose powder in the overheated zones adjacent to the meltpool, were the major causes of high surface roughnesses in down-facing surfaces; this can be clearly seen in Figure 12.

Another observation made is that the 35° down-facing surfaces exhibited higher surface roughnesses than the 45° down-facing surfaces. It can be concluded that this occurred because, even though both surfaces were printed on top of loose powder, the 45° down-facing surface had a less steep slope when compared to the build platform, which resulted in a smaller overheating zone as some of the heat was conducted away through the solid bulk of the part. However, for the 35° down-facing surface a larger amount of energy was transferred into the powder, causing a larger meltpool and larger dross formation.

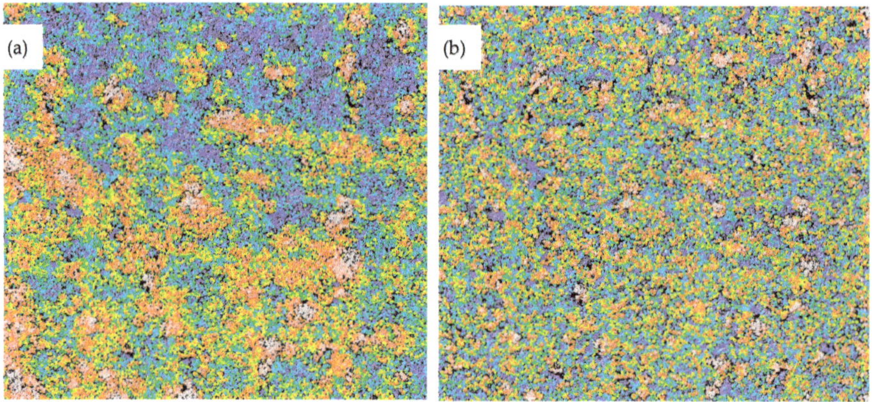

Figure 12. Microscopic images of the down-facing surfaces depicting the large presence of partially melted powder on the surface which thereby created surfaces with high roughness. (**a**) Unfiltered image, (**b**) image after filtering for waviness and roughness.

Additionally, while visually examining the samples it was found that some of the samples that showed relatively low surface roughness values in fact exhibited a significantly large dimensional deviation from the CAD of the samples. The down-facing surface was thicker than intended for many of the samples. This clearly indicated a large dross formation on the surface. This could be attributed to the larger energy input on some samples, resulting in the formation of larger meltpools. As a result, this could cause large dross with less measured roughness, as the larger meltpools had better wettability and would cause more interconnected melt pools which cause a uniform dross, and cannot be detected just by measuring roughness. Therefore, factors such as line energy, which could relate to the size of the meltpool formed, could lead to a better understanding of optimum overlap parameters in order to optimize the process for minimizing dross formation and surface roughness.

Dimensional accuracy tests are also required in order to investigate this phenomenon of formation of low roughness and large dross. It clearly becomes evident that, in the SLM process, the process parameters display significant degrees of interdependency that affects the final quality of printed parts.

5. Conclusions

This paper has presented a method to aid in the understanding of the effects of various SLM printing parameters on the surface roughness of down-facing surfaces. Clear trends were presented on the interaction effects of the laser power, scan speed and scan spacing. Additionally, a predictive process model has also been presented for the suggestion of process parameters in order to attain a certain *Sa* value for 45° and 35° down-facing surfaces. Current works-in-progress are aimed at further improving this model by including more process data and also including other quality marks such as dimensional accuracy.

The experimental trials with the various parameter sets and replications showed that the process is stable and therefore made it possible to obtain a process model based on the three selected predictors and which uses *Sa* as the quality mark. This model was also used to optimize the process by suggesting process parameters with the objective of minimizing the *Sa*. The model was also tested by comparing the measured and predicted values; average errors of 5% (for 45° samples) and 6.3% (for 35° samples) were obtained. The continuous development of this model is currently taking place to include even more experimental data and quality marks. Process modelling by using artificial neural networks (ANN) is also an option and is being considered in order to further improve the process understanding and to minimize further the errors in prediction.

Such a model shows promise and is a step towards achieving a level of closed loop feedback control for the SLM process. By integrating it with existing pre-processing software, as well as other in-process monitoring tools, it will theoretically be able to assess the print quality during printing and make necessary changes as per the user's quality requirements. This system integration is planned for future work.

Author Contributions: Conceptualization, A.C., L.T., S.S.; methodology, A.C., A.E., L.T., S.S.; software, A.C., A.E.; validation, A.C., A.E., L.T.; formal analysis, A.C., A.E.; investigation, A.C., A.E.; resources, V.H., L.T., S.S.; data curation, A.C., A.E.; writing—original draft preparation, A.C.; writing—review and editing, A.C., A.E., L.T., V.H. and S.S.; visualization, A.C., A.E.; supervision, L.T., V.H. and S.S.; project administration, S.S.; funding acquisition, S.S.

Funding: This work was done in the H2020-MSCA-ITN-2016 project PAM^2, Precision Additive Metal Manufacturing, which is funded by The EU Framework Programme for Research and Innovation—Grant Agreement No. 721383.

Acknowledgments: The support by the Karlsruhe Nano Micro Facility (KNMF-LMP, http://www.knmf.kit.edu/) a Helmholtz research infrastructure at KIT, is gratefully acknowledged. The authors also acknowledge support by Deutsche Forschungsgemeinschaft and open access publishing fund of Karlsruhe Institute of Technology. The authors gratefully acknowledge the technical help provided by Fabian Grinschek from the Institute for Micro Process Engineering (IMVT) at Karlsruhe Institute of Technology, Germany.

Conflicts of Interest: The authors declare no conflict of interest.

References

1. Eyers, D.R.; Potter, A.T. Industrial Additive Manufacturing: A manufacturing systems perspective. *Comput. Ind.* **2017**, *92*, 208–218. [CrossRef]
2. Berman, B. 3-D printing: The new industrial revolution. *Bus. Horiz.* **2012**, *55*, 155–162. [CrossRef]
3. Brans, K. 3D Printing, A Maturing Technology. *IFAC Proc.* **2013**, *46*, 468–472. [CrossRef]
4. Kannan, G.B.; Rajendran, D.K. A Review on Status of Research in Metal Additive Manufacturing. In *Advances in 3D Printing & Additive Manufacturing Technologies*; Wimpenny, D.I., Pandey, P.M., Kumar, L.J., Eds.; Springer: Berlin/Heidelberg, Germany, 2017; pp. 95–100.
5. Clare, A.T.; Chalker, P.; Davies, S.; Sutcliffe, C.; Tsopanos, S. Selective laser melting of high aspect ratio 3D nickel–titanium structures two way trained for MEMS applications. *Int. J. Mech. Mater.* **2008**, *4*, 2. [CrossRef]
6. Rochus, P.; Plesseria, J.Y.; Van Elsen, M.; Kruth, J.P.; Carrus, R.; Dormal, T. New applications of rapid prototyping and rapid manufacturing (RP/RM) technologies for space instrumentation. *Acta Astronaut.* **2007**, *61*, 1–6. [CrossRef]
7. Guo, N.; Leu, M.C. Additive manufacturing: Technology, applications and research needs. *Front. Mech. Eng.* **2013**, *8*, 215–243. [CrossRef]
8. Bourell, D.; Kruth, J.P.; Leu, M.; Levy, G.; Rosen, D.; Beese, A.M.; Clare, A.T. Materials for additive manufacturing. *CIRP Ann. Manuf. Technol.* **2017**, *66*, 659–681. [CrossRef]

9. Thijs, L.; Verhaeghe, F.; Craeghs, T.; Humbeeck, J.V.; Kruth, J.P. A study of the microstructural evaluation during selective laser melting of Ti-6Al-4V. *Acta Mater.* **2010**, *58*, 3303–3312. [CrossRef]
10. Triantaphyllou, A.; Giusca, C.L.; Macaulay, G.D.; Roerig, F.; Hoebel, M.; Leach, R.K.; Tomita, B.; Milne, K.A. Surface texture measurement for additive manufacturing. *Surf. Topogr. Metrol. Prop.* **2015**, *3*, 024002. [CrossRef]
11. Sufiiarov, V.S.; Popovich, A.A.; Borisov, E.V.; Polozov, I.A.; Masaylo, D.V.; Orlov, A.V. The Effect of Layer Thickness at Selective Laser Melting. *Procedia Eng.* **2017**, *174*, 126–134. [CrossRef]
12. Wang, X.; Keya, T.; Chou, K. Build Height Effect on the Inconel 718 Parts Fabricated by Selective Laser Melting. *Procedia Manuf.* **2016**, *5*, 1006–1017. [CrossRef]
13. Shipley, H.; McDonnell, D.; Culleton, M.; Coull, R.; Lupoi, R.; O'Donnell, G.; Trimble, D. Optimisation of process parameters to address fundamental challenges during selective laser melting of Ti-6Al-4V: A review. *Int. J. Mach. Tools Manuf.* **2018**, *128*, 1–20. [CrossRef]
14. Elkaseer, A.; Mueller, T.; Charles, A.; Scholz, S. Digital Detection and Correction of Errors in As-built Parts: A Step Towards Automated Quality Control of Additive Manufacturing. In Proceedings of the World Congress on Micro and Nano Manufacturing (WCMNM), Portorož, Slovenia, 18–20 September; pp. 389–392.
15. Tian, Y.; Tomus, D.; Rometsch, P.; Wu, X. Influences of processing parameters on surface roughness of Hastelloy X produced by selective laser melting. *Addit. Manuf.* **2017**, *13*, 103–112. [CrossRef]
16. Robinson, J.; Ashton, I.; Fox, P.; Jones, E.; Sutcliffe, C. Determination of the effect of scan strategy on residual stress in laser powder bed fusion additive manufacturing. *Addit. Manuf.* **2018**, *23*, 13–24. [CrossRef]
17. Kruth, J.-P.; Deckers, J.; Yasa, E.; Wauthlé, R. Assessing and comparing influencing factors of residual stresses in selective laser melting using a novel analysis method. *Proc. Inst. Mech. Eng. Part B J. Eng. Manuf.* **2012**, *226*, 980–991. [CrossRef]
18. Fox, J.C.; Moylan, S.P.; Lane, B.M. Effect of process parameters on the surface roughness of overhanging structures in laser powder bed fusion additive manufacturing. *Procedia CIRP* **2016**, *45* (Suppl. C), 131–134. [CrossRef]
19. Charles, A.; Elkaseer, A.; Mueller, T.; Thijs, L.; Torge, M.; Hagenmeyer, V.; Scholz, S. A Study of the Factors Influencing Generated Surface Roughness of Downfacing Surfaces in Selective Laser Melting. In Proceedings of the World Congress on Micro and Nano Manufacturing (WCMNM), Portorož, Slovenia, 18–20 September; pp. 327–330.
20. Khorasani, A.; Gibson, I.; Awan, U.S.; Ghaderi, A. The effect of SLM process parameters on density, hardness, tensile strength and surface quality of Ti-6Al-4V. *Addit. Manuf.* **2019**, *25*, 176–186. [CrossRef]
21. Shi, X.; Ma, S.; Liu, C.; Wu, Q. Parameter optimization for Ti-47Al-2Cr-2Nb in selective laser melting based on geometric characteristics of single scan tracks. *Opt. Laser Technol.* **2017**, *90*, 71–79. [CrossRef]

applied sciences

Article

Spatial Uncertainty Modeling for Surface Roughness of Additively Manufactured Microstructures via Image Segmentation

Namjung Kim [1],*, Chen Yang [2], Howon Lee [2],* and Narayana R. Aluru [1]

[1] Department of Mechanical Science and Engineering, University of Illinois at Urbana-Champaign, 1206 W. Green ST., Urbana, IL 61801, USA; aluru@illinois.edu

[2] Department of Mechanical and Aerospace Engineering, Rutgers University–New Brunswick, 98 Brett Rd., Piscataway, NJ 08854, USA; chen.yang@rutgers.edu

* Correspondence: kim847@illinois.edu (N.K.); howon.lee@rutgers.edu (H.L.); Tel.: +1-848-445-2213 (H.L.)

Received: 6 February 2019; Accepted: 11 March 2019; Published: 15 March 2019

Featured Application: The proposed modeling framework helps to generate a mathematical spatial roughness model including the staircase side profile as well as spatial uncertainty of additively manufactured parts, which might significantly affect the mechanical behavior of printed parts. This general approach can be applied to any complex AM parts which have spatial roughness that can be obtained via optical images.

Abstract: Despite recent advances in additive manufacturing (AM) that shifts the paradigm of modern manufacturing by its fast, flexible, and affordable manufacturing method, the achievement of high-dimensional accuracy in AM to ensure product consistency and reliability is still an unmet challenge. This study suggests a general method to establish a mathematical spatial uncertainty model based on the measured geometry of AM microstructures. Spatial uncertainty is specified as the deviation between the planned and the actual AM geometries of a model structure, high-aspect-ratio struts. The detailed steps of quantifying spatial uncertainties in the AM geometry are as follows: (1) image segmentation to extract the sidewall profiles of AM geometry; (2) variability-based sampling; (3) Gaussian process modeling for spatial uncertainty. The modeled spatial uncertainty is superimposed in the CAD geometry and finite element analysis is performed to quantify its effect on the mechanical behavior of AM struts with different printing angles under compressive loading conditions. The results indicate that the stiffness of AM struts with spatial uncertainty is reduced to 70% of the stiffness of CAD geometry and the maximum von Mises stress under compressive loading is significantly increased by the spatial uncertainties. The proposed modeling framework enables the high fidelity of computer-based predictive tools by seamlessly incorporating spatial uncertainties from digital images of AM parts into a traditional finite element model. It can also be applied to parts produced by other manufacturing processes as well as other AM techniques.

Keywords: spatial uncertainty modeling; additive manufacturing; uncertainty quantification; Image segmentation; gaussian process modeling

---

1. Introduction

Additive manufacturing (AM) or 3D printing refers to a set of manufacturing processes that can fabricate a three-dimensional (3D) physical object from a digital computer-aided design (CAD) model by directly joining materials in a layer-by-layer fashion. In contrast, traditional manufacturing processes are subtractive since a part is mainly produced by removing unnecessary parts from a bulk material, which typically increases material waste, and thereby production cost. Additionally, the manufacturing time of a subtractive process is highly dependent on the geometrical complexity of products. However, AM builds a product in a layer-by-layer fashion, making it possible to construct products in a complex geometry without increasing manufacturing time or cost. Given these advantages, AM is increasingly used to produce a wide range of parts and products and replace traditional manufacturing processes [1,2]. Furthermore, AM even allows for the creation of sophisticated geometries that would not be possible to realize otherwise. For example, AM is used to fabricate lightweight cellular structures with an unprecedented capability to sustain high external loads while maintaining an extremely low density [3,4]. Highly ordered cellular structures that comprise spatially arranged thin struts result in high stiffness and toughness per unit weight and high surface-to-volume ratios. Several applications of cellular structures are suggested, including energy absorbing systems [5–7], thermal applications [8,9], and biomimetic materials [10].

Despite significant advances in AM techniques, distinctive geometrical deviations continue to exist between a CAD model and a final printed product due to various sources, such as digitization of 3D models, non-uniform material forming, and defects from the printing processes [11,12]. The staircase sidewall profile is one of the most prominent problems observed in most AM parts. Given that the use of AM gradually shifts from creating prototyping to manufacturing of functional end products, the accurate prediction of mechanical properties of AM parts is increasingly important. Recent studies indicate that dimensional accuracy, geometrical alignment, and surface roughness of micro and nanostructures fabricated by AM techniques can lead to high deviations in the mechanical properties [13,14]. In particular, the impact of geometrical uncertainties on mechanical performance is increasingly pronounced when the length scale of the AM parts approaches the spatial deviations caused by the manufacturing process. There has been a previous study in which the stiffness variations of additively manufactured microlattice structures resulting from geometrical uncertainties were quantified [13]. However, the geometrical difference investigated is limited to in-plane radial deviations, and the effects from the staircase sidewall profile and roughness are not considered. In addition, previous studies do not consider the actual deviation between planned and manufactured geometries to investigate the effect of topological uncertainties on the mechanical performance of AM parts.

Figure 1a shows a schematic illustration of additive manufacturing process of a 3D microlattice using a mask projection stereolithography technique. The manufactured microlattice has a hierarchical structure of repeating unit cells which consist of thin struts arranged in multiple directions. In the study, we present a probabilistic modeling framework that can predict the spatial variations of printed geometry via digital images of AM parts. We limit our interest to the spatial variations of a single strut given that it is a unit building block of various types of 3D microlattices and that the load-bearing characteristics of a single strut determine the mechanical behavior of the overall lattice structure. Figure 1b shows the schematic of the proposed framework to quantify the spatial uncertainty in AM parts and incorporate spatial uncertainty in a computer simulation model to better predict the mechanical performance of the AM parts. A strut is initially additively manufactured using a mask projection stereolithography technique [15]. The sidewall profile caused by the layer-by-layer manufacturing process is extracted from a microscope image, based on which a probabilistic surface model is generated. The details of the three steps involved in generating the probabilistic model are given in the following sections.

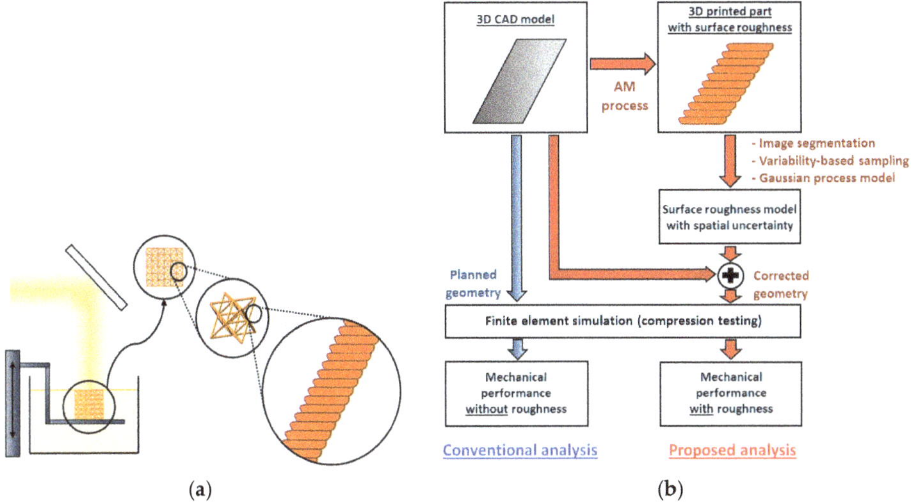

(a)   (b)

Figure 1. Schematic description of probabilistic modeling framework for an additively manufactured part: (**a**) Manufacturing process of a microlattice structure: a hierarchical structure of a unit cell consisting of multiple struts; (**b**) A probabilistic modeling framework for surface roughness estimation and mechanical behavior of an additively manufactured geometry.

2. Materials and Methods

2.1. Sample Fabrication

All materials including the liquid polymer, photo initiator (PI), and photo absorber (PA) are purchased from Sigma-Aldrich (St. Louis, MO, USA) and used in the as-received condition. Poly (ethylene glycol) diacrylate (PEGDA) (Mn 250) is used as a base polymer. Phenylbis (2,4,6-trimethylbenzoyl) phosphine oxide as PI and Sudan I as PA are added into the polymer at concentrations of 2 wt.% and 0.02 wt.%, respectively.

To examine the sidewall profile of AM part, CAD models of struts with angles of $60°$, $75°$, and $90°$ with respect to a horizontal plane were designed in SolidWorks. All struts are 6 mm in height and exhibit a diameter of 800 μm. Each 3D model is digitally sliced into a series of two-dimensional (2D) bitmap images with a layer thickness of 100 μm.

We used a custom-built mask projection stereolithography AM system in the study [15,16]. An ultraviolet (UV) digital projector (CEL5500, Digital Light innovations) generates spatially patterned light for each cross-sectional image, and this is projected through a 2× microscope objective lens and focused on the top surface of the liquid polymer mixture. When a layer is photopolymerized with an exposure time of 20 s, the sample holder on which the object rests is lowered by a linear stage (MTS50-Z8, Thorlabs) by a layer thickness of 100 μm. The subsequent cross-sectional image is projected to photopolymerize the next layer on top of the preceding layer. The process is repeated until all layers are completed. Each sample is individually fabricated at the center of the build area for consistency. Process parameters of the AM process, including light intensity, layer thickness, and curing time, are listed in Table 1.

Table 1. Mask projection stereolithography process parameters.

Total Number of Layer	60
Layer thickness (μm)	100
Curing time per layer (sec)	20
Light intensity (mW/cm^2)	1

Fabricated samples are rinsed in ethanol to remove the uncured polymer from the object. The prepared samples were optically inspected via a microscope. Images of the side profile of the samples were captured using a digital camera attached to the microscope.

2.2. Image Segmentation by Using a Level Set Method

The accuracy of data-driven stochastic model for spatial uncertainties depends on the experimental characterization data. Although profilometry is a fast and simple method to measure the spatial profiles, the usage of profilometry is strictly limited by its dimensional accessibility [17]. Therefore, we employ an image-based technique to extract the spatial variation of AM products. The method involves collecting sidewall profile data of additively manufactured struts from multiple sample images and extracting the data set that can be used to quantify the spatial variations by using image segmentation. The image-based technique can be applied to any spatially varying field obtained from any imaging technique.

Image segmentation is a partitioning process that groups clusters of pixels of a digital image based on parameters including their colors, intensities, and textures to extract useful information from the image [18]. Various techniques are proposed for segmenting images [18–20]. Among these, thresholding is one of the simplest and most widely used techniques for image segmentation. Thresholding separates pixels in which the grayscale level exceeds a critical value from the background, and the critical value is defined by the image gray-level histogram. However, a few disadvantages of thresholding are also reported, including susceptibility to pixel noise and the possibility that the extract profile is potentially not a closed contour due to lack of consideration of spatial characteristics [21]. In this study, we use the level set method to extract the profiles of objects in the image [22]. In the level set method, a higher dimensional function termed as level set function (ϕ) is defined to describe the contour. The contour evolves toward the object's boundaries in the image based on the predefined constraints such as difference or gradient of pixel intensities. With respect to the evolving level set function, the contours are given by the zero-level set, $C = \{(x,y)|\phi(x,y) = 0\}$, which evolves toward the object's boundaries.

We assume that $\Omega \subset R^2$ is the domain inside an image, and $I(x,y)$ denotes the intensity of pixel at the pixel's location (x, y). The general energy function E in the level set method is shown as follows [23]:

$$E = \lambda_o \int_\Omega |I(x,y) - u_o|^2 H(\phi(x,y))dxdy$$
$$+\lambda_b \int_\Omega |I(x,y) - u_b|^2 (1 - H(\phi(x,y)))dxdy + \gamma \int_\Omega g(\phi(x,y))|\nabla\phi(x,y)|dxdy, \tag{1}$$

where u_o and u_b denote the pixel intensity averages inside the contour and in the background region, respectively, and λ_o, λ_b, and γ are non-negative weighting factors. A smoothing function $g(x) = (1 + \exp((x - \xi)/L))^{-1}$ is selected for the robustness of the algorithm. L defines the smoothing length, and ξ locates the center of the smoothing function. The Heaviside function (H) is used to differentiate the object Ω_o and the background Ω_b. The physical meaning of each term in Equation (1) is as follows: The first term matches the average pixel intensity inside the contour with the pixel intensity of objects by minimizing its mean-squared error. Similarly, the second term matches the pixel intensity of background in the image. The third term regulates the smoothness of the contour, and thus a higher weight of the term implies the smoothing of the contour. The initial condition of the level set function is given by the threshold pixel intensities. With the fixed level set function, u_o and u_b are obtained as follows:

$$u_o = \frac{\int_\Omega I(x,y)H(\phi(x,y))dxdy}{\int_\Omega H(\phi(x,y))dxdy}, \tag{2}$$

$$u_b = \frac{\int_\Omega I(x,y)(1 - H(\phi(x,y)))dxdy}{\int_\Omega (1 - H(\phi(x,y)))dxdy}, \tag{3}$$

Given u_o and u_b, the energy function E is minimized by a standard gradient descent algorithm [23]. This two-step energy minimization is repeated until the converged level set function is obtained. As a result, the zero level set of the converged level set function is matched with the object's boundaries in the image. Further details regarding the level set method is described in [23].

2.3. Variability-Based Sampling and Bisection Method

The extracted boundaries from image segmentation correspond to a uniformly distributed data set that shows the edge profile. The data can be used for the Gaussian process model. However, the number of data points in the set is proportional to the number of hyperparameters to be fitted, and thus an efficient sampling of the data set is required to reduce the computational cost and improve the convergence rate. To determine the sampling locations along the profile $y(x)$, the local variability, σ_y^2, is defined to calculate the probability of sampling at each location, $\Pr(x)$, as follows [24,25]:

$$\Pr(x) \propto \sigma_y^2(x) = \frac{1}{n_x - 1} \sum_{x' \in R(x)} \left(\frac{y(x') - \bar{y}(x)}{(x' - x)} \right)^2, \tag{4}$$

in which x' denotes a location within the region around x, $\bar{y}$ denotes the local mean of the profile values in the region $R(x)$, and n_x denotes the number of measurement locations in the region. The numerator, $(y(x') - \bar{y}(x))$, measures the geometrical distance between x and x' and the denominator, $(x' - x)$, makes x' near the x more influential to the variability. $1/(n_x - 1)$ averages the calculated variability. The intent behind the variability-based sampling is that the regions in which the profile varies more frequently should include more sampling locations, and thus the details within the region can be captured properly. The probability distribution for sampling as defined in Equation (4) is used, and there is a higher probability of selecting the sampling locations with higher local variability as the initial measurements.

In addition to the variability-based sampling, we also employ an iterative bisection algorithm as follows: if a certain region between two selected measurement locations exhibits a high deviation between the estimated mean and the mean from realizations, then we add an additional measurement point in the middle of the region to improve the match. The process is repeated until the deviations in the entire region are lower than the threshold value. The bisection method ensures that every detail that might be missed in the variability-based sampling is included in the sampling domain.

2.4. Gaussian Process Modeling

In the study, we employ mask projection stereolithography technique as a model AM process [15,16]. The layer-by-layer nature of the process inevitably yields a staircase sidewall profile of the part, leading to geometric difference between the manufactured part and the planned geometry (or the 3D computer model). Furthermore, when high-aspect-ratio structures, such as thin struts or beams, are manufactured, the geometric difference can also be affected by the angle between the substrate and the strut [26]. This problem is universal and found in most AM processes. In order to quantify the deviation, we consider the sidewall profiles of the AM part captured using a digital camera attached to a microscope. We assume that the extracted edge roughness profiles of AM parts result from the following two parts: (1) **staircase profile**: consistent curing behavior from the printing apparatus (consistent deviation between the planned geometry and actual printed geometry) and (2) **spatial uncertainty** (small variations): stochastic roughness variations from the AM process arising from inherent uncertainties, such as light scattering, mask projection not perfectly focused on the curing plane, perturbation on the liquid resin surface, and uncontrolled external environmental factors including fluctuations of temperature or oxygen level. After separating the effects from both the aforementioned factors, we focus on the second part, namely the spatially varying stochastic roughness uncertainty that is not a function of the printed angle.

Spatial uncertainty is often represented by a random process that is defined as a collection of random variables in space or time. In the study, we assume that the random process behind the spatial roughness uncertainty is represented by a Gaussian process [27]. The advantage of using a Gaussian process is that the random process is fully described by the mean and covariance functions, and thus it is possible to obtain the full probabilistic prediction as well as the estimation of uncertainty in the prediction.

We define the spatial uncertainty in the sidewall profile as the roughness d and assume that this uncertainty is the realization of a Gaussian process (f). The Gaussian process is a statistical model that produces a set of random variables; and the finite selection of the random values within the set follows a multivariate Gaussian distribution [27]. The Gaussian process is fully described in terms of its mean M and covariance C, and its realization is also fully specified by $M(X)$ and $C(X,X)$ in which X denotes the domain of the realization [28]. The stationary Gaussian process model for roughness spatial uncertainty is defined as follows:

$$
\begin{aligned}
f|M,C &\sim GP(M,C),\\
d|M,C &\sim N(M(X),C(X,X)),\\
M &: X,\alpha,t_M \to \textstyle\sum_i \alpha_i B_{i,k}(X;t_M),\\
C &: X,X',\nu,\phi,\theta \to Matern(X,X',\nu,\phi,\theta),
\end{aligned}
\tag{5}
$$

where GP and N stand for the Gaussian process and normal distribution, respectively. $B_{i,k}$ denotes the B-spline basis function of order k, α_i denotes the corresponding weight factor at i^{th} location, and t_M denotes the knot vector for the mean function [29]. The B-spline representation is used as the mean function to model the local variability in roughness. This is especially useful when the shape of the mean function is not known a priori. The Matérn covariance is used because it represents a diverse class of covariance functions that belong to the exponential family, which is commonly used to model physical stochastic processes [27]. The Matérn covariance includes three hyperparameters, namely ν, ϕ, and θ that control its differentiability, correlation length, and amplitude, respectively.

In order to determine the stationary Gaussian process properly, we applied the Bayesian inference approach to estimate the probability density functions (PDFs) of unknown parameters. It follows the Bayes' theorem as follows [30]:

$$
\Pr(\Phi|d) = \frac{\Pr(d|\Phi) \times \Pr(\Phi)}{\Pr(d)} \propto \Pr(d|\Phi) \times \Pr(\Phi),
\tag{6}
$$

where $\Phi = \{\alpha_i, \nu, \phi, \theta\}$ is a set of unknown parameters in the Gaussian process and $\Pr(\Phi)$ and $\Pr(\Phi|d)$ are the prior and posterior distributions of Φ, respectively. If prior information about the distribution of the unknown parameters exists, it can be incorporated into the prior probability density functions (PDFs) of the corresponding unknown parameters [31]. Conversely, if prior information on the parameters is absent, the uniform PDFs in the certain range can be included. If each parameter is statistically independent to the others, $\Pr(\Phi)$ can be expressed as a multiplication of prior distributions of each parameter. The $\Pr(\Phi|d)$ is the inferred distribution of unknown parameters as a result of having the observed data. $\Pr(d|\Phi)$ is the probability density of the observed data given a model parameterized with parameters Φ. This is called likelihood. $\Pr(d)$ is a normalizing term and it is often ignored because it does not affect the posterior PDFs and the computational cost for calculating this term is quite demanding. In practice, the posterior PDFs of unknown parameters are proportional to the multiplication of the prior PDFs and the likelihood. In this work, we used the multivariate normal distribution as the likelihood and the uniform distribution for prior PDFs of unknown parameters. The sampled data from the variability-based sampling and bisection algorithm is used as the observed data, d.

Since it is not always possible to estimate the closed form for the posterior PDFs of unknown parameters and calculation of the posterior PDFs involves high dimensional integration. Therefore, the sampling-based approach, called the Markov Chain Monte Carlo (MCMC) algorithm, is used to estimate the posterior PDFs of unknown parameters in this study [30,32]. The convergence of posterior PDFs of unknown parameters is determined by inspecting the trace of each estimating parameter from the sampling. A high value is chosen as the number of iteration of MCMC sampling, so that it ensures the trace of each estimating parameter shows an asymptotic behavior. The asymptotic behavior stands for the state that the mean and the variance of sample stay unchanged. The open-source Bayesian analysis package PyMC [32] is used to perform Monte Carlo sampling to estimate the posterior PDFs of unknown parameters.

3. Results

3.1. Gaussian Process Modeling and Sampling Algorithm Verification

We present a test problem to illustrate and verify our Gaussian process modeling with the variability-based sampling and bisection algorithm. Equation (7) is the mathematical form of the given mean function q and is expressed as follows:

$$q(x) = \sin\left(2(x - 0.9)^4\right)\cos\left(x - 0.2\right) + \frac{1}{2}(x - 0.5)^3, \tag{7}$$

The linear slope between the starting and ending points of the function is removed since locally varying spatial variation is the main focus. We decide to use Equation (7) that has a higher spatial variation frequency in the region $x \in [-1,\ 0]$ when compared to that in region $[0,\ 1]$ to test whether our algorithm captures the correct behavior of spatial roughness variation when its characteristics are not uniform over the entire domain. A Matérn covariance function with the parameters: $v = 2$, $\phi = 2$, $\theta = 0.4$ is used to generate the spatial variation. We generate realizations of the random process and sample these realizations at a total of 61 measurement locations. The sampled data is used as the input for estimation, wherein we try to reconstruct the parameters of the actual stochastic process from which the data originated. We assign the following prior PDFs for the unknown parameters:

$$\begin{aligned}
\alpha_i &\sim Uniform\ [-10,\ 10], \\
v &\sim Uniform\ [1,\ 3], \\
\phi &\sim Exponential\ (1), \\
\theta &\sim Exponential\ (1),
\end{aligned} \tag{8}$$

The prior PDFs can be specifically selected to incorporate any knowledge regarding the values of the unknown parameters. Otherwise, they can be set as uniform distribution in the absence of the aforementioned information. We use 500 realizations of the given random process that are sampled at the given locations to generate the joint posterior distribution of the parameters and select 1000 samples from it using the Markov chain Monte Carlo method with a burn-in of 30,000 and a thinning factor of 10 [17].

Figure 2a shows 10 realizations from the given stochastic process. The sampled points in the 1st (circles), 2nd (triangles), and 3rd (rectangles) runs of non-uniform sampling are also shown in Figure 2b. Fourteen points from the 1st sampling based on the local variability effectively capture the high frequency spatial variation near the left sidewall. However, they miss a few details near the left sidewall as well as near the right sidewall. The lack of sampling points is compensated in the 2nd and 3rd bisection steps by placing additional sampling points and reducing the deviation between the mean function calculated from all realizations and the estimated mean function. Total sampled points in the 2nd and 3rd sampling from the bisection method are 20 and 31, respectively.

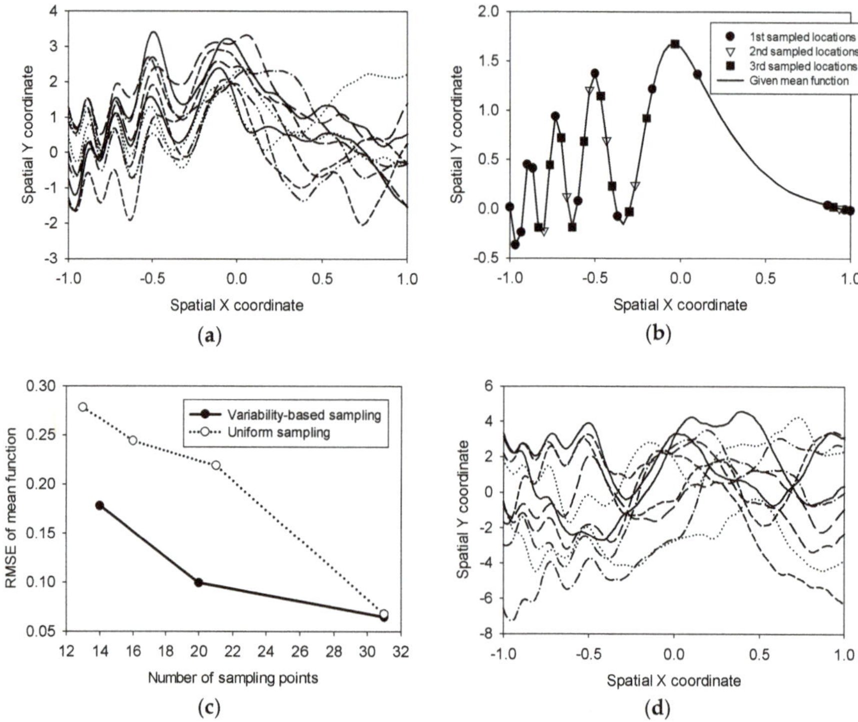

Figure 2. Given and estimated stochastic processes: (**a**) 10 realizations from the given model; (**b**) Given mean function (solid line) and sampled points in 1st (circle), 2nd (triangle), and 3rd (rectangle) runs after variability-based sampling algorithm; (**c**) root-mean-square error (RMSE) of the mean function from uniform and variability-based sampling algorithm; (**d**) 10 sampled realizations from the estimated Gaussian process model.

To visualize the effectiveness of variability-based sampling and the bisection algorithm, the root-mean-square error (RMSE) between the estimated mean and the mean from realizations is calculated and compared in two different sampling cases: proposed variability-based sampling and uniform sampling. The result is shown in Figure 2c. The x-axis denotes the number of sampling points and the y-axis denotes the RMSE value. Given the efficient placement of measurement locations, the RMSE in non-uniform sampling is considerably lower than that in the uniform sampling result with an equal number of sampling points. Additionally, while the RMSE is reduced with increases in the sampling locations, the RMSE decreases significantly faster in the variability-based method. The RMSEs in both methods are identical when the sampling locations correspond to almost half of the total measurement locations. Therefore, we conclude that variability-based sampling and the bisection algorithm is a more attractive option for the effective allocation of computing power and resources. Figure 2d shows 10 sampled realizations from the estimated Gaussian process model.

In addition to the comparison of the mean function, the estimated parameters in the Matérn covariance function with respect to the number of sampling points are shown in Table 2. There are three parameters in the Matérn covariance function, and the given values are denoted in the parentheses. In contrast to the deviations in the mean function, the estimated parameters do not significantly depend (within a 5% error) on the number of sampling points. It can be speculated that the sampled points from the variability-based sampling gives enough information to accurately estimate the hyperparameters in the covariance function even though it missed a few local variabilities in the mean function. Figure 3 shows the prior and the posterior PDFs of three unknown parameters in the Matérn covariance function. The prior PDF from Equation (8) is displayed with a blue line at the displayed region and the blue histogram is from the posterior PDF obtained by histogram of the trace values from MCMC sampling. The red line is a fitted normal distribution to the histogram. Figure 3 clearly visualizes the updating process of Bayesian inference by incorporating the observed data. It is important to emphasize that since the estimated parameters are obtained as PDFs instead of single values, the Gaussian process described by the estimated parameters is a set of multiple Gaussian processes defined by hyperparameters realized from the posterior PDFs. This helps us to model a wide range of uncertainties in the observed data while keeping the practical advantages of Gaussian process.

Table 2. Estimated hyperparameters in the covariance function for the test problem. The given values are denoted in the parentheses.

	ν (2)	ϕ (2)	θ (0.4)
1st sampled locations (14 points)	2.011 ± 0.027	2.052 ± 0.028	0.416 ± 0.012
2nd sampled locations (20 points)	1.978 ± 0.025	2.063 ± 0.030	0.418 ± 0.009
3rd sampled locations (31 points)	1.971 ± 0.016	2.055 ± 0.029	0.415 ± 0.007

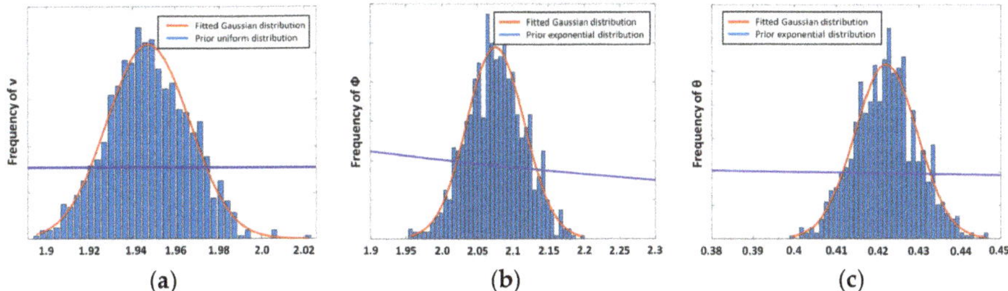

Figure 3. Probability density functions (PDFs) of the estimated parameters in Matérn covariance function: (a) ν, (b) ϕ, and (c) θ. The blue line is a part of the given prior PDF at the displayed region and blue histogram is the frequency of sampled parameters from the posterior PDF. The red curve is a fitted normal distribution to the histogram. The histograms and the prior PDFs are not in the same scale.

3.2. Extracting Profiles from the Images of Additively Manufactured Struts.

Gaussian process modeling with the non-uniform sampling algorithm is applied to the sidewall profiles of AM fabricated struts. Figure 4a–c shows the CAD geometries of struts in 3 different printing angles ($90°$, $75°$, and $60°$) and printing directions. The layer thickness corresponds to 80 pixel units. The sidewall profiles of multiple AM fabricated struts with different angles ($90°$, $75°$, and $60°$) are shown in Figure 4d–f. The sidewall profile consists of multiple layers printed repeatedly layer-by-layer, and we assume that the spatial uncertainty in each printed layer is generated from an independent event, so that is not related to the neighboring layers. There are three parameters in Equation (1) that are reduced by considering the two ratios, λ_b/λ_0 and γ/λ_0. We manually segmented one printed layer in the image and varied the two ratios to determine the optimized parameter values to obtain the optimal match with respect to the manually segmented edge. After calibrating two ratios in the level set method, the same algorithm is applied to the remaining layers to extract the edge information from the image.

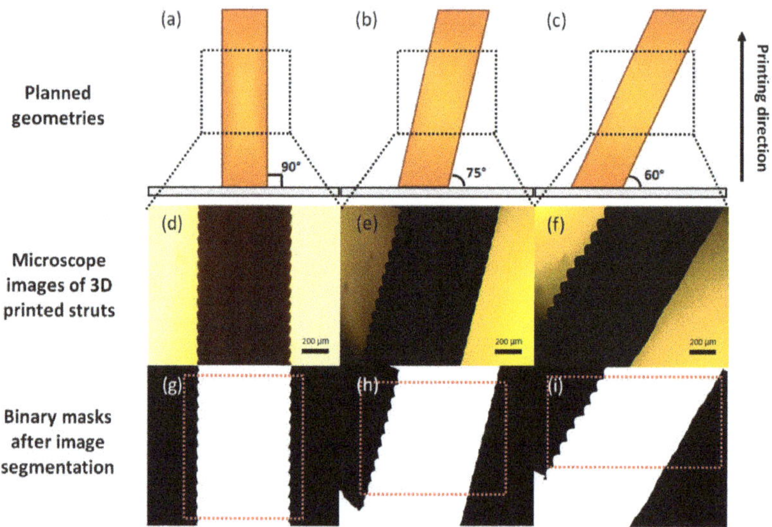

Figure 4. Microscope images and corresponding binary images of additively manufactured struts:
(**a–c**) CAD geometries of struts in different printing angles (90°, 75°, and 60°); (**d–f**) Microscope images
of additively manufactured struts (90°, 75°, and 60°); (**g–i**) Binary masks after image segmentation
(90°, 75°, and 60°). The sidewall profiles in the red box (dotted line) are used for characterizing the
spatial uncertainty.

Figure 4g–i shows the binary masks of the images after segmentation in which the white pixels
represent the segmented region of the strut and the black pixels denote the background region. Given
the smooth intensity gradient from the center to the edge of the image, the segmentation at the edges
of the images is inaccurate in the 75° and 60° cases in Figure 4h,i. The artifact from the background
inhomogeneity can be easily removed by measuring and compensating the intensity gradient in the
background. The compensation is not applied in the study to minimize the manual intervention.
A more systematic approach to reduce the effect from the local inhomogeneity is given in [33].

The edge profile is extracted by collecting the outermost pixels in the left and right sidewalls
of binary masks. Figure 5a,b shows the extracted edge profiles in the left and right sidewalls of
each printed layer, respectively. It is clear that the side profile of all layers show consistent trend
depending on the printing angle. This characteristic staircase profile is caused by the light absorption
and scattering in the resin vat and light penetration during the curing of upper layer. In a manner
similar to the test case, the mean of all observations, which represents the staircase profile, is subtracted
from the sidewall profile; thus, the roughness spatial uncertainty is independent of the printed
angle. The mean profile is obtained by simple summation of all side profiles in each printing angles.
The summation averages out the local roughness from the mean profile. The mean trends extracted
from all observations and the roughness extracted spatial uncertainties are shown in Figure 5c–f,
respectively. In total, 37 roughness profiles are collected and the roughness is independent of printing
angle as well as direction of side profile as shown in the Figure. This will be used for estimating
hyperparameters in the Gaussian process model.

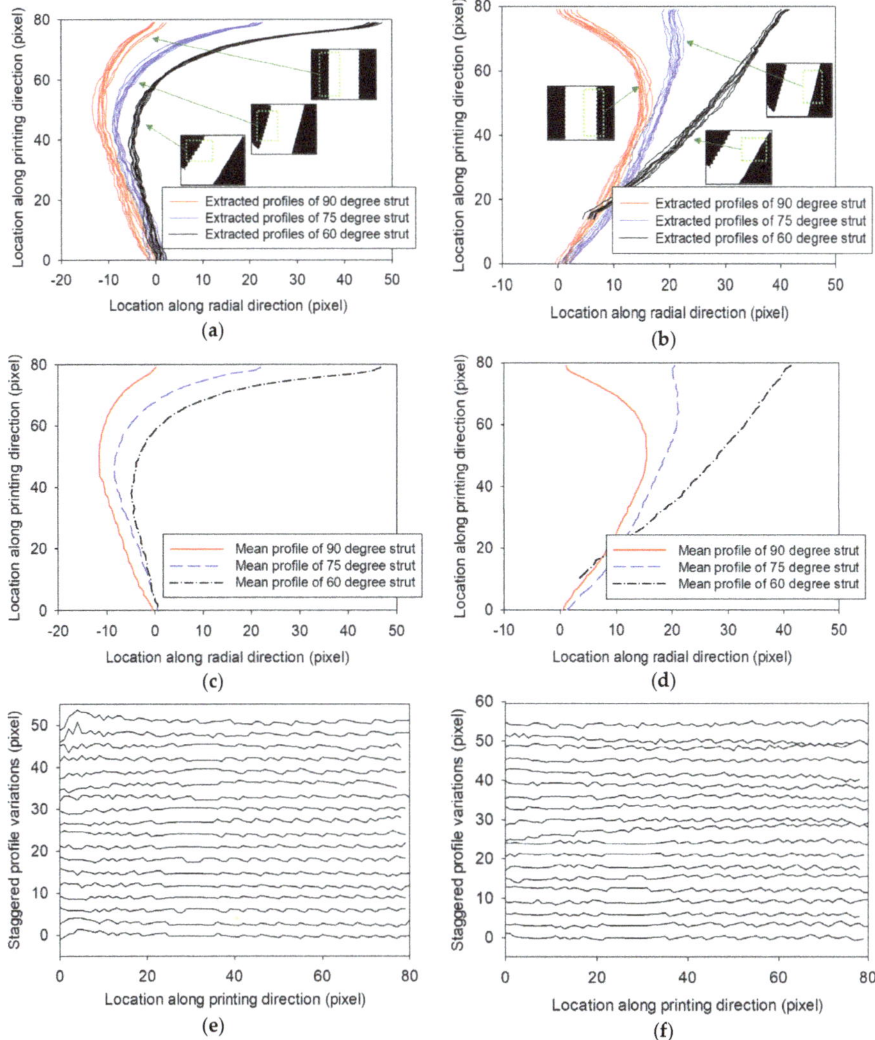

Figure 5. Extracted sidewall profiles of additively manufactured struts: (**a**) The extracted left-sidewall profiles in different angles (90°, 75°, and 60°); (**b**) The extracted right-sidewall profiles in different angles (90°, 75°, and 60°); (**c**) Extracted left-sidewall mean profiles in different angles (90°, 75°, and 60°); (**d**) Extracted right-sidewall mean profiles in different angles (90°, 75°, and 60°); (**e**) Extracted left-sidewall roughness profiles after clean-up; (**f**) Extracted right-sidewall roughness profiles after clean-up.

In contrast to the test case, we possess prior information on the distribution of the amplitude and scaling parameters of covariance function from the geometrical constraints of the printed struts. Based on the roughness spatial variation amplitude from observations, the upper limit on the prior distribution of the amplitude parameter in the covariance function is 5. Additionally, the maximum correlation length is limited up to the height of a printed layer. Equation (9) describes the prior information of the hyperparameters.

$$\begin{aligned}
\alpha_i &\sim Uniform\ [-10,\ 10], \\
\nu &\sim Uniform\ [1,\ 3], \\
\phi &\sim Uniform\ [0,\ 5], \\
\theta &\sim Uniform\ [0,\ 80],
\end{aligned} \tag{9}$$

Twenty realizations from the estimated stochastic model are shown in Figure 6. In a manner similar to the observations from the extracted profiles of actual struts, the fluctuations are distributed uniformly over the domain, and this indicates the characteristics of the stationary Gaussian process. This reproduced roughness will be superimposed in the CAD geometry with the mean profile to quantify the effect of side profile on the mechanical behavior of AM parts.

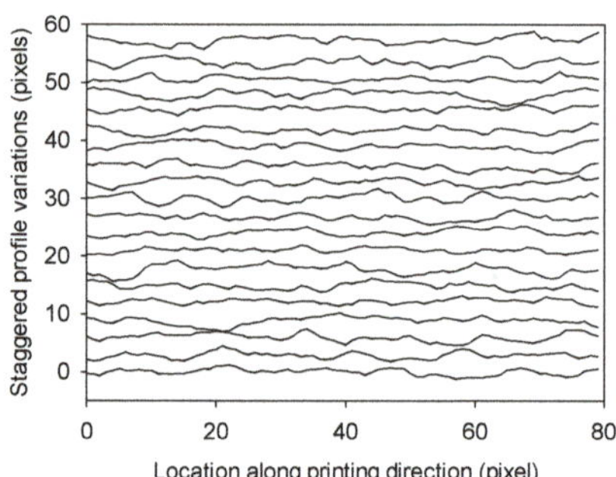

Figure 6. Twenty sampled realizations from the estimated Gaussian process model.

3.3. Effect of Spatial Uncertainty on the Mechanical Behavior of Additively Manufactured Struts

In the study, we consider simple struts with three different printing angles (90°, 75°, and 60°). In order to clearly visualize the effect from the spatial uncertainty, the struts in FE analysis are simplified as 2D struts. To investigate the effect of spatial uncertainty on the mechanical properties of additively manufactured struts, three computational models with different strut geometries are first prepared as follows: (1) **nominal strut**: desired CAD geometry with a smooth side surface (planned geometry without any staircase profile as well as spatial uncertainty), (2) **predicted AM strut**: geometry with predicted roughness in which the staircase profile and the spatial uncertainty are reproduced from our estimated Gaussian process model, and (3) **volume-matched strut**: strut also with a smooth side surface, having a thickness adjusted by smoothing the ridges and valleys of the rough side profile such that the volume is matched with the predicted AM strut. The analysis of (3) is necessary for fair comparison because the thickness of the manufactured struts is slightly lower than that of the planned geometry. By including (3), the difference in the mechanical properties that are caused by the volume change can be investigated. The thickness of the strut in the CAD model is 800 μm, and the thicknesses of struts in the volume-matched geometries extracted from the sample images are approximately 754 μm, 754 μm, and 760 μm for 90°, 75°, and 60°, respectively. For the computational efficiency, one third of the height of printed strut (2000 μm) is modeled in the simulation. The three different aforementioned geometries are created for each printing angle, namely 90°, 75°, and 60°.

A linear static structural analysis was performed to calculate the mechanical property of struts in three different printing angles (90°, 75°, and 60°) under compressional loading conditions. A commercial FE software package, COMSOL Multiphysics™, running on the personal computer (Windows 10 64 bit, Intel i7-4790 CPU, 16 GB RAM), was used for all numerical simulations. The governing equation for the struts is as follows [34]:

$$0 = \nabla \cdot (FS)^{T} + Fv$$
$$\sigma = J^{-1}FSF^{T}$$
$$\varepsilon = \tfrac{1}{2}(F^{T}F - I) \tag{10}$$
$$S = \frac{\partial W_S}{\partial \varepsilon}$$
$$W_S = \tfrac{\mu}{2}(\bar{I}_1 - 3)$$

where F is the deformation gradient, S is the second Piola–Kirchhoff stress, v is the left stretch tensor. The Cauchy stress tensor, σ, is the multiplication of F, S, the transpose of F, and the volume ratio, J, which is the determinant of F. The Lagragian–Green strain, ε, is a function of F and the identity tensor, I. The second Piola–Kirchhoff stress, S, is the gradient of the strain energy-density function, W_s with respect to ε. $\bar{I}_1$ is the scalar invariant of the left Cauchy–Green deformation tensor. The struts are made of a photo-polymerizable poly(ethylene glycol) diacrylate (PEGDA) (molecular weight 250). PEGDA is assumed to be a hyperelastic, incompressible, and isotropic material and modeled by using neo-Hookean model [34]. The uniaxial compression test is performed to obtain the material property of PEGDA. The stress–strain curve obtained from the experiment is fitted to the neo-Hookean model to determine the shear modulus, μ. The fitted value of μ is approximately 36.1 MPa for the PEGDA.

The interfacial boundaries of each layer were merged together with its neighboring layers using "union assembly" feature in COMSOL software, assuming the perfect binding condition between layers. Zero-displacement was imposed on the bottom of the strut, and the controlled displacement was applied on the top surface of the strut in the vertical direction as described in Equation (11).

$$u_{bottom} = 0$$
$$u_{top,vertical} = u_0 \tag{11}$$

where u is the displacement vector in the computational domain and u_{bottom} is the displacement vector of elements at the bottom of strut. $u_{top,vertical}$ is the vertical displacement of elements at the top of the strut and u_0 is the controlled displacement that was applied as a boundary condition.

Figure 7a shows the loading direction and the boundary condition in the three struts: nominal, predicted AM, and volume matched struts. The applied strain was limited up to 0.03 to avoid the nonlinear deformation of the strut. Since the strut is considered as the building block of unit cell structure, the strain is defined as the actual deformation of the strut divided by the height of the strut. With respect to the deformation of the strut, the resulting stress was calculated as the average of total pressure over the bottom surface of the strut. Triangular elements as well as the adaptive mesh size control were adopted to make sure that it captures all details of spatial uncertainties at the edge of struts. The mesh size convergence test was performed to ensure that the simulation result is not sensible to the mesh size. The number of elements in three different geometries, giving the consistent simulation results, are 10,040, 831,944, and 9474 for the nominal strut, predicted AM strut, and volume-matched strut, respectively. The number of elements for predicted AM strut is significantly higher than the others because of the geometrical details of spatial uncertainty at the edge of strut.

Figure 7b shows the stress–strain curves of 9 different examples obtained from the FE simulation. The slope of the stress–strain curve denotes the stiffness of the strut. The slopes of all stress–strain curves are grouped into three categories based on its printing angle, and a slope deviation exists given the existence of the spatial deviation. The decrease in slope is qualitatively observed in the predicted AM strut and volume-matched strut. To quantitatively understand the effect of spatial deviation on the stiffness of the struts, the stiffness of the volume-matched and predicted AM struts are normalized with the stiffness of the nominal struts as shown in Figure 7c. The change in the effective stiffness of the volume-matched geometry is not significant (less than 7% of the stiffness of the nominal) when the printing angle changes from 90° to 60°. However, when the geometry exhibits roughness including the staircase profile as well as the spatial uncertainty, the stiffness is reduced to 70% of the nominal geometry when the printing angle is 60°. The reduction in stiffness is expected since the roughness decreases the cross-sectional area of each strut and decreases the stiffness of the strut. However, it is observed that the roughness effect is more pronounced when compared the reduction in volume-matched case to the predicted AM case. This result indicates that the necessity of details of surface roughness becomes more important as the printing angle reduces. Specifically, the angled strut is frequently used in the lattice structures, which is the assembly of multiple unit cells, consisting of angled struts. Therefore, modeling the surface roughness including spatial uncertainty is important for the accurate estimation of mechanical properties of the lattice structure.

Additionally, we compare the maximum value of von Mises stress in the three cases. Table 3 shows the maximum von Mises stresses of the volume-matched geometry strut and the predicted AM strut normalized with respect to the maximum von Mises stress of the nominal strut. The von Mises stress distribution and the location of the maximum value is displayed in Figure 7d. The stress distribution in Figure 7d shows distinctive difference from the stress distribution in the vertical strut because of two major factors: (1) bending and shear stresses and (2) surface roughness. When a strut is loaded axially, it only develops compressive normal stress. When a strut is loaded with an angle, bending and shear stresses begin to arise, and it becomes more likely that the strut undergoes a large deformation and becomes more susceptible to developing a large stress as shown in Figure 7d. In addition to the effect from bending and shear stresses, local stress concentrations are observed near the side profiles in the predicted AM, which is not observed in the volume-matched strut or the nominal strut in which the sidewall profiles are smooth. The geometric discontinuities caused by the spatial uncertainties change the stress distribution in the strut, and the amount of stress concentration is thrice that in the smooth cases. Thus, it is anticipated that the total stiffness of the additively manufactured strut is affected by the change in stress distribution. Therefore, considering spatial uncertainties in the geometry to accurately predict the mechanical behavior of struts using FE analysis is essential.

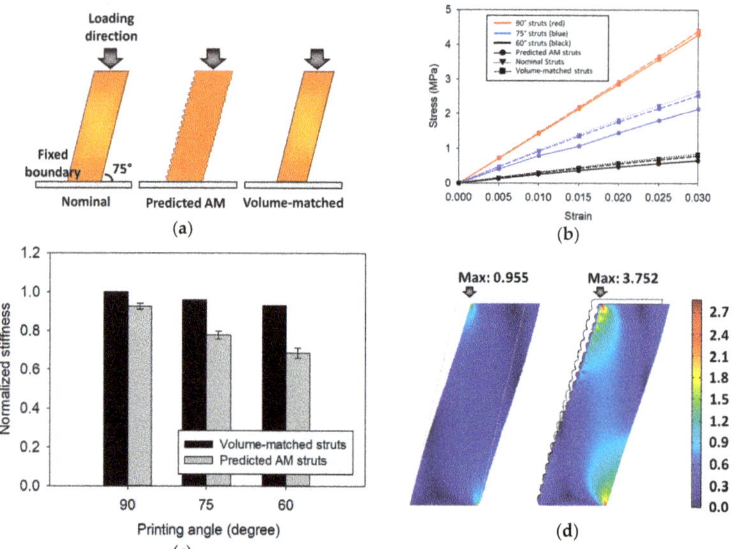

Figure 7. Compression test results of additively manufactured struts: (**a**) Loading and boundary conditions of three struts: nominal, predicted AM, and volume matched struts; (**b**) Stress–strain curve of nominal, predicted AM, and volume-matched struts in different printing angles (90°, 75°, and 60°); (**c**) Normalized effective stiffness of volume-matched geometry and predicted AM struts in different printing angles (90°, 75°, and 60°); (**d**) Normalized von Mises stress distributions in a predicted AM strut and a volume-matched strut.

Table 3. Maximum von Mises stress of the volume-matched struts and the predicted AM struts normalized with respect to those of the nominal struts in different printed angles.

Struts \ Angles	90°	75°	60°
Volume-matched struts	0.975	0.955	0.947
Predicted AM struts	3.410	3.752	3.273

The proposed modeling framework is also applicable to complex AM parts with inner structures. The spatial uncertainties can be estimated by the proposed modeling framework if the inner structure of the complex AM parts is expected to exhibit spatial uncertainties that are statistically similar to the outer structure. Even if the inner structures of the complex AM parts are expected to exhibit statistically different spatial uncertainties, the proposed modeling framework can be still applied to the complex AM parts with the exception of the image segmentation sub-step. As opposed to the use of optical images, there are multiple techniques to measure the inner structure profile of the complex parts, such as X-ray computed tomography and profilometry. Following the extraction of the inner structure profile, it is possible to apply the remaining sub-steps of the proposed modeling framework, namely variability-based sampling and Gaussian process modeling, to the problem.

4. Conclusions

The study presents a general method to model spatial variations of additively manufactured products from their sidewall profile images. Specifically, this can be achieved using three approaches: (1) Gaussian process modeling using Bayesian framework, (2) level set method for image segmentation, and (3) variability-based sampling algorithm. The method is verified via a test problem and applied to the real AM products fabricated by mask projection stereolithography. In order to verify the necessity of the proposed framework, the mechanical properties of printed high-aspect-ratio struts with and without roughness is simulated under compression loading via FE simulation. The FE simulation results show that the staircase profile with the spatial uncertainty reduces the stiffness of printed strut to 70% when the printing angle is 60°. It is also shown that the effect from the spatial roughness increases as the printing angle decreases. This indicates that the accurate estimation of mechanical behavior of AM products as well as lattice structures consisting of multiple angled struts requires to consider the spatial roughness. In addition, the influence from the spatial roughness significantly affects the stress distribution under compressive loading conditions that could result in the location of failure in the structure. It is also expected that the effect from roughness is more significant when the dimension of AM structures approaches the magnitude of roughness. The proposed modeling framework can be applied to complex AM parts which have spatial roughness that can be measured through optical images. The proposed method is not limited to AM products, and it can be applied to other problems in which spatial variations exist and its mechanical behavior is expected to be affected by its spatial roughness.

Author Contributions: Conceptualization, N.K., H.L. and N.A.; Data curation, N.K. and C.Y.; Formal analysis, N.K.; Funding acquisition, H.L. and N.A.; Investigation, N.K.; Methodology, N.K. and C.Y.; Supervision, H.L. and N.A.; Visualization, C.Y.; Writing–original draft, N.K.; Writing–review & editing, C.Y., H.L. and N.A.

Funding: This research was funded by the National Science Foundation under grant number 1420882. We also acknowledge the support of DMDII. Additionally, we acknowledge support from the Agency for Defense Development under contract number UD150032GD.

Conflicts of Interest: The authors declare no conflict of interest.

References

1. Melchels, F.P.W.; Feijen, J.; Grijpma, D.W. A review on stereolithography and its applications in biomedical engineering. *Biomaterials* **2010**, *31*, 6121–6130. [CrossRef] [PubMed]
2. Alapan, Y.; Hasan, M.N.; Shen, R.; Gurkan, U.A. Three-dimensional printing based hybrid manufacturing of microfluidic devices. *J. Nanotechnol. Eng. Med.* **2015**, *6*, 021007. [CrossRef] [PubMed]
3. Zheng, X.; Lee, H.; Weisgraber, T.H.; Shusteff, M.; DeOtte, J.; Douss, E.B.; Kuntz, J.D.; Biener, M.M.; Ge, Q.; Jackson, J.A. Ultralight, ultrastiff mechanical metamaterials. *Science* **2014**, *344*, 1373–1377. [CrossRef] [PubMed]
4. Bauer, J.; Meza, L.R.; Schaedler, T.A.; Schwaiger, R.; Zheng, X.; Valdevit, L. Nanolattices: An emerging class of mechanical metamaterials. *Adv. Mater.* **2017**, *29*, 1701850. [CrossRef] [PubMed]
5. Schaedler, T.A.; Ro, C.J.; Sorensen, A.E.; Eckel, Z.; Yang, S.S.; Carter, W.B.; Jacobsen, A.J. Designing metallic microlattices for energy absorber applications. *Adv. Eng. Mater.* **2014**, *16*, 276–283. [CrossRef]
6. Frenzl, T.; Findeisen, C.; Kadic, M.; Gumbsch, P.; Wegener, M. Tailored buckling microlattices as reusable light-weight shock absorbers. *Adv. Mater.* **2016**, *28*, 5865–5870. [CrossRef] [PubMed]
7. Dejean, T.T.; Spierings, A.B.; Mohr, D. Additively-manufactured metallic micro-lattice Materials for high specific energy absorption under static and dynamic loading. *Acta Materialia* **2016**, *116*, 14–28. [CrossRef]
8. Roper, C.S.; Fink, K.D.; Lee, S.T.; Kolodzijeska, J.A.; Jacobsen, A.J. Anisotropic convective heat transfer in microlattice materials. *AIChE J.* **2013**, *59*, 622–629. [CrossRef]
9. Maloney, K.J.; Fink, K.D.; Schaedler, T.A.; Kolodzijeska, J.A.; Jacobsen, A.J.; Roper, C.S. Multifunctional heat exchangers derived from three-dimensional micro-lattice structures. *Int. J. Heat Mass Tranf.* **2012**, *55*, 2486–2493. [CrossRef]
10. Warnke, P.H.; Doyglas, T.; Wolny, P.; Sherry, E.; Steiner, M.; Galonska, S.; Becker, S.T.; Springer, I.N.; Wiltfang, J.; Sivananthan, S. Rapid prototyping: Porous titanium alloy scaffolds produced by selective laser melting for bone tissue engineering. *Tissue Eng. Part C Methods* **2008**, *15*, 115–124. [CrossRef]
11. Ahn, D.; Kim, H.; Lee, S. Surface roughness prediction using measured data and interpolation in layered manufacturing. *J. Mater. Process. Technol.* **2009**, *209*, 664–671. [CrossRef]
12. Ancau, M.; Caizar, C. The optimization of surface quality in rapid prototyping. In Proceedings of the WSEAS International Conference on Engineering Mechanics, Structures, Engineering Geology, Heraklion, Crete Island, Greece, 22–24 July 2008.
13. Hasan, R. Progressive Collapse of Titanium Alloy Micro-Lattice Structures Manufactured Using Selective Laser Melting. Ph.D. Thesis, University of Liverpool, Liverpool, UK, 2013.
14. Meza, L.R.; Greer, J.R. Mechanical characterization of hollow ceramic nanolattices. *J. Mater. Sci.* **2014**, *49*, 2496–2508. [CrossRef]
15. Zheng, X.; Deotte, J.; Alomso, M.P.; Farquar, G.R.; Weisgraber, T.H.; Gemberling, S.; Lee, H.; Fang, N.; Spadaccini, C.M. Design and optimization of a light-emitting diode projection micro-stereolithography three-dimensional manufacturing system. *Rev. Sci. Inst.* **2012**, *83*, 125001. [CrossRef]
16. Sun, C.; Fang, C.; Wu, D.M.; Zhang, X. Projection micro-stereolithography using digital micro-mirror dynamic mask. *Sens. Actuators A Phys.* **2005**, *121*, 113–120. [CrossRef]
17. Alwan, A.; Aluru, N.R. Data-driven stochastic models for spatial uncertainties in micromechanical systems. *J. Micromech. Microeng.* **2015**, *25*, 115009. [CrossRef]
18. Pal, N.R.; Pal, S.K. A review on image segmentation techniques. *Pattern Recognit.* **1993**, *26*, 1277–1294. [CrossRef]
19. Shapiro, L.; Stockman, G.C. *Computer Vision*; Prentice Hall: New Jersey, NJ, USA, 2001.
20. Cremers, D.; Rousson, M.; Deriche, R. A review of statistical approaches to level set segmentation: Integrating color, texture, motion and shape. *Int. J. Comput. Vis.* **2007**, *72*, 195–215. [CrossRef]
21. Sahoo, P.K.; Soltani, S.; Wong, A.K.C. A survey of thresholding techniques. *Comp. Vis. Graph. Image Process.* **1988**, *41*, 233–260. [CrossRef]
22. Osher, S.; Sethian, J.A. Fronts propagating with curvature-dependent speed: Algorithms based on hamilton-jacobi formulations. *J. Comput. Phys.* **1988**, *79*, 12–49. [CrossRef]
23. Chan, T.F.; Vese, L.A. Active contours without edges. *IEEE Trans. Image Process.* **2001**, *10*, 266–277. [CrossRef]
24. Anderson, A.B.; Wang, G.; Gertner, G. Local variability based sampling for mapping a soil erosion cover factor by co-simulation with landsat tm images. *Int. J. Remote Sens.* **2006**, *27*, 2423–2447. [CrossRef]

25. Jin, R.; Chang, C.J.; Shi, J. Sequential measurement strategy for wafer geometric profile estimation. *IIE Trans.* **2012**, *44*, 1–12. [CrossRef]

26. Kim, N.; Bhalerao, I.; Han, D.; Yang, C.; Lee, H. Improving surface roughness of additively manufactured parts using a photopolymerization model and multi-objective particle swarm optimization. *Appl. Sci.* **2019**, *9*, 151. [CrossRef]

27. Rasmussen, C.E. Gaussian processes for machine learning. In *Lecture Notes in Computer Science*; Springer: Berlin/Heidelberg, Germany, 2006; pp. 63–71.

28. Cressie, N. *Statistics for Spatial Data*; Wiley Series in Probability and Statistics; John Wiley and Sons Ltd.: New York, NY, USA, 1993.

29. De Boor, C. *A Practical Guide to Splines*; Springer: New York, NY, USA, 1978; Volume 27.

30. Bishop, C.M. *Pattern Recognition and Machine Learning*; Springer: New York, NY, USA, 2006.

31. Alwan, A.; Aluru, N.R. A nonstationary covariance function model for spatial uncertainties in electrostatically actuated microsystems. *Int. J. Uncertain. Quantif.* **2015**, *5*. [CrossRef]

32. Patil, A.; Huard, D.; Fonnesbeck, C.J. Pymc: Bayesian stochastic modelling in python. *J. Stat. Softw.* **2010**, *35*, 1. [CrossRef]

33. Wang, X.F.; Huang, D.S.; Xu, H. An efficient local chan–vese model for image segmentation. *Pattern Recognit.* **2010**, *43*, 603–618. [CrossRef]

34. Asaro, R.; Lubarda, V. *Mechanics of Solids and Materials*; Cambridge University Press: Cambrige, UK, 2006.

 *applied sciences*

Article

Fabrication of Multiscale-Structure Wafer-Level Microlens Array Mold

Shuping Xie [1,2], **Xinjun Wan** [1,2,*] **and Xiaoxiao Wei** [1,2]

[1] Engineering Research Center of Optical Instrument and System, The Ministry of Education, University of Shanghai for Science and Technology, Shanghai 200093, China; 161390011@st.usst.edu.cn (S.X.); weixx@usst.edu.cn (X.W.)

[2] Shanghai Key Laboratory of Modern Optical System, University of Shanghai for Science and Technology, Shanghai 200093, China

* Correspondence: xinjun.wan@usst.edu.cn; Tel.: +86-136-5192-1672

Received: 3 January 2019; Accepted: 28 January 2019; Published: 31 January 2019

Abstract: The design and manufacture of cost-effective miniaturized optics at wafer level, using advanced semiconductor-like techniques, enables the production of reduced form-factor camera modules for optical devices. However, suppressing the Fresnel reflection of wafer-level microlenses is a major challenge. Moth-eye nanostructures not only satisfy the antireflection requirement of microlens arrays, but also overcome the problem of coating fracture. This novel fabrication process, based on a precision wafer-level microlens array mold, is designed to meet the demand for small form factors, high resolution, and cost effectiveness. In this study, three different kinds of aluminum material, namely 6061-T6 aluminum alloy, high-purity polycrystalline aluminum, and pure nanocrystalline aluminum were used to fabricate microlens array molds with uniform nanostructures. Of these three materials, the pure nanocrystalline aluminum microlens array mold exhibited a uniform nanostructure and met the optical requirements. This study lays a solid foundation for the industrial acceptation of novel and functional multiscale-structure wafer-level microlens arrays and provides a practical method for the low-cost manufacture of large, high-quality wafer-level molds.

Keywords: wafer-level optics; antireflection nanostructure; microlens array mold; ultraprecision machining; anodic aluminum oxide

---

1. Introduction

Microlens arrays (MLA) are fundamental micro-optical elements composed of a series of lenslets, with diameters ranging from several micrometers to several millimeters, that are arranged in a certain configuration. They are used in diverse applications, such as micro-optical collimation, diffusion lighting, three-dimensional (3D) imaging, light homogenization, and wavefront sensing [1–6]. Wafer-level MLA fabrication is of special interest in the manufacture of mobile phone or augmented reality eye-tracking cameras, in which the wafer-stacking process is used to achieve the parallel fabrication of thousands of compact camera modules [7,8]. However, Fresnel surface reflection loss is still an issue in wafer-level MLA applications. Various antireflection (AR) technologies have been used to suppress surface reflection, including single or multilayer dielectric coatings and subwavelength textured surfaces [9]. Dielectric coatings are currently used in wafer-level MLAs; however, 4−8″ diameter wafer-level MLAs are only several hundred microns thick, giving a high probability of wafer warpage during the coating deposition process [10]. Wafer warpage may have a considerable negative impact on wafer registration. Moreover, microcamera modules are required to withstand high-temperature reflow processes for subsequent soldering onto circuit boards. Dielectric AR coatings are unsuitable for this process, since the coefficient of thermal expansion (CTE) mismatch between the AR coating and the base microlens material can lead to coating fracture [6,9,10]. There are previous studies that report the fabrication of AR moth-eye nanostructures on a single lens or on flexible material-based samples [11–13]. In this paper, we have proposed the fabrication of AR moth-eye nanostructures onto wafer-level MLAs; these nanostructures not only satisfy the AR requirement of the wafer MLA, but also overcome the problem of coating fracture during thermal reflow processes.

For economic reasons, MLAs are usually manufactured by melting photoresist, ultraviolet (UV) imprinting, injection molding, two-photon polymerization, or other batch processes [14–19]. Melting of the photoresist layer provides a simple and practical method for MLA fabrication, but this conventional process is limited by the high relief depth of the MLA. In addition, it is difficult to control the surface-shape error of microlenses [20]. In contrast, a high-precision MLA mold is used in the UV-imprinting process, which is followed by UV-curing the photoresist layer [16], and in the injection-molding process, in which soft optical resin material is injected into the mold to obtain MLAs with the desired optical properties. Since the width to thickness ratio of the wafer MLA is very large, the mold-based UV-imprint process is currently used in industrial wafer MLA manufacturing. Using high-precision MLA molds provides simple-processing, low-cost, and easier mass production of complex aspherical MLAs. Ultraprecision machining is capable of directly producing metal optical surface molds [21–25] in the different shapes that are required for various wafer-level MLA applications, such as spherical, aspherical, or freeform microlens surfaces.

In order to combine moth-eye nanostructures with the wafer-level MLA-mold fabrication method, a multiscale-structure wafer-level MLA mold is proposed, which encompasses hundreds of nanometer-level structured arrays for AR construction, hundreds of micrometer-level MLAs for camera imaging, and finally hundreds of millimeter-level wafer substrates for wafer-level camera manufacture.

In this study, three different materials, namely 6061-T6 aluminum alloy, high-purity polycrystalline aluminum, and pure nanocrystalline aluminum were used as substrates to fabricate MLA molds with an aspherical design, using ultraprecision machining. Then, each mold was treated with a self-assembly process based on anodic aluminum oxidation (AAO) to generate nanostructures over the mold surface. Optical and scanning electron microscopy (SEM) measurements were performed on the three types of multiscale molds. The multiscale mold fabricated using pure nanocrystalline aluminum material exhibited uniform nanostructures, and the optical measurements demonstrated that both the profile peak-to-valley (PV) error of the lenslet and the surface roughness met the optical requirements.

2. Materials and Methods

2.1. Fabrication Process

The wafer-level MLA is a thin, 4–8″ diameter wafer, distributed with thousands of microlenses. As shown in Figure 1, the UV-curable resin-based imprint process transfers the microlens shape on the metal mold to a glass wafer substrate, which is free of high pressure, and then the wafer-level microlens array is released from the mold after curing. Several MLA wafers can be stack-assembled and cut into thousands of multisurface lens modules, resulting in very compact microcameras or camera arrays [26]. Currently, the wafer-level MLA process is mainly used to fabricate compact lens modules for low-resolution near-infrared (NIR) eye-ball tracking cameras or disposable endoscope modules. The general requirements of MLA surface characteristics are a PV error <0.5 μm with a sharp edge, a surface roughness (Sa) <20 nm, and a Fresnel reflectivity <1% to increase sensitivity and avoid ghost images. The wafer-level MLA mold is one of the core components in the whole process. Here, the fabrication of wafer-level MLA molds is followed by an AAO-based self-assembly process to provide periodically arranged nanostructuring.

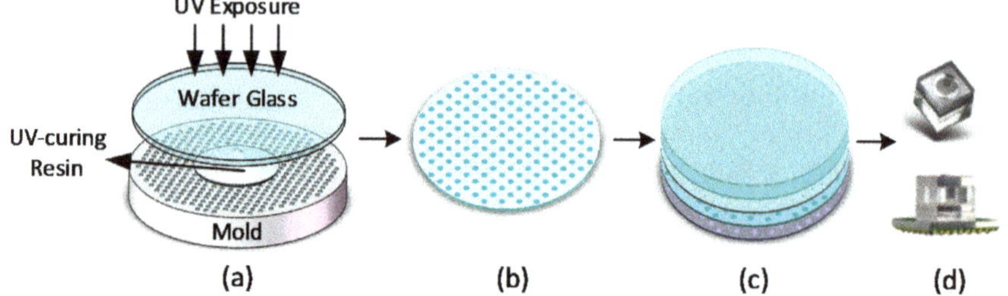

Figure 1. Wafer-level camera fabrication and assembly process, which creates extremely thin camera modules: (**a**) replication of microlens array using a mold, (**b**) wafer-level microlens array, (**c**) wafer-to-wafer bonding, and (**d**) dicing of wafer stack and bonding of the lens module to the sensor. UV: ultraviolet.

The common ultraprecision machining methods used to generate precision MLA metal molds include fast tool servo (FTS), micromilling, and slow slide servo (SSS) processes [21,22,27]. In this study, micromilling and SSS diamond-machining processes were adopted to fabricate the MLA metal molds, using an ultraprecision machine (350FG, Moore Nanotechnology, Swanzey, NH, USA). For low-density arrays, the SSS-machining process was selected, as its implementation requires only 3 axes of freedom. For high-density MLA molds, the micromilling process was selected, as it allows more flexibility in large-slope microlens design.

The key factors that decide the quality of the MLA mold include the lenslet profile accuracy, mold surface roughness, and the lenslet positioning accuracy, which are closely related to the metal mold material, the diamond-machining performance, the machine-tool lifetime, and the machining process parameters. The mold material is the most crucial influencing factor. Metal materials that are used conventionally in diamond machining are usually high-ductility metals, such as aluminum and its alloys, copper alloys, and electroless nickel. Here, considering the further fabrication of nanostructures, various aluminum materials were selected as candidates for mold materials, namely 6061-T6 aluminum alloy, high-purity polycrystalline aluminum, and pure nanocrystalline aluminum.

Wafer-level MLA trial molds based on these aluminum materials were diamond-machined and evaluated according to the mold requirements. For the following machining tests, we designed the MLA aspherical shape with the cavity aperture at 1.0 mm and the sag at 85 μm. 6061-T6 aluminum alloy is the most widely diamond-machined material owing to its relatively high ductility and strength. A 6061-T6 4″ MLA mold with a 20×20 array was machined via the SSS process. The nose radius of the single-point diamond tool was 0.3 mm, the spindle rotational speed was ~100 rpm, and the diamond tool traveled from the periphery to the center of the mold at a speed of 1 mm/min. The depth of the cut for a rough cut was 70 μm, the semifinished depth of the cut was 8 μm, and the finished depth of the cut was 2 μm. Next a high-purity polycrystalline aluminum pin mold was machined via the traditional diamond-turning method. Finally, we adopted the micromilling process for the 2″ pure nanocrystalline aluminum MLA mold to create a dense lenslet array.

To generate a uniform layer of nanostructures on the MLA mold, we adopted the two-step AAO process, which is widely used for nanoporous template applications [28,29]. In the AAO experiments, we investigated characteristics, such as pore size, interpore distance, and nanolayer thickness, by varying the applied voltage, electrolyte concentration, and reaction time. In the first anodization step, the clean MLA mold was anodized in oxalic acid solution at 25 °C, the voltage was set at 40–50 V, and then a mixed solution of H_3PO_4 and H_2CrO_4 was used to remove the oxidized layer. Then, a pore-widening process was carried out using H_3PO_4 for 30 min. After optimizing the parameters of the anodization process, the nanostructured mold was finally fabricated with an average pore diameter of ~100 nm and a spatial period of ~120 nm. However, it needs to be clarified whether the AAO process is effective on curved array surfaces and whether the process will destroy the surface accuracy and roughness of mirror-quality precision molds.

2.2. Characterization

The proposed multiscale wafer-level MLA mold needed to satisfy the requirements for the lens mold profile, the surface roughness, and the nanostructure quality of the curved lens mold. The surface profiles of the MLA mold trial samples were obtained using a phase grating interferometer (PGI) Dimension stylus profiler (Taylor Hobson, Leicester, UK). The surface roughness was measured using a white light interference microscope (GTK-contour, Bruker, Billica, MA, USA), based on regional topography (Sa). Finally, the surface nanostructure morphology of the molds was characterized using a Zeiss SEM (Carl Zeiss Microscopy GmbH, Jena, Germany), after the AAO process.

3. Results and Discussion

3.1. 6061-T6 Aluminum Alloy MLA Mold

Figure 2a shows a photograph of the MLA mold fabricated with 6061-T6 aluminum alloy. Visual inspection did not indicate any surface artifacts, such as tool marks or haze, indicating that a true optical surface finish was achieved. Figure 2b demonstrates that the boundary of the lenslet has a sharp appearance with neither tool marks nor a tool tracking error. Figure 2c shows the surface 3D microtopology of the lenslet apex region. Curvature and tilt were removed from the measurement results to better represent the high spatial frequency of the surface texture. The surface roughness of the mold was 4.932 nm—this value meets the requirements for optical surface applications.

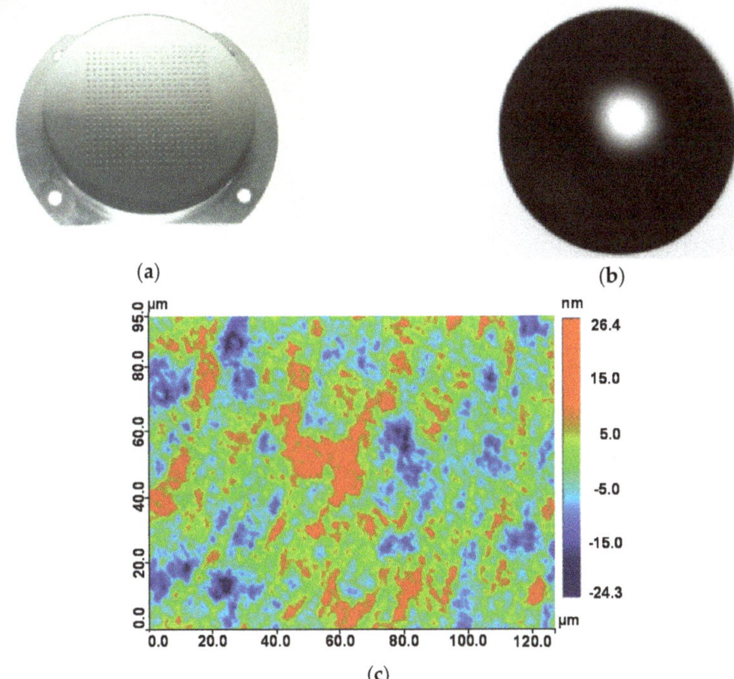

Figure 2. Finished 6061-T6 aluminum alloy mold: (**a**) photograph of the processed sample of the dense microlens array (MLA) mold (20 × 20 array), (**b**) edge profile of the processed sample, and (**c**) surface roughness of the aluminum alloy MLA mold (Sa = 4.932 nm).

We then applied the AAO process to generate a uniform layer of nanostructures on the aluminum alloy MLA mold, as mentioned above. 6061-T6 aluminum alloy contains a variety of alloying elements, including magnesium and silicon. During the anodizing process, these alloying elements react to form nonuniform random nanopores and partial defects, as shown in Figure 3. At the same time, the nanopores are not uniform in either shape or depth. The nanostructured surface thus turns into a nonspecular surface. Therefore, the 6061-T6 MLA mold can meet the requirements of an optical mold, but it is not suitable for fabricating nanostructured MLAs.

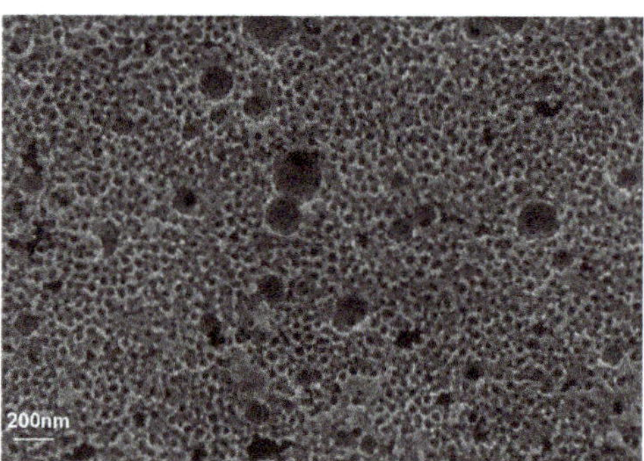

Figure 3. Scanning electron microscopy (SEM) image of the nanostructured 6061-T6 MLA mold.

3.2. High-Purity Polycrystalline Aluminum Pin Mold

To avoid the formation of nonuniform nanostructures due to alloy impurities, high-purity polycrystalline aluminum (with a purity of over 99.99%) was used as the mold material in the subsequent experiment. However, during the single-point diamond SSS-turning process, the cutting edge of the diamond tool was severely damaged, because the pure polycrystalline aluminum was too soft for single-point SSS turning. Hence, a wafer-level MLA mold meeting the optical requirements could not be obtained. Therefore, a pure polycrystalline aluminum optical mold was fabricated via the pin-mold mode, as shown in Figure 4a. Pin molds can be used to fabricate wafer MLAs via a slower step and flash process. Since each pin mold contains only a single lens cavity, the traditional high-speed single-point diamond-turning method can be used. Figure 4b shows the edge profile of the pin mold—it has a sharp appearance. The surface roughness of the lenslet region was 3.632 nm (Figure 4c), which meets the requirements for optical applications.

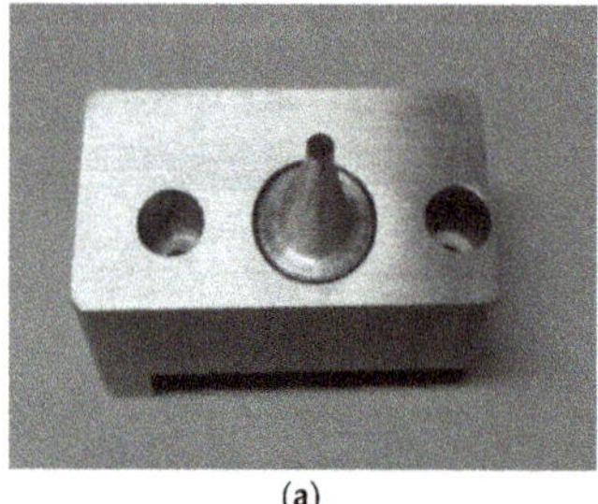

(a)

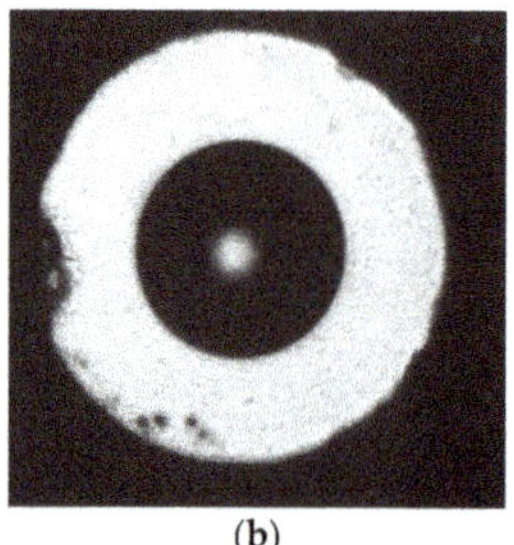

(b)

Figure 4. *Cont.*

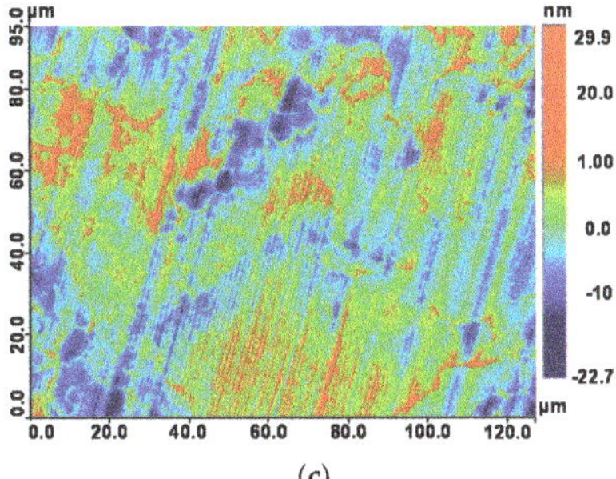

(c)

Figure 4. Finished high-purity polycrystalline aluminum pin mold: (**a**) photograph of the processed sample of the pin mold, (**b**) edge profile of the processed sample, and (**c**) surface roughness of the polycrystalline aluminum pin mold (Sa = 3.632 nm).

The high-purity polycrystalline aluminum pin mold was treated with the AAO process. Figure 5a shows a SEM image of the lenslet apex region. A hexagonal lattice with well-defined nanopores was formed as desired, exhibiting a fundamental pitch of 120 nm and a pore diameter of ~90 nm. The results of the surface roughness measurements are shown in Figure 5b, which demonstrate that the pin mold meets the requirements for optical applications (Sa = 9.161 nm). However, it also indicates that the polycrystalline aluminum surface is composed of multiple microsized grains. Figure 5c shows the 3D microtopology of the lens mold after one replication, which clearly indicates a surface profile deformation due to grain interfacial sliding under the demolding force. Consequently, the poor strength and shape stability of the pure crystalline pin mold restrict its application in wafer MLA manufacturing.

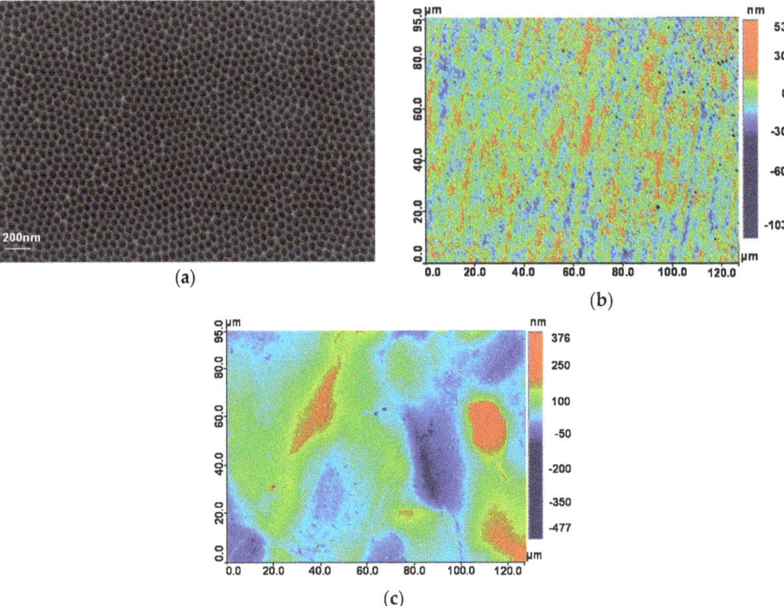

Figure 5. (**a**) SEM image of the lenslet mold apex region. (**b**) Surface roughness of the nanostructured pin mold (Sa = 9.161 nm). (**c**) The lens mold 3D microtopology after one replication (Sa = 48.939 nm).

3.3. Pure Nanocrystalline Aluminum MLA Mold with Dense Array

After numerous trials to identify the right material, such as single crystal/polycrystalline pure aluminum and different aluminum alloys, an aluminum alloy (6061-T6) substrate electroplated with ~0.3 mm-thick pure aluminum consisting of nanosized grains (AlumiPlate, Coon Rapids, MN, US) was finally selected as the mold material, by considering the requirements of both the ultraprecision machining process and the subsequent AAO process. The strength of the 6061-T6 alloy substrate assures the shape stability of the mold, while the pure nanocrystalline-plated layer simultaneously supports the generation of a uniform nanopore layer without impurity defects. A 2″ wafer mold was fabricated to validate the process. First, a flat optical surface was created on the aluminum substrate by a diamond-turning process, as shown in Figure 6a. Next, a concave MLA with a dense array was machined by micromilling using a high-speed air-bearing spindle. The measured results demonstrated that the boundary of the lenslets had a sharp appearance, as shown in Figure 6b. The surface roughness of the lenslet was 3.362 nm, as shown in Figure 6c. We then adopted the AAO process to obtain the pure nanocrystalline aluminum-dense MLA mold with nanostructures, as shown in Figure 6d.

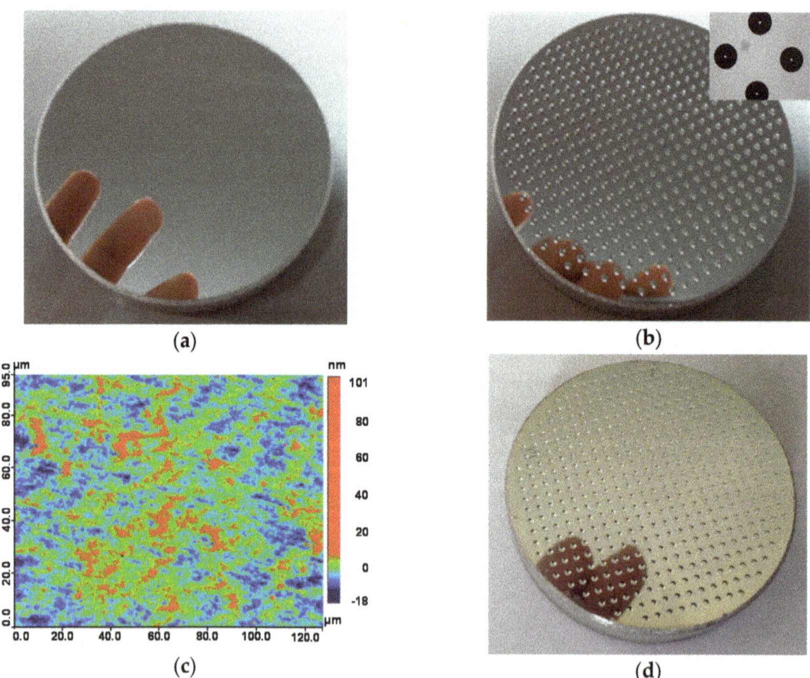

Figure 6. Different stages in the fabrication of the hybrid-structure microlens array (MLA) metal wafer mold: photographs of (**a**) the flat mirror surface obtained by diamond-turning and (**b**) the micromilled MLA mold cavities on the substrate and edge profile of the processed sample, (**c**) the MLA mold 3D surface finish of the lenslet apex region without nanostructures (Sa = 3.362 nm), and (**d**) the final MLA mold with antireflection nanostructures, which was fabricated using the anodic aluminum oxidation (AAO) process.

The SEM images of the mold nanostructure, in Figure 7, show the uniformity of the nanoporous layer on the MLA mold cavity, both at the apex flat region and the lens boundary region. Cross-sectional SEM images of the thick metal mold were difficult to record, so Figure 7d shows a SEM image of a nanostructured sheet plate sample that was fabricated using the same process parameters as those of the metal mold. The nanostructures exhibited a V-shaped cross section with an average pore diameter of ~100 nm and a spatial period of ~120 nm, as shown in Figure 7d. Figure 8a shows the lens profile error of the lenslet mold with AR nanostructures—the PV error is ~0.098 µm, which indicates that the AAO process has little negative impact on the mold profile. The surface roughness of the nanostructured mold was measured to be 15.683 nm, as shown in Figure 8b, and the mold surface displays almost the same visual mirror quality as before the AAO treatment. The results of the surface profile and the surface roughness measurements showed that the MLA mold of pure nanocrystalline aluminum meets the optical requirements even after the nanostructuring treatment, and it paves the way for a further study to replicate a multiscale wafer-level MLA with an inherent AR capability.

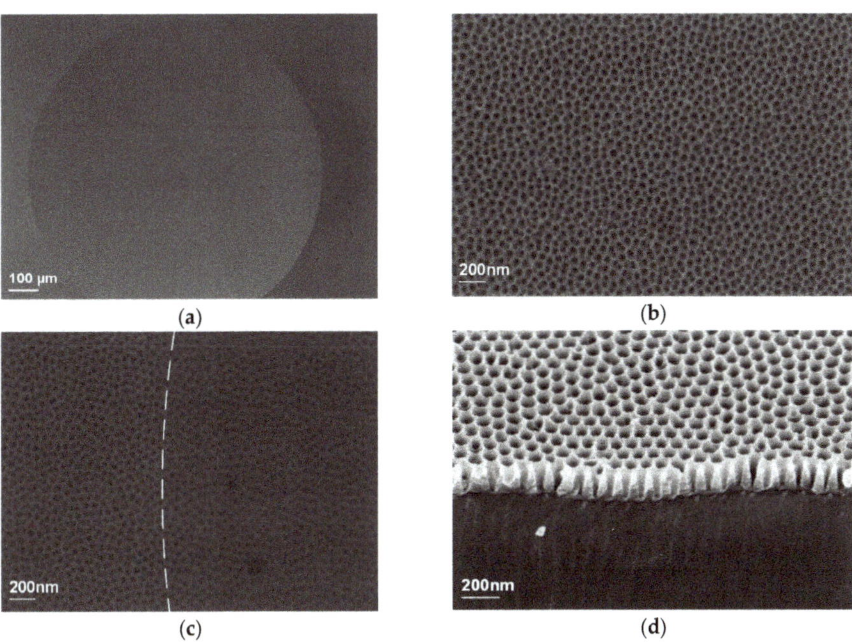

Figure 7. (**a**) SEM images of the nanostructured MLA mold. (**b**) High-magnification SEM image of the lenslet apex region. (**c**) High-magnification SEM image of the lenslet boundary region, with the boundary being indicated by the dashed line. (**d**) Cross-sectional SEM image of the nanopores.

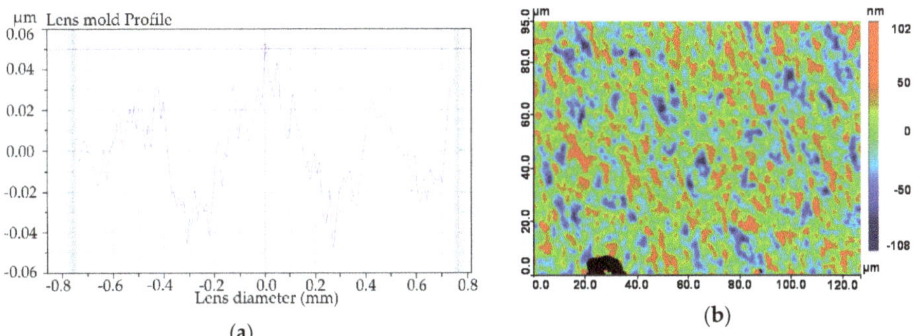

Figure 8. (**a**) Stylus profile measurement results of the lenslet mold with nanostructures (peak-to-valley (PV) error = 0.0985 µm). (**b**) Surface roughness of the MLA mold with nanostructures (Sa = 15.683 nm).

4. Conclusions

In this study, three different aluminum materials were selected to fabricate multiscale wafer-level MLA molds with uniform nanostructures. After the trials with different materials, a 6061-T6 substrate, electroplated with pure aluminum consisting of nanosized grains (US proprietary), was identified to satisfy the requirements of both lens mold quality and uniform nanostructure. We demonstrated the fabrication of a high-quality, multiscale-structure wafer-level MLA mold using ultraprecision machining and self-assembling AAO processes, which produced hundreds of accurate aspherical microlens cavities and a uniform layer of AR nanostructures on the mold surface. The surface measurements indicated that the fabricated mold met the requirements for precision optical applications.

Compared with the former reports of AR nanostructured planar or flexible material molds, for the first time, this study generated AR nanostructures on a hard mold with densely arrayed lens surfaces and quantitatively evaluated its quality based on lens requirements. This study is expected to benefit the industrial production of novel and functional multiscale-structure wafer-level MLAs, as it provides a practical method for the manufacture of large wafer-level molds with high-quality and uniform nanostructures.

Author Contributions: Conceptualization and design of experiments, S.X., X.W. (Xinjun Wan), and X.W. (Xiaoxiao Wei); fabrication, S.X.; scanning electron microscopy (SEM) measurements, S.X.; data analysis, S.X., X.W. (Xinjun Wan), and X.W. (Xiaoxiao Wei); writing—original draft preparation, S.X.; writing—review and editing, X.W. (Xinjun Wan).

Funding: This work was funded by the National Natural Science Foundation of China under 61505107, the Capacity Building of Local Institutions of Shanghai Science and Technology Commission under 18060502500 and the National Key Scientific Instrument and Equipment Development Projects under 2012YQ170004.

Acknowledgments: The authors gratefully acknowledge the assistance of Yun Zhu from Shangmu Tech. who participated in the fabrication of the nanostructures.

Conflicts of Interest: The authors declare no conflict of interest.

References

1. Cho, M.; Daneshpanah, M.; Moon, I.; Javidi, B. Three-Dimensional Optical Sensing and Visualization Using Integral Imaging. *Proc. IEEE* **2011**, *99*, 556–575.
2. Tanida, J.; Kitamura, Y.; Yamada, K.; Miyatake, S.; Miyamoto, M.; Morimoto, T.; Masaki, Y.; Kondou, N.; Miyazaki, D.; Ichioka, Y. Compact image capturing system based on compound imaging and digital reconstruction. *Proc. SPIE* **2001**, *4455*, 34–41.
3. Aldalali, B.; Li, C.; Zhang, L.; Jiang, H. Micro Cameras Capable of Multiple Viewpoint Imaging Utilizing Photoresist Microlens Arrays. *J. Microelectromech. Syst.* **2012**, *21*, 945–952. [CrossRef]
4. Buettner, A.; Zeitner, U.D. Wave optical analysis of light-emitting diode beam shaping using microlens arrays. *Opt. Eng.* **2002**, *41*, 2393–2401. [CrossRef]
5. Seifert, L.; Liesener, J.; Tiziani, H.J. The adaptive Shack–Hartmann sensor. *Opt. Commun.* **2003**, *216*, 313–319. [CrossRef]
6. Wippermann, F.; Zeitner, U.D.; Dannberg, P.; Bräuer, A.; Sinzinger, S. Beam homogenizers based on chirped microlens arrays. *Opt. Express* **2007**, *15*, 6218–6231. [CrossRef] [PubMed]
7. Buschbeck, E.; Ehmer, B.; Hoy, R. Chunk versus point sampling: Visual imaging in a small insect. *Science* **1999**, *286*, 1178–1180. [CrossRef]
8. Andreas, B.; Jacques, D.; Robert, L.; Peter, D.; Andreas, B.U.; Andreas, T. Thin wafer-level camera lenses inspired by insect compound eyes. *Opt. Express* **2010**, *18*, 24379–24394.
9. Raut, H.K.; Ganesh, V.A.; Nair, A.S.; Ramakrishna, S. Anti-Reflective Coatings: A Critical, In-Depth Review. *Energy Environ. Sci.* **2011**, *4*, 3779–3804. [CrossRef]
10. Walheim, S.; Schäffer, E.; Mlynek, J.; Steiner, U. Nanophase-Separated Polymer Films as High-Performance Antireflection Coatings. *Science* **1999**, *283*, 520–522. [CrossRef]
11. Mano, I.; Uchida, T.; Taniguchi, J. Fabrication of the antireflection structure on aspheric lens surface and lens holder. *Microelectron. Eng.* **2018**, *191*, 97–103. [CrossRef]

12. Bae, B.J.; Hong, S.H.; Kwak, S.U.; Lee, H. Fabrication of Moth-Eye Pattern on a Lens Using Nano Imprint Lithography and PVA Template. *J. Korean Inst. Surf. Eng.* **2009**, *42*, 41–173. [CrossRef]

13. Tan, G.; Lee, J.H.; Lan, Y.H.; Wei, M.K.; Peng, L.H.; Cheng, I.; Wu, S.T. Broadband antireflection film with moth-eye-like structure for flexible display applications. *Optica* **2017**, *4*, 678. [CrossRef]

14. Hung, S.Y.; Chang, T.Y.; Shen, M.H.; Yang, H. Tilted microlens fabrication method using two photoresists with different melting temperatures. *J. Micromech. Microeng.* **2014**, *24*, 68–75. [CrossRef]

15. Kim, S.; Kim, H.; Kang, S. Development of an ultraviolet imprinting process for integrating a microlens array onto an image sensor. *Opt. Lett.* **2006**, *31*, 2710–2712. [CrossRef] [PubMed]

16. Chen, L.; Kirchberg, S.; Jiang, B.Y.; Xie, L.; Jia, Y.L.; Sun, L.L. Fabrication of long-focal-length plano-convex microlens array by combining the micro-milling and injection molding processes. *Appl. Opt.* **2014**, *53*, 7369–7380. [CrossRef] [PubMed]

17. Zhang, J.; Shen, S.; Dong, X.X.; Chen, L.S. Low-cost fabrication of large area sub-wavelength anti-reflective structures on polymer film using a soft PUA mold. *Opt. Express* **2014**, *22*, 1842–1851. [CrossRef]

18. Chidambaram, N.; Kirchner, R.; Altana, M.; Schift, H. High fidelity 3D thermal nanoimprint with UV curable polydimethyl siloxane stamps. *J. Vac. Sci. Technol. B* **2016**, *34*, 6K401. [CrossRef]

19. Guo, R.; Xiao, S.; Zhai, X.; Li, J.; Xia, A.; Huang, W. Micro lens fabrication by means of femtosecond two photon photopolymerization. *Opt. Express* **2006**, *14*, 810–816. [CrossRef]

20. Yan, J.H.; Ou, W.; Ou, Y.; Zhao, L.J. Design and fabrication of novel microlens–micromirrors array for infrared focal plane array. *Microw. Opt. Technol. Lett.* **2012**, *54*, 879–884.

21. Yi, A.Y.; Li, L. Design and fabrication of a microlens array by use of a slow tool servo. *Opt. Lett.* **2005**, *30*, 1707–1709. [CrossRef] [PubMed]

22. Yi, A.Y.; Huang, C.; Klocke, F.; Brecher, C.; Pongs, G.; Winterschladen, M.; Demmer, A.; Lange, S.; Bergs, T.; Merz, M.; et al. Development of a compression molding process for three-dimensional tailored free-form glass optics. *Appl. Opt.* **2006**, *45*, 6511–6518. [CrossRef] [PubMed]

23. Mukaida, M.; Yan, J. Fabrication of Hexagonal Microlens Arrays on Single-Crystal Silicon Using the Tool-Servo Driven Segment Turning Method. *Micromachines* **2017**, *8*, 323. [CrossRef] [PubMed]

24. Chang, X.; Xu, K.; Xie, D.; Luo, S.; Shu, X.; Ding, H.; Zheng, K.; Li, B. Microforging technique for fabrication of spherical lens array mold. *Int. J. Adv. Manuf. Technol.* **2018**, *96*, 3843–3850. [CrossRef]

25. Gao, P.; Liang, Z.; Wang, X.; Zhou, T.; Xie, J.; Li, S.; Shen, W. Fabrication of a Micro-Lens Array Mold by Micro Ball End-Milling and Its Hot Embossing. *Micromachines* **2018**, *9*, 96. [CrossRef] [PubMed]

26. Feldman, M.; America, T.N. Wafer-Level Camera Technologies Shrink Camera Phone Handsets. *Photonic Spectra* **2009**, *41*, 58–60.

27. Zhan, Z.; Liang, N.; Bian, R.; Zhao, M. Design and Manufacturing of Ultra-Hard Micro-Milling Tool. *Trans. Tianjin Univ.* **2014**, *20*, 415–421. [CrossRef]

28. Huangfu, Y.; Zhan, W.; Hong, X.; Fang, X.; Ding, G.; Ye, H. Heteroepitaxy of Ge on Si(001) with pits and windows transferred from free-standing porous alumina mask. *Nanotechnology* **2013**, *24*, 185302. [CrossRef] [PubMed]

29. Ding, J.; Zhu, Y.; Yuan, N.; Ding, G. Reduction of nanoparticle deposition during fabrication of porous anodic alumina. *Thin Solid Film* **2012**, *520*, 4321–4325. [CrossRef]

 *applied sciences*

Article

Miniaturized Optical Encoder with Micro Structured Encoder Disc

Jonathan Seybold [1,*], André Bülau [1], Karl-Peter Fritz [1], Alexander Frank [2], Cor Scherjon [2], Joachim Burghartz [2] and André Zimmermann [1,3]

[1] Hahn-Schickard, 70569 Stuttgart, Germany; Andre.Buelau@Hahn-Schickard.de (A.B.); Karl-Peter.Fritz@Hahn-Schickard.de (K.-P.F.); zimmermann@ifm.uni-stuttgart.de (A.Z.)
[2] Institut für Mikroelektronik Stuttgart, 70569 Stuttgart, Germany; frank@ims-chips.de (A.F.); scherjon@ims-chips.de (C.S.); burgh@ims-chips.de (J.B.)
[3] Institute for Micro Integration (IFM), University of Stuttgart, 70569 Stuttgart, Germany
* Correspondence: Jonathan.Seybold@Hahn-Schickard.de

Received: 30 November 2018; Accepted: 23 January 2019; Published: 29 January 2019

Abstract: A novel optical incremental and absolute encoder based on an optical application-specific integrated circuit (opto-ASIC) and an encoder disc carrying micro manufactured structures is presented. The physical basis of the encoder is the diffraction of light using a reflective phase grating. The opto-ASIC contains a ring of photodiodes that represents the encryption of the encoder. It also includes the analog signal conditioning, the signal acquisition, and the control of a light source, as well as the digital position processing. The development and fabrication of the opto-ASIC is also described in this work. A laser diode was assembled in the center on top of the opto-ASIC, together with a micro manufactured polymer lens. The latter was fabricated using ultra-precision machining. The encoder disc was fabricated using micro injection molding and contains micro structures forming a blazed grating. This way, a 10-bit optical encoder with a form factor of only 1 cm^3 was realized and tested successfully.

Keywords: optical encoder; grating; blaze; injection molding; micro assembly; active alignment; opto-ASIC

---

1. Introduction

Rotary encoders are widely used to detect the rotation angle of motors and gear shafts. State-of-the-art encoders mostly operate on potentiometric, capacitive, magnetic, or optical principles [1]. Magnetic and optical encoders are most common in use with an annual growth rate of 11%, whereby the optical technology dominates the global industrial encoder market. The market of encoders is segmented into automotive (35%), electronics (28%), machines (21%), and others (16%) [2]. It is expected that, by 2021, the global market size of the encoders will reach $760 million [3].

The common approach of magnetic encoders is to use a magnetic rotor and to detect the position of the rotor with a hall sensor. In contrast, the common structure of optical encoders is to use a chrome structured encoder disc made of glass. The position of the rotor is then detected with a light barrier. Both operating principles are shown in Figure 1. The advantages of magnetic encoders are their high robustness and low fabrication costs, whereas optical encoders show high accuracy and high dynamics. A distinction is made between incremental and absolute encoders. While absolute encoders deliver the angular position immediately after switching on the encoder, incremental encoders firstly need a reference run. Otherwise only relative position measurement is possible. This article deals solely with absolute optical encoders. As a result of the extensive topic which comprises optical design, injection molding, ultra-precision machining, application-specific integrated circuit (ASIC) design and micro assembly, not all details can be presented in this article.

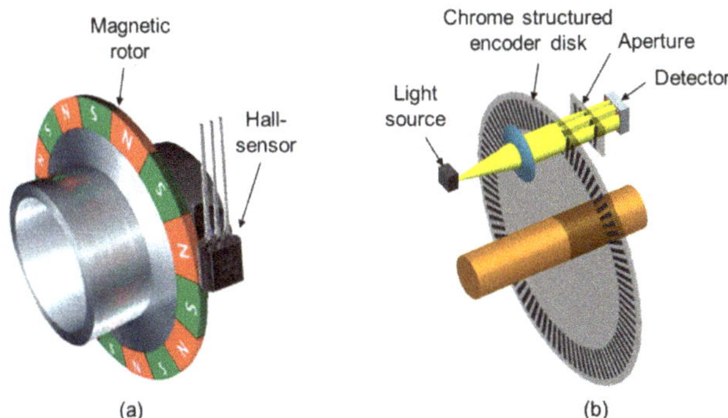

Figure 1. (**a**) Operating principle of a magnetic encoder with a magnetic rotor and hall sensor; (**b**) operating principle of an optical encoder with a light source, encoder disc, aperture, and detector.

Absolute encoder discs contain the angle information typically as a Gray code pattern, which is read out in accordance with a transmission principle. Every position of an absolute encoder is unique. Figure 2a shows the Gray code pattern for an encoder disc with $n = 6$ tracks (6-bit resolution) and its optical path. This results in $2^n = 64$ positions per revolution of the encoder disc. This means that, for a higher resolution, there are more tracks necessary. The tracks consist of opaque and transparent parts. Each track is read out optically with a light barrier as shown in Figure 2b. This means that, for a resolution of n bits, there are n tracks with n light barriers necessary. As a result, the dimensions of optical encoders increase with their resolution, and the effort for assembling increases due to elaborate alignment processes between the optical device and the encoder disc [4].

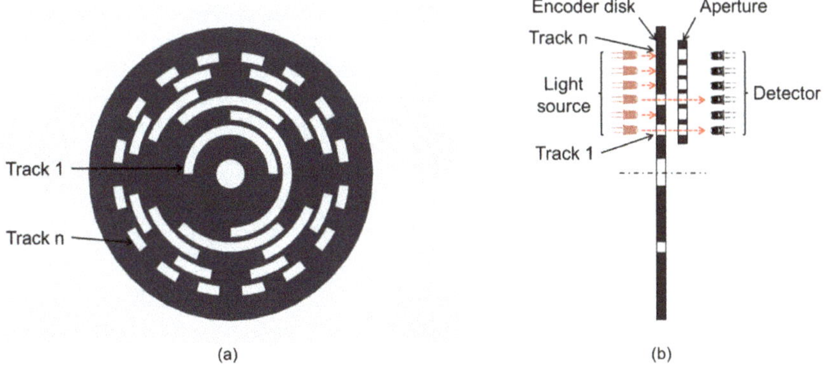

Figure 2. (**a**) Encoder disc of an absolute optical encoder with Gray code patterns of opaque and transparent parts; (**b**) cross-section of the optical path.

However, the smaller the encoder is, the lower its resolution and the higher its costs will be, because, on small discs, there is less space for the Gray code patterns and the associated light barriers; additionally, the expense for the assembly increases due to the complicated fine mechanical–optical system [5,6]. For example, an optical high-end encoder with a diameter of 13 mm delivers a resolution of only 256 increments, while the costs are on the order of several hundreds of euros [7,8]. Also, the accuracy decreases with smaller diameters because the influence of the angular error due to eccentricity between encoder disc and encoder shaft increases for smaller diameters of the encoder disc [9]. Table 1 shows two examples of high-end state-of-the-art-encoders with comparatively very small dimensions. Especially for small devices such as finger prostheses, miniaturized drives, or miniaturized robots, the need for small encoders with high resolution is obvious. Therefore, the motivation of this work was to develop a small absolute optical low-cost encoder with a significant higher resolution, such as state-of-the-art encoders with comparable small dimensions.

Table 1. Examples of high-end state-of-the-art encoders.

Feature/Series	CUI MAS10	CUI MES6
Method	Absolute	Incremental
Resolution	256 positions	500 positions
Speed	6000 rpm	6000 rpm
Interface	Digital	Digital
Temperature	$0 \ldots +60\,°C$	$0 \ldots +60\,°C$
Size	Ø 13 mm × 16 mm	Ø 8 mm × 11 mm

To overcome the physical limitations for optical state-of-the-art encoders, a new approach was developed using a reflective blazed grating as an encoder disc and a specific photo detector with a circular array of photodiodes. In this way, the diameter of the encoder disc was no longer the limitation for the resolution because, on the encoder disc, there are no longer any Gray code patterns. In this way, a centric optical scanning of the encoder disc was possible. This achieved a higher integration density because the optical device could also be arranged centrically. Of course, the circular array of photodiodes was now the limitation factor. By comparison, state-of-the-art encoders need an eccentric arrangement of the optical device and, therefore, more construction space. With this novel approach, an optical encoder with a resolution of 1024 increments within only 1 cm^3 was developed. A demonstrator was built and tested successfully.

The main advances of the proposed encoder are the small dimensions with a comparatively high resolution and an adjustment-free assembling of the encoder disc. Another advance is the manufacturing method of the encoder disc itself. While traditional optical encoders with high resolution need expensive encoder discs made of glass with chromium patterns and an additional adjustment process of the encoder disc, the proposed encoder needs only an injection-molded encoder disc with a simple blazed grating. The optical set-up is also relatively simple because the optical path is in a reflection order and there is no additional adjustment process necessary. A further advantage is the modular design concept. The same optical module and grating can be used for different encoder diameters. In summary, this results in a low-cost encoder with high resolution and small dimensions.

2. A New Approach for Optical Encoders

2.1. Main Principle of the Encoder

The physical basis of the optical encoder was the diffraction of light using a reflective phase grating. The approach was to use a reflective phase grating as an encoder disc for an optical encoder. Figure 3 shows the mechanism of diffraction schematically. After reflection of the light at the phase grating, destructive and constructive interference occurs because of the path difference Δ of the light, which is caused by the grating. If Δ is equal to λ, constructive interference occurs and, if Δ is equal to 0.5λ, destructive interference occurs.

The relationship between the grating constant and the angle of diffraction is described by the so-called grating equation as follows:

$$-\sin \alpha_e - \sin \alpha_m = \frac{m\lambda}{g},$$

where α_e is the angle of the incident light and α_m is the angle of the diffracted light. With a given wavelength λ and grating constant g, the angle of the diffracted light can be calculated for each diffraction order m ($m = 0, \pm1, \pm2, \ldots$) using this equation.

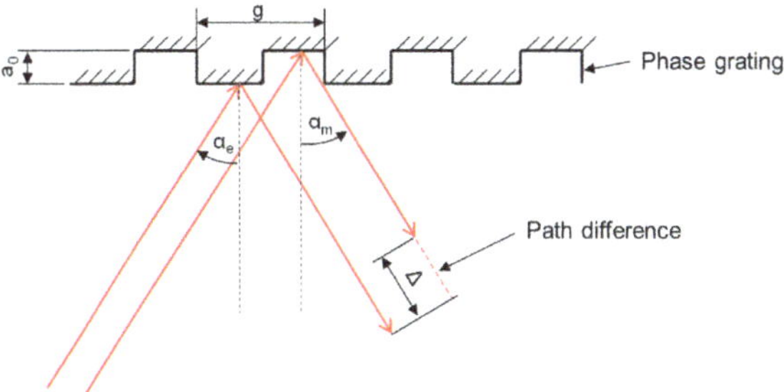

Figure 3. Diffraction of light by a reflective phase grating.

The concept for the optical encoder was to use a light source on top of a photo detector that transmits light through a lens to the encoder disc. The disc contains micro structures forming the grating. The light is then diffracted at the encoder disc and reaches the photo detector. While laminar gratings result in two beams of equal intensity with the zero order of the beam being reflected into the light source [10–15], a blazed grating diffracts two beams of different intensities. This is due to the fact, that the zero order is diffracted into the +1st order beam.

When the disc rotates, the two diffracted light spots also rotate on the photo detector. In this way, it is possible to detect the rotational movement of the encoder disc. This circulation was detected by a ring of photodiodes on an opto-ASIC. While the beams were of different intensity, an unambiguous assignment of the disc angle over 360° was accomplished. Figure 4 shows the functional principle of the optical encoder and the underlying concept for the arrangement of the individual components [16,17].

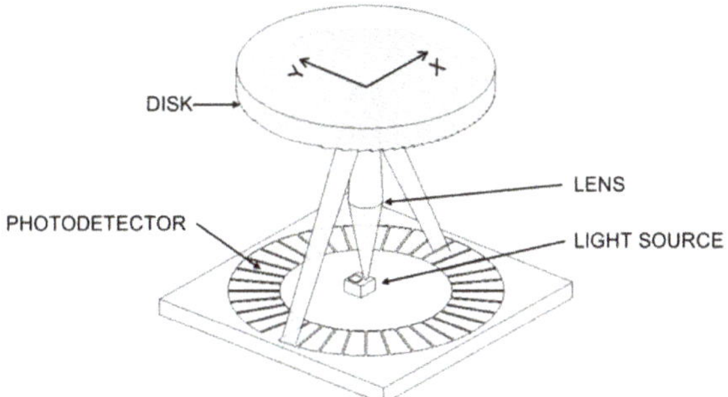

Figure 4. Functional principle of the optical encoder.

To detect the rotational angle of the diffracted beams on the photodiode ring, different concepts for the photodiodes were investigated and designed. The first concept consists of a segmented arrangement of photodiodes as shown in Figure 5a. This approach leads to a simple focusing lens to form circular spots, but requires 1024 photodiodes on the opto-ASIC for an encoder resolution of 10 bits. Thus, the signal processing unit had to be designed to read out 1024 photodiodes.

The second concept consists of a Gray code arrangement of the photodiodes as shown in Figure 5b. This approach requires a number of tracks of photodiodes of significantly bigger size, but a more complex lens, as the diffracted beam has to be additionally expanded in the radial direction to illuminate all photodiode tracks. For 10-bit resolution, 10 tracks of photodiodes with the most significant bit (MSB) on the inner track and the least significant bit (LSB) on the outer track are required.

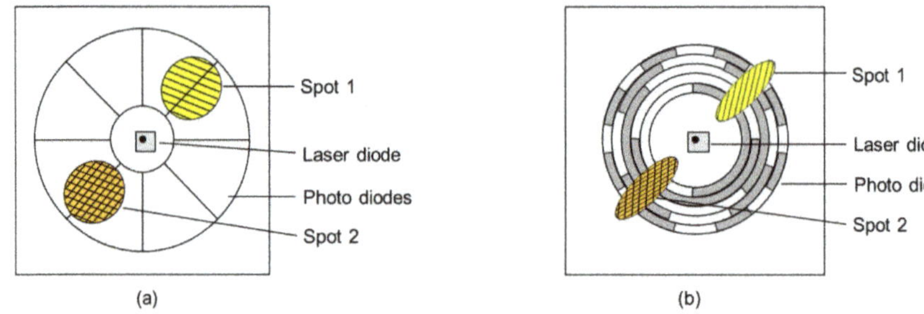

Figure 5. Two different layouts for the photodetector of the optical encoder. (**a**) Segmented arrangement of the photodiodes; (**b**) Gray code arrangement of the photodiodes.

By using photodiodes for the dark and bright parts of the Gray code, differential signals can be obtained leading to a higher signal-to-noise ratio (SNR) and offset compensation. An additional outer track with an offset by half a diode was provided. Hence, the interpolation of the expected sin/cos signals from the last two outer LSB tracks was possible, allowing increased relative resolution of the rotational angle between the two photodiodes.

The two lenses (one for the segmented and another one for the Gray code detector) were designed by ray-tracing, using the optical design software OSLO® (Optics Software for Layout and Optimization). Because of the two different layouts of the photo detector, two different designs for the optical path were necessary. Figure 6 shows the optical path for both versions. For the first photo detector with the segmented photodiodes, a centric lens to focus the light onto the photo detector was needed. For the second photo detector with Gray code arrangement of the photodiodes, an additional lens was necessary to expand the diffracted spot in the radial direction and to illuminate all photodiode tracks. This led of course to a more complex optical path.

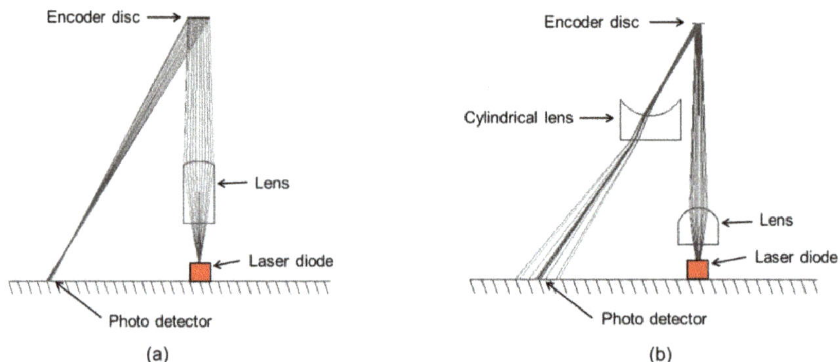

Figure 6. Design of the optical path for the encoder. (**a**) Optical path for the segmented photo detector; (**b**) optical path for the photo detector with Gray code arrangement of the photodiodes.

The main advantage of the novel approach for optical encoders is that a lateral displacement or eccentricity of the disc does not influence the optical path. The encoder disc can be illuminated in its center, as well as on its outer diameter. Thus, hollow shaft encoders can be realized as well.

A further advantage is that the functional principle of the encoder can be used to develop a modular design concept for optical encoders. When the design for the optical module is defined, the same optical module can also be used for different encoder diameters. Even the grating remains unchanged. Only the diameter of the disc needs to be adapted. A further aspect of this concept is the small dimensions with a comparatively high resolution. Thus, a low-cost encoder is possible with this novel approach.

2.2. Proof of Principle

To prove the optical principle, a first set-up was realized using a High Dynamic Range complementary metal-oxide semiconductor (CMOS) (HDRC®) image sensor and a glass lid on top of it. Figure 7 shows the demonstrator for the proof of principle. To mount the laser diode with a lens holder on it, the lid was processed with traces made of Cr/Au using a laser-cut mask and physical vapor deposition (PVD). The laser diode and the lens were assembled directly on the glass lid. The glass lid itself was assembled on top of the image sensor. The grating with a grating constant of 1.7 μm was attached at the rotatable shaft with a specific mounting. After joining the housing with the shaft and the image sensor, the characteristics of the spots could be observed using the image sensor.

For fabrication of the lenses, ultra-precision machining (UPM) was used. This allowed fabricating custom lenses in polymethyl methacrylate (PMMA) using a diamond tool. In this way, a dimensional accuracy of 1–2 μm and a surface roughness of 10 nm were achieved for the lens. A commercially available blazed grating [18] was used and integrated into a ball-bearing encoder chassis that was mounted on top of the image sensor.

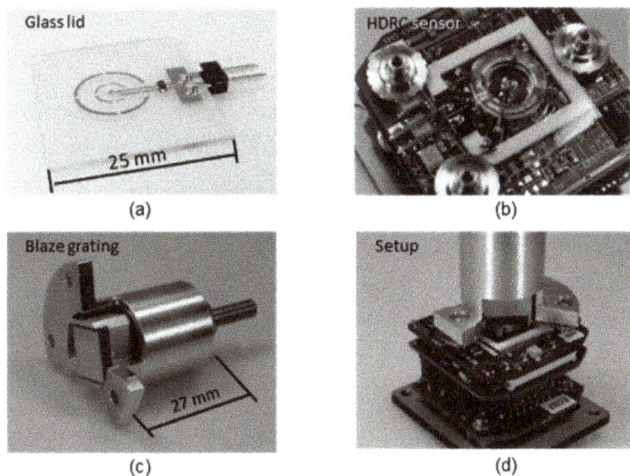

Figure 7. Demonstrator for proof of principle. (**a**) Glass lid with laser diode; (**b**) High Dynamic Range complementary metal-oxide semiconductor (CMOS) (HDRC®) image sensor with the glass lid and lens on top of it; (**c**) blazed grating on a rotatable shaft; (**d**) fully assembled demonstrator.

With this equipment, it was possible to visualize and to investigate the spot shape of the two concepts with two different lenses. Measurements of this arrangement with the two lenses were performed—the first one for the segmented photo detector with two elliptical spots with different intensity, and the second one for the Gray code pattern. With the additional toroid lens, it was possible to expand the diffracted beam in the radial direction to realize two linear spots. The measured spot shapes are shown in Figure 8a,b for both lenses. Figure 8c shows the relative intensity of the spots from the second lens.

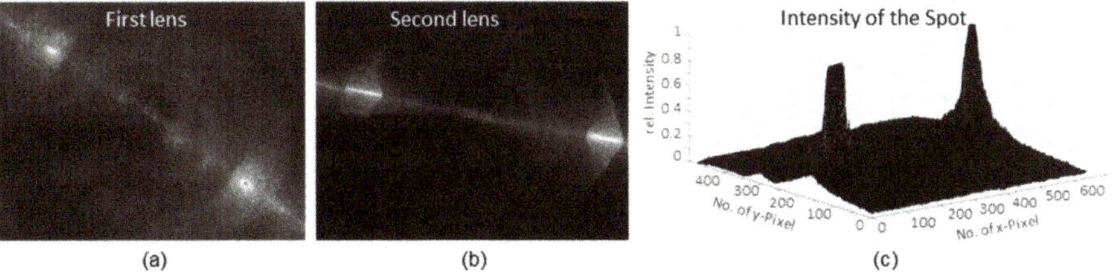

Figure 8. Spot shape measured with log-scale HDRC® image sensor.

Analysis of the images obtained using the log-scale HDRC® image sensor showed good agreement within the theoretically calculated optical laser spots. Figure 9a shows the simulated spot for the segmented photo detector, while Figure 9b shows the spot for the Gray code detector. For the simulation, the optical ray-tracing software OSLO® was used. For the light source, a Gaussian beam profile was assumed. By turning the encoder shaft, the laser spots of the diffracted beams circulated around the laser diode. That means that both lenses were suitable for the optical path with each characteristic spot shape.

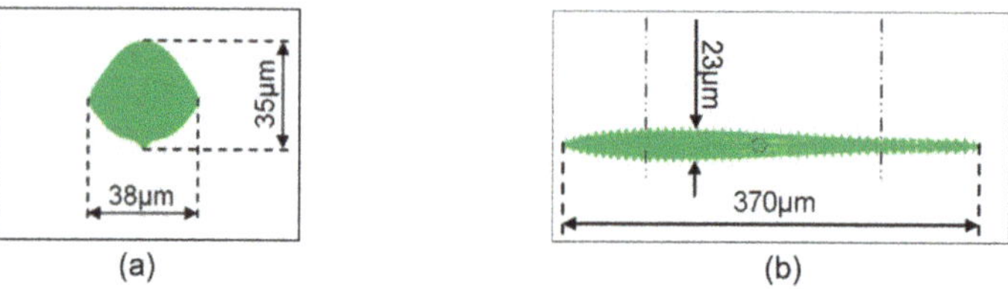

Figure 9. Spot shape measured with log-scale HDRC® image sensor.

2.3. Photodiode Readout Circuits of the Opto-ASIC

In order to convert the photodiode current into voltage, transimpedance amplifiers were integrated inside the opto-ASIC. The photodiodes of the Gray code structure were read out in a differential mode, as can be seen in Figure 10. Therefore, all dark photodiodes (PD_{dark}) of one ring were shorted, together as well as the bright photodiodes (PD_{bright}). The output voltage of the transimpedance amplifiers of those two pairs was compared using a comparator and translated into a logical signal at once. The 10 bit positions of the spot were achieved by implementing ten similar circuit structures and were read out in parallel. A following digital block calculated the spot location and output the result via a serial interface.

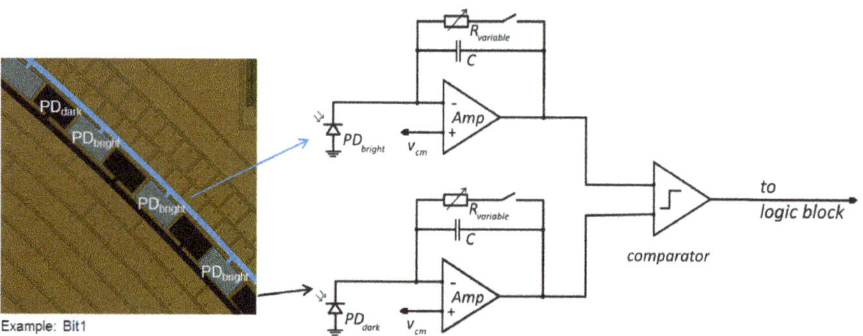

Figure 10. Read-out circuit of the Gray code structure (photodiode ring bit 1).

The simplified readout circuit of the segmented photodiode structure can be seen in Figure 11. In comparison to the Gray code structure, each of the 1024 photodiodes had to be read out separately to get the spot position. To overcome the limited chip array, speed, and the power consumption, 128 segmented photodiodes were connected to one read-out circuit consequently via a multiplexer structure, followed by a transimpedance amplifier and a comparator for a sequential selection. This yielded eight independent read-out blocks, which could be controlled and read out in parallel by the digital block. The threshold reference voltages of the comparator could be separately adjusted for each block, also allowing a compensation of the photodiode process variation. The comparator decisions of all eight blocks were connected to a digital logic block, which calculated the spot position.

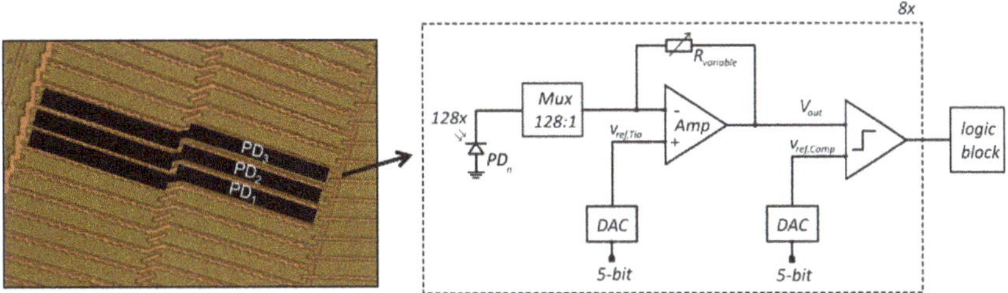

Figure 11. Read-out circuit of the segmented photodiode an optical application-specific integrated circuit (opto-ASIC).

2.4. Design Concept of the Demonstrator for the Miniaturized Optical Encoder

Parallel to the development of the opto-ASIC, the demonstrator for a miniaturized optical encoder was designed. Figure 12 shows a cross-section of the demonstrator.

Figure 12. Cross-section of the demonstrator for the miniaturized optical encoder.

The main emphasis was the miniaturized assembly and the development of the assembling concept based on chip-on-board technology. The dimensions of the demonstrator were about 10 mm in length and width, and 10 mm in height, measured without the connector and without the encoder shaft.

The assembly concept for the optical module was a combination of chip-on-board and chip-on-chip technology. Figure 13 shows the assembling concept of the demonstrator. Firstly, the opto-ASIC was assembled onto a board. After this, the laser diode was assembled directly onto the opto-ASIC. It was important to assemble the laser diode concentric to the ring of photodiodes, as an eccentric arrangement would lead to an incorrectly measured angle during operation. After wire bonding of the opto-ASIC and laser diode, the lens was assembled on top of the opto-ASIC. Finally, the assembly of the housing, the encoder disc, the bearing, and the encoder shaft followed to complete the encoder.

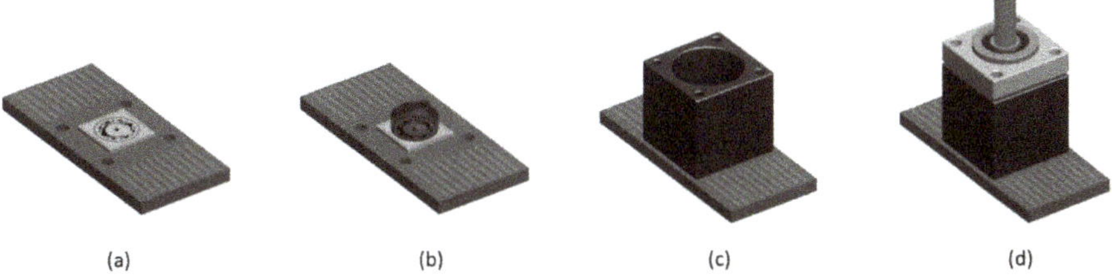

Figure 13. Assembly concept of the demonstrator. (**a**) Assembly of the opto-ASIC onto the printed circuit board (PCB) and the laser diode onto the opto-ASIC; (**b**) assembly of the lens onto the opto-ASIC; (**c**) assembly of the housing; (**d**) assembly of the encoder disc, the bearing, and the shaft.

3. Results

3.1. Opto-ASIC

Both encoding concepts of the photodiodes have their pros and cons. While the segmented concept has a simple optical path but complex signal processing, the concept with the Gray code structure has simple signal processing but a more complex optical path. Both opto-ASICs with the two different photodiode structures are shown in Figure 14.

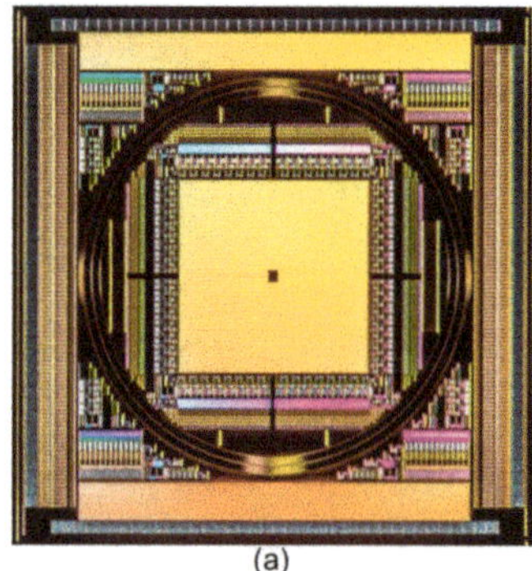

Figure 14. Microscope images of the opto-ASICs. (**a**) Opto-ASIC with segmented photodiodes; (**b**) opto-ASIC with Gray code photodiode structure.

Opto-ASICs for both variants were designed and fabricated using a mask programmable process. This approach allowed for a pre-fabricated, so-called master wafer containing analog and digital structures. The final wiring of its function was realized by the last two metal layers. Thus, a batch of master wafers was processed, and the different circuit variants were personalized by the metallization layers. In addition to the photodiode rings, the opto-ASIC includes transimpedance amplifiers, programmable comparators for digitizing the signals [19], and the logic to analyze the position of the spots on the opto-ASIC. It has a total size of $4.7 \times 4.7 \times 0.3$ mm^3 and detects and tracks the spot position. It is directly mounted on a printed circuit board (PCB) with necessary passives for the system.

3.2. Light Source

For the light source, a vertical-cavity surface-emitting laser (VCSEL) with a feed size of 200 μm and a wavelength of 850 nm was chosen, because this wavelength fit best to the grating. The laser diode was mounted on top of the opto-ASIC by gluing and bonding it to the surface of the chip. The laser diode cavity was placed in the center of the photodiode rings (see Figure 15), whereby the opto-ASIC itself was mounted onto a PCB.

Figure 15. Opto-ASIC mounted onto PCB with assembled laser diode on the additional metal top layer, in comparison to Figure 12.

Due to the temperature dependence of the laser, the light intensity was controlled by the opto-ASIC using the signal from the photodiodes in a closed-loop fashion. In this manner, a constant intensity of the laser light and, therefore, stable photodiode signals over temperature and time were guaranteed.

3.3. Micro Lens

For the miniaturized optical encoder, an additional micro lens was designed to fit on top of the opto-ASIC, focusing the light to the disc with a spot size of ~40 μm. It had a total diameter of 4.2 mm, while the diameter of the lens itself was only 1 mm. For the Gray code encoder, an additional toroid lens was included to expand the diffracted beam in the radial direction. The lenses were fabricated in PMMA using UPM and glued on top of the opto-ASIC. Figure 16 shows the assembled lens on top of the opto-ASIC.

Figure 16. Lens mounted on top of the opto-ASIC.

3.4. Manufacturing of the Micro Structured Encoder Discs

Because of the miniaturized dimensions of the optical encoder, small encoder discs were necessary. The encoder discs were fabricated using precision injection molding of polycarbonate (Makrolon® OD2015). Therefore, an injection mold was constructed. Figure 17 shows the design concept of the injection mold. The cavity of the encoder disc was located at the inner position of the mold and the marked stamper carrying the grating was necessary to form the micro structures of the blazed grating into the surface of the disc. The commercially available blazed grating used for the first demonstrator for proof of principle of the encoder was used to fabricate nickel stampers with a thickness of 0.3 mm using the galvanic reproduction technique. The nickel stamper was clamped inside the cavity of the injection mold. This way, encoder discs were fabricated, while this process allows for a scalable production at low manufacturing costs.

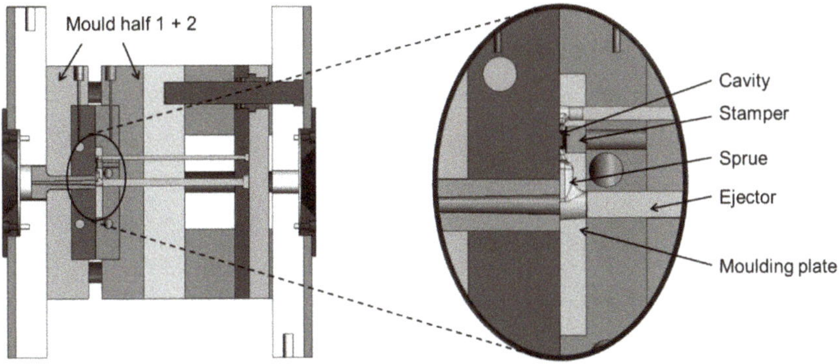

Figure 17. Design of the injection mold for encoder discs.

As the blazed grating was a continuous shape, an alignment of the stamper inside the cavity was not required. Figure 18 shows the molded encoder disc directly after its fabrication. The picture shows the encoder disc with the sprue still present to the left and the right of the encoder disc. To create a reflective surface, the discs were coated with gold using a PVD process. The quality of the gold depended on the sputtering target. The used target had a purity level of 99.999%. That means that the gold layer had a purity level slightly lower, because, during the sputtering process, there is the possibility that foreign atoms are incorporated into the gold crystal lattice. The gold layer had a thickness of 50 nm.

Figure 18. Injection-molded encoder disc with blazed grating.

With a total diameter of 4.2 mm, a height of 1.2 mm, and a grating constant of 1.7 μm, the encoder discs can be duplicated economically. The discs were assembled to the shaft of the encoder using a suitable adhesive. The size of the disc toward smaller diameters is only limited for practical reasons such as handling. The challenge is probably to find a feasible handling concept for very small discs. However, bigger discs are also possible with this concept, such as hollow shaft encoders.

4. Characterization of the Miniaturized Encoder

With the fabricated components, the demonstrator samples for the novel optical encoders were assembled and characterized. Figure 19 shows the completely assembled demonstrator for the miniaturized optical encoder. Thereby, the PCB acts also as the connector of the encoder. The part of the PCB which is taller than the encoder chassis is suitable for a conventional board connector [20].

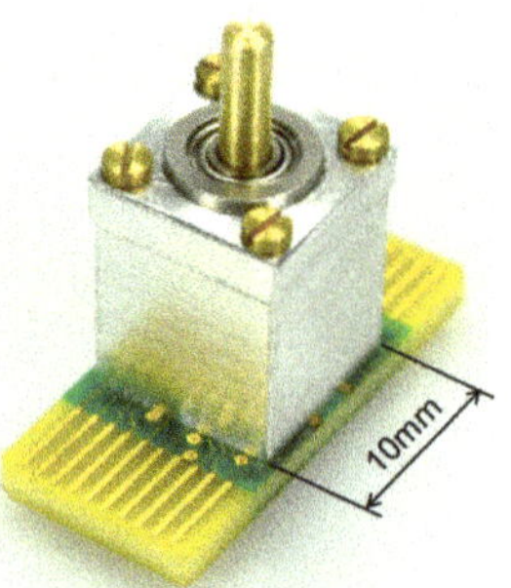

Figure 19. Completely assembled demonstrator for the miniaturized optical encoder.

For characterization, a customized test bench for encoders was used. Figure 20 presents the signal of bit 6 of the Gray code encoder as an example over one revolution (360°). Because of the layout of the photodiodes, it was possible to generate differential signals. This automatically led to more stable signals with twice the amplitude. As shown, each angular position delivered a signal for the bright spot (bit 6) and the dark spot (bit 6n) on the opto-ASIC. By measuring the difference between them, a defined distinction of cases was possible using comparators instead of analog-to-digital converters.

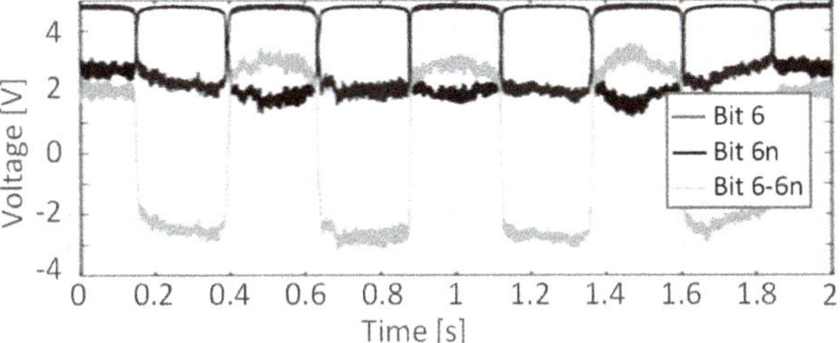

Figure 20. Example of the Gray code encoder signals (bit 6) for one revolution (360°).

To analyze the performance of the Gray code encoder, the diagram in Figure 21 shows the differential signals over a small angular range. Displayed are all 10 tracks from bit 0 to bit 9, whereby bit 0 is the bit with the finest scale and, therefore, has the most oscillations. It can be seen that all bits delivered the correct signals and that they could be digitized at a constant threshold, for example, at 0 V. Furthermore, it can be seen that the signals were still not perfect because a drift could be observed during rotation of the encoder disc. This effect could probably be improved after a design review of the system with a better assembly process of the lens, whereby the optical system could maybe be improved in respect of a more tolerant optical path.

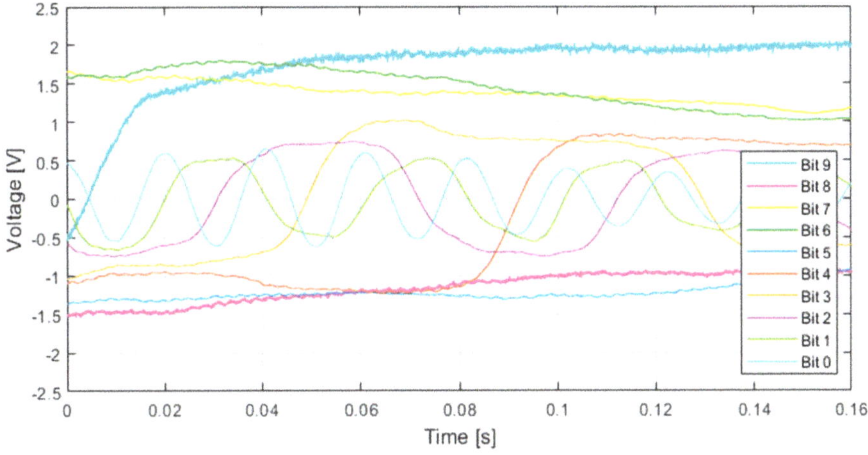

Figure 21. Signals of the Gray code encoder for all bits over a small angular range.

The encoder with the segmented photodiodes provided only digital signals; thus, no analog signals can be shown here. To demonstrate the functional principle, Figure 22a shows the laser spot on the segmented photo detector. The position of the laser spot could be calculated with a software tool by evaluating all 1024 photodiodes of the opto-ASIC. Figure 22b shows the user interface for the evaluation of the position of the laser spot with a Serial Peripheral Interface (SPI). Because the Gray code and the segmented encoder have two different evaluation principles, the results were not directly comparable. For further investigations of the segmented encoder, it should be enabled to also have access to the analog signals.

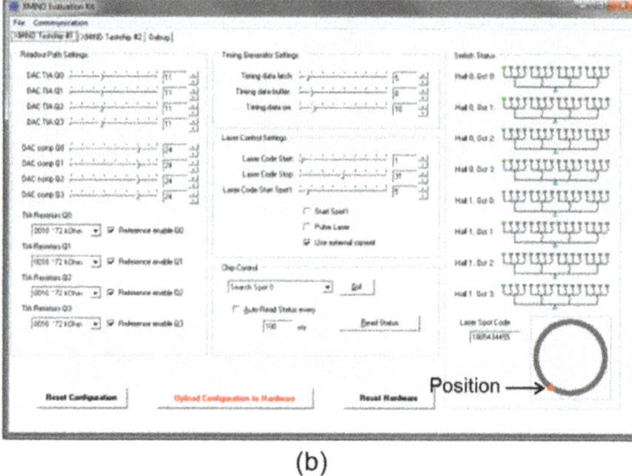

(a)   (b)

Figure 22. Encoder with the segmented photodiodes. (**a**) Laser spot on the photo detector; (**b**) user interface for evaluation of the position of the laser spot.

The most important features of the two optical encoders are summarized in Table 2.

Table 2. The most important features of the encoder; SPI—Serial Peripheral Interface.

Feature	
Method	Incremental and absolute
Resolution	1024 positions
Speed	12,000 rpm
Interface	Digital and SPI
Temperature	$-40\dots +85\,^{\circ}\mathrm{C}$
Size	$10 \times 10 \times 10\ \mathrm{mm}^3$

Further investigations were performed to study the influence of the required assembly accuracy of the lens and the optical components to determine the allowed tolerances. As expected, it became apparent that the assembly accuracy of the lens has an important role in determining the accuracy of the whole encoder. Therefore, special attention is needed during assembly of the lens. The successful method was to build an active alignment tool and use an active alignment process. This means that, during the assembling process of the lens, the laser diode was switched on and the required beam profile was controlled with an integrated image sensor inside this tool. The misalignment of the lens from the ideal position was calculated automatically. If necessary, the position of the lens was aligned automatically with an additional integrated piezo actuator with an accuracy of 80 nm [21,22].

5. Discussion

A demonstrator for a miniaturized optical encoder with a resolution of 10 bits was presented in this article. In comparison to available absolute optical encoders on the market with comparable small dimensions, this represents a significant fourfold improvement. The MAS 10 series from CUI with a housing diameter of 13 mm only achieves 8-bit resolution [2].

Attention was paid to a relatively simple assembly concept. Once the optical module was built, the final assembly of the encoder was possible without any further alignment process. In the case of conventional optical encoders, an alignment process between the encoder disc and the optical module is essential. Therefore, it is expected that, in conjunction with the injection-molded encoder disc, the proposed approach for the new optical encoder is even suitable for low-cost applications. Currently, the opto-ASIC is the most expensive part of the encoder. That means further research projects should focus on smaller opto-ASICs with a smaller silicon surface area to reduce costs. For mass production of the presented encoder, an automatic assembly of lenses using active alignment has yet to be implemented into assembly lines.

Another topic is the encoder signals. As shown in Figure 20, the signals were very stable; however, they can obviously be further improved. Slight fluctuations over one revolution could be observed. Firstly, the reasons for these fluctuations should be analyzed. A small misalignment of the lens can possibly cause such effects. Once the signals are improved, the resolution of the encoder can be increased using interpolation. The interpolation can be integrated into the opto-ASIC. Therefore, specific interpolation algorithms need to be developed, such as a correction of the analog sine and cosine signals. In this way, the usual conversion function arc tangent can be used, such that the phase angle can directly be determined from the sine and cosine voltage, thereby resulting in a linear correlation between angular position and the encoder signal. Thus, a discretization of the angular position is possible [23]. Initial assessments prove the expectation that a resolution of up to 15 bits is achievable.

6. Conclusions and Outlook

A new approach for miniaturized encoders was presented, overcoming the current physical limitations in the resolution of common optical encoders, because the solid measure was transformed from the encoder disc onto an opto-ASIC. The encoder achieved a resolution of 10 bits both for absolute and incremental values with a size of only $10 \times 10 \times 10$ mm^3.

The encoder has the potential to be manufactured as a low-cost encoder due to the simple but robust design and the reduced number of components. The optical components such as the lens and encoder disc can be manufactured by injection molding, and costs in the single-digit euro range are expected for these components for higher-volume manufacturing. The most expensive component of the encoder is the circular photo detector array, because the silicon area of the opto-ASIC is relatively large with 5×5 mm^2. Costs in the lower two-digit range are expected here. The other components including the laser diode, bearing, shaft, housing, and connector contain no special parts and, therefore, no cost drivers, because state-of-the-art encoders need such parts too. Cost benefits are expected as a result of the simple structure of the encoder during assembly and the removal of the adjustment process. Thus, production costs <80 € in total should be achieved easily for the encoder in the case of higher-volume manufacturing.

Furthermore, the encoder can be integrated into specific miniaturized products; moreover, there is no encoder enclosure necessary, and smaller form factors are possible. The encoder has also the potential as a modular encoder concept for different diameters or hollow-shaft encoders, when the opto-ASIC is located eccentric to the encoder shaft, whereby the same optical module and the same grating can be used. Further investigations will be directed toward increasing the resolution of the presented encoder.

Author Contributions: Conceptualization, J.S., A.B., A.F., and C.S.; methodology, J.S., A.B., A.F., and C.S.; validation, J.S., A.B., A.F., and C.S.; formal analysis, J.S., A.B., A.F., and C.S.; investigation, J.S. and A.F.; resources, A.Z. and J.B.; writing—original draft preparation, J.S., A.B., and A.F.; writing—review and editing, K.-P.F. and C.S.; visualization, J.S.; supervision, A.Z. and J.B.; project administration, J.S. and A.F.; funding acquisition, J.S., A.B., K.-P.F., A.F., and C.S.

Funding: This research was funded by the Federal Ministry for Economic Affairs and Energy on the basis of a decision by the German Bundestag, grant number IGF-project 17898 N.

Conflicts of Interest: The authors declare no conflict of interest.

References

1. Eitel, E. Basics of Rotary Encoders: Overview and New Technologies. Machine Design, 2014. Available online: www.machinedesign.com (accessed on 10 January 2019).
2. Market Research Report. Global Industrial Encoder Market 2017–2021. 2017. Available online: www.technavio.com (accessed on 10 January 2019).

3. Global Rotary Encoder Market Research Report 2017. 2018. Available online: www.marketresearchstore.com (accessed on 10 January 2019).
4. Incremental Optical Encoder. Contactless Position Sensors General Information, Presentation. 2016. Available online: www.exxelia.com/uploads/PDF/ie09-v1.pdf (accessed on 7 January 2019).
5. Monari, G. Understanding Resolution In Optical And Magnetic Encoders. 2013. Available online: www.electronicdesign.com (accessed on 7 January 2019).
6. Seybold, J. *Untersuchungen zur Industrialisierung von Miniaturisierten Optischen Drehwinkelsensoren mit Diffraktiver Kodierscheibe aus Kunststoff*; Dissertation Universität Stuttgart; Verlag Dr. Hut: München, Germany, 2013.
7. CUI Inc. Absolute, Optical Shaft Encoder. Datasheet of the Series MAS10-256G. 2018. Available online: www.cui.com (accessed on 26 October 2018).
8. Trupke, T. Optoelektronische Miniatur-Drehgeber, Einsätze in der minimalinvasiven Chirurgie. In *Medizin&Technik*; Konradin Verlag R. Kohlhammer GmbH: Leinfelden-Echterdingen, Germany, 2012.
9. Qin, S.; Huang, Z.; Wang, X. Optical Angular Encoder Installation Error Measurement and Calibration by Ring Laser Gyroscope. *IEEE Trans. Instrum. Meas.* **2010**, *59*, 506–511. [CrossRef]
10. Mayer, V. *Untersuchungen zu Optischen Drehgebern mit Mikrostrukturierten Maßverkörperungen aus Kunststoff*; Dissertation Universität Stuttgart; Verlag Dr. Hut: München, Germany, 2009.
11. Hopp, D. Inkrementale und Absolute Kodierung von Positionssignalen Diffraktiver Optischer Drehgeber. Ph.D. Thesis, Universität Stuttgart, Stuttgart, Germany, 2012.
12. Wibbing, D.; Binder, J.; Schinköthe, W.; Pauly, Ch.; Gachot, C.; Mücklich, F. Optical Absolute Position-Encoder by Single-Track, q-ary Pseudo-Random-Sequences for Miniature Linear Motors. In Proceedings of the Sensor + Test Conference—OPTO, Nürnberg, Germany, 7–9 June 2011.
13. Hopp, D.; Pruss, C.; Osten, W.; Seybold, J.; Mayer, V.; Kück, H. Optischer inkrementaler Drehgeber in Low-Cost-Bauweise. *Tm Technisches Messen* **2010**, *77*, 358–363. [CrossRef]
14. Mayer, V. A new concept for an absolutely encoded angular resolver. In Proceedings of the 4M 2007 Conference on Multi-Material Micro Manufacture, Borovets, Bulgaria, 3–5 October 2007.
15. Seybold, J.; Mayer, V.; Kück, H.; Hopp, D.; Pruss, Ch.; Osten, W. Hochauflösender optischer Drehgeber mit MID-Optikmodul. In Proceedings of the 6th Paderborner Workshop, Entwurf mechatronischer Systeme, Paderborn, Germany, 2–3 April 2009.
16. Seybold, J.; Fritz, K.-P.; Mayer, V.; Kück, H. Miniature Optical Encoder for Reflow Solder Mounting. In Proceedings of the 8th International Conference on Multi-Material Micro Manufacture, Stuttgart, Germany, 8–10 November 2011.
17. Seybold, J.; Scherjon, C. Untersuchungen zu extrem miniaturisierten optischen Drehgebern (X-MIND). IGF-Vorhaben 17898 N, Abschlussbericht. Available online: www.hahn-schickard.de (accessed on 20 December 2016).
18. Edmund Optics, Ruled Holo Grating 600x800n. Available online: www.edmundoptics.de (accessed on 18 July 2015).
19. Craubner, S.; Graf, H.-G.; Harendt, C.; Platz, W.; Schubert, M. *Hochleistungs-Front-End-Elektronik. Endbericht Phase 2, Entwicklung "Integrierter Detektorsysteme"*; BMBF FKZ 50TT9724; BMBF: Bonn, Germany, 2002.
20. Bülau, A.; Seybold, J.; Fritz, K.-P.; Frank, A.; Scherjon, C.; Burghartz, J.; Zimmermann, A. Miniaturized optical encoder with micro structured encoder disk. In Proceedings of the WCMNM-Conference, World Congress on Micro and Nano Manufacturing, Portorož, Slovenia, 18–20 September 2018.
21. Höhn, M. *Sensorgeführte Montage Hybrider Mikrosysteme*; Herbert Utz Verlag: München, Germany, 2001.
22. Martin, C. *Commissioning and Optimization of a Pick-up Tool with Integrated Beam Profiling for Precise Assembling of Lenses*; Student Research Project; University of Stuttgart: Stuttgart, Germany, 2015.
23. High-Precision Sine/Cosine Interpolation. White Paper. 2014. Available online: www.ichaus.de (accessed on 17 September 2018).

Article

Fabrication of a Novel Culture Dish Adapter with a Small Recess Structure for Flow Control in a Closed Environment

Reiko Yasuda [1,2], Shungo Adachi [3], Atsuhito Okonogi [2], Youhei Anzai [2], Tadataka Kamiyama [2], Keiji Katano [2], Nobuhiko Hoshi [1], Tohru Natsume [3] and Katsuo Mogi [3,*]

[1]   Laboratory of Animal Molecular Morphology, Department of Animal Science,
       Graduate School of Agricultural Science, Kobe University, 1-1 Rokkodai, Nada-ku, Kobe 657-8501, Japan;
       r.yasuda@icomes.co.jp (R.Y.); nobhoshi@kobe-u.ac.jp (N.H.)
[2]   Icomes Lab Co., Ltd., 1-8-25 Kitaiioka, Morioka-shi, Iwate 020-0857, Japan; a.okonogi@icomes.co.jp (O.A.);
       y.anzai@icomes.co.jp (Y.A.); t.kamiyama@icomes.co.jp (T.K.); katano@icomes.co.jp (K.K.)
[3]   Molecular Profiling Research Centre for Drug Discovery, National Institute of Advanced Industrial Science
       and Technology (AIST), 2-4-7 Aomi, Koto-ku, Tokyo 135-0064, Japan; s.adachi@aist.go.jp (S.A.);
       t-natsume@aist.go.jp (T.N.)
*   Correspondence: mogi.k@aist.go.jp; Tel.: +81-3-3599-8251

Received: 3 December 2018; Accepted: 4 January 2019; Published: 14 January 2019

Featured Application: The developed adapter makes it possible to spatially control flow in a commercially available culture dish in a manner previously only possible in microchannel devices. It will also be possible to create an arbitrary concentration gradient in a culture dish. Potential applications in a very wide range of fields include simultaneously reacting multiple single cells with various reagents in a culture dish and conducting drug discovery screening using very delicate concentration gradients.

Abstract: Cell culture medium replacement is necessary to replenish nutrients and remove waste products, and perfusion and batch media exchange methods are available. The former can establish an environment similar to that in vivo, and microfluidic devices are frequently used. However, these methods are hampered by incompatibility with commercially available circular culture dishes and the difficulty in controlling liquid flow. Here, we fabricated a culture dish adapter using polydimethylsiloxane that has a small recess structure for flow control compatible with commercially available culture dishes. We designed U-shaped and I-shaped recess structure adapters and we examined the effects of groove structure on medium flow using simulation. We found that the U-shaped and I-shaped structures allowed a uniform and uneven flow of medium, respectively. We then applied these adaptors to 293T cell culture and examined the effects of recess structures on cell proliferation. As expected, cell proliferation was similar in each area of a dish in the U-shaped structure adapter, whereas in the early flow area in the I-shaped structure adapter, it was significantly higher. In summary, we succeeded in controlling liquid flow in culture dishes with the fabricated adapter, as well as in applying the modulation of culture medium flow to control cell culture.

Keywords: flow control; culture dish adapter; small recess structure; closed environment; perfusion culture

---

1. Introduction

The basic unit of an organism is the cell and an individual is formed by cells taking a more complicated structure to become tissues, organs, and organ systems. Cell culture is a technique for growing these cells in an artificial environment outside the body. One of the major advantages of cell culture is that, in the environment that the cells grow in, physicochemical characteristics such as temperature, pH, and shear stress, as well as physiological components such as hormone and waste concentration, can be adjusted [1–3]. Moreover, it also makes it possible to analyze biological phenomena using a simpler experimental system compared to cellular analysis in vivo. Cell culture is a technology still in development: from culture dishes, a conventional technique, to perfusion culture in micro-channels for gene therapy and regenerative medicine and application in single-cell analysis and drug discovery screening [4–6].

Various conditions are required for culturing cells, but in terms of culture vessels, commercially available culture dishes are the most commonly used. In fact, when culturing cells, it is necessary to adjust the external environmental conditions, such as pH, CO_2 concentration, and temperature of the culture medium, for cells to grow in an environment most suitable for them, and these conditions will vary depending on cell type and experimental system. It is necessary to select the cell culture vessel, such as culture dishes or flasks, according to the desired purpose [7,8]. However, what is widely used in laboratories are the commercially available culture dishes sold by manufacturers [9]. Each laboratory has a detailed know-how of each product, such as adhesion of cells to culture dishes, and even among commercially available culture dishes, each researcher has his or her own preference.

Advances in microfluidic device technology have led to progress in research on perfusion culture [10], and it is becoming clear that liquid flow has a large influence on cells. Presently, owing to the progress in microfabrication technology, it is possible to create microscale fluidic devices that are very complex and have a high difficulty of fabrication [11], and that allow reproducible cell culture experiments within microchannels [12–14]. In fact, studies using microfluidic devices showed that stem cells and iPS cells/ES cells are influenced by shear stress, and this promotes differentiation [15–17]. Therefore, the influence of culture medium flow during cell culture is very large. If this flow can be controlled, researchers can modulate the influence of flow on cells, particularly shear stress. However, in culture experiments using microfluidic devices, a skilled technician is required to deliver liquid to the microchannel, and contamination from connection points in the complex structure of the device is often a problem [18]. Thus, unsolved problems remain in culture experiments using microfluidic devices for practical applications owing to the complexity of the equipment.

An efficient perfusion culture equipment using culture dishes remains to be developed [19], and it is difficult to control the flow in currently used culture dishes. In perfusion culture experiments using commercially available culture dishes, problems such as flow rate accuracy of the tube pump and meniscus in the culture dish remain to be solved, and a method for controlling the flow in the culture dish has not been established. In typical tube pumps, errors in the inner diameter of the tube affect the amount of liquid delivered [20], and when injection and discharge are controlled using a tube pump, it is difficult to keep the injection and discharge amounts constant, leading to stagnant liquid or leakage in the culture dish. In addition, because the culture dish itself is an open type chamber, a meniscus occurs between the dish and the medium, which results in a faster flow of culture medium along the edge of the dish owing to the property that liquid flows more easily toward the higher liquid bulk volume, and thus it is very difficult to control flow in a culture dish [21].

If these problems are resolved, it becomes possible to use, in combination with a commercially available culture dish, a device manufactured using laboratory equipment that arbitrarily controls the flow in a culture dish, which is difficult to realize with a commercialized device [19]. In addition, it will be possible to assemble and manufacture a simple handling device and evaluate the influence of fine flow in a macro culture environment that is difficult with microfluidic devices [9–11,22]. In other words, it becomes possible to spatially control flow in a commercially available culture dish in a manner previously only possible in microchannel devices. It will also be possible to create an arbitrary concentration gradient in a culture dish (Figure 1). Potential applications in a very wide range of fields include simultaneously reacting multiple single cells with various reagents in a culture dish and conducting drug discovery screening using very delicate concentration gradients.

In this study, we fabricated a polydimethylsiloxane (PDMS) culture dish adapter (CD-adapter) that enables fluid control with a small recess structure, and also investigated whether flow control affects cultured cells.

2. Fabrication of the PDMS CD-Adapter

We fabricated an adapter with a small recess structure to control the flow of liquid in a culture dish in a closed environment. To fabricate a PDMS adapter that can be attached to a commercially available culture dish without gaps, a method using a culture dish equipped with a negative structure of a small recess structure as a mold was adopted. A U-shaped small recess structure (U-shaped structure) and an I-shaped small recess structure (I-shaped structure) as shown in Figure 1 were provided in a CD-adapter corresponding to a 35-mm culture dish. A semicircular arc with outer and inner diameters of 32 mm and 30 mm, respectively, and a rectangle with a width and length of 2 mm and 34 mm, respectively, were cut out from a 500-μm thick polyethylene sheet (HRHG711303, KYOWA). The circular arc was pasted on the bottom of a 35-mm culture dish with a 2-mm gap from the edge, and the rectangle was stuck on a line bisecting the culture dish (Figure 2A). The outer periphery of the 35-mm culture dish with a mold structure was filled with epoxy resin (STYCAST 126J PTA, Henkel) for the bearing surface of the adapter. Therefore, as shown in Figure 2A, a 60-mm culture dish was used as a tray, being attached centered under the 35-mm culture dish. Epoxy resin was poured into the gap between both culture dishes, which were left for 9 h to stand at 25 °C to cure the epoxy resin (Figure 2B). After checking that the epoxy resin had cured, 2 mL of PDMS (CAT-106F, Shin-Etsu Chemical) was injected into the 35-mm culture dish and left to stand at 25 °C for 6 h to cure the PDMS (Figure 2C). The reason for setting the volume of PDMS to 2 mL is to make the thickness of the bottom membrane of the CD-adapter 2 mm. Because the bottom area of the 35-mm culture dish is 9 cm^2, the volume of PDMS required to make the bottom membrane thickness 2 mm is 1.8 mL. In this study, we set it to 2 mL for easy preparation of the experiments. After confirming that the PDMS had cured, the 35-mm culture dish and the center of a cylinder (aluminum cylinder, Showa Denko, Tokyo, Japan) with a diameter of 25 mm was placed on the PDMS membrane (Figure 2D), and PDMS was poured in a 60-mm culture dish so as not to overflow the edge (Figure 2E). After leaving it at 25 °C for 9 h to cure the PDMS, the aluminum cylinder and culture dish were peeled off, and the molded PDMS taken out. Using a biopsy punch (BP-L20K, Kai, Tokyo, Japan), a 2-mm diameter hole was drilled at the intersection between the circular arc of the small recess structure and the Y-Y' part of the fabricated PDMS adapter, and at a linear small recess structure 2 mm away from the edge of the adapter, to serve as an inlet. Similar holes were made on each opposite side to form an outlet (Figure 2F,G). The completed sealed adapter was a cylindrical-shaped PDMS with a 2 mm wide, 500 μm deep recess structure on a 2 mm thick bottom membrane, and an inlet and outlet with an inner diameter of 2 mm (Figure 2G).

The fabricated CD-adapter had a very high transparency and was composed of PDMS only, and thus it has good oxygen permeability and a very suitable structure for cell culture. Furthermore, because the central membrane portion was a thin transparent membrane with a thickness of 2 mm, it was also possible to observe the culture dish with the adapter attached on it with a microscope (Figure 3A). The fabricated U-shaped and I-shaped structures are shown in Figure 3B,C, respectively.

As shown in Figure 3D, a transparent framework made of acrylic resin was used as a holding jig to prevent liquid leakage. It has a circular shape so that a constant pressure can be applied to the side of the culture dish for tight sealing. In addition, the center of the culture jig contains a hole with a diameter of 10 mm through which cells can be observed using a microscope. The holding jig has a height of 5 mm and an outer diameter of 27.0 mm. Furthermore, to adjust the amount of medium in the culture dish, a ring spacer was placed between the CD-adapter and the culture dish to raise the adapter. Because the CD-adapter was molded from the culture dish, when the CD-adapter is attached to it, the bottom surface of the CD-adapter touches the bottom of the culture dish. In this study, to set the amount of medium in the culture dish to 2 mL, the ring spacer was made of silicone with a thickness of 2 mm and the CD-adapter was lifted 2 mm from the bottom of the culture dish.

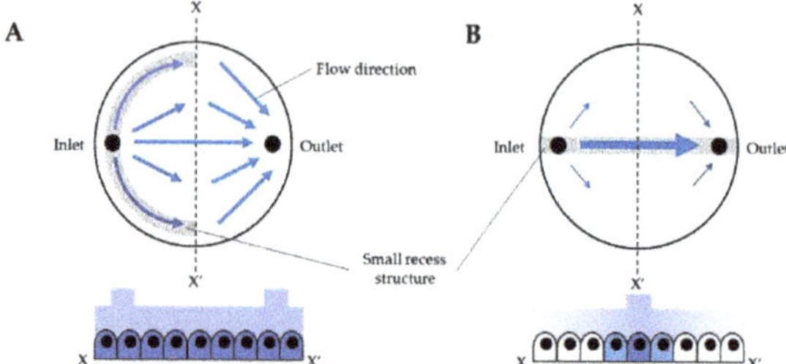

Figure 1. Flow control in the closed environment of a culture dish. By providing a structure that can control flow in a culture dish by adapter selection, it is possible to arbitrarily modulate the liquid flow in the closed environment without changing the culture medium, culture dish, or injection rate. (**A**) Delivered liquid flows uniformly from the inlet towards the outlet at an almost constant speed irrespective of the position in the culture dish. Cells in culture dishes can be stably cultured under the same conditions. (**B**) Delivered liquid flows linearly from the inlet to the outlet. A concentration gradient occurs in the culture dish along the flow of the delivered liquid, and, as a result, it is possible to modulate its influence on cells by adapter selection.

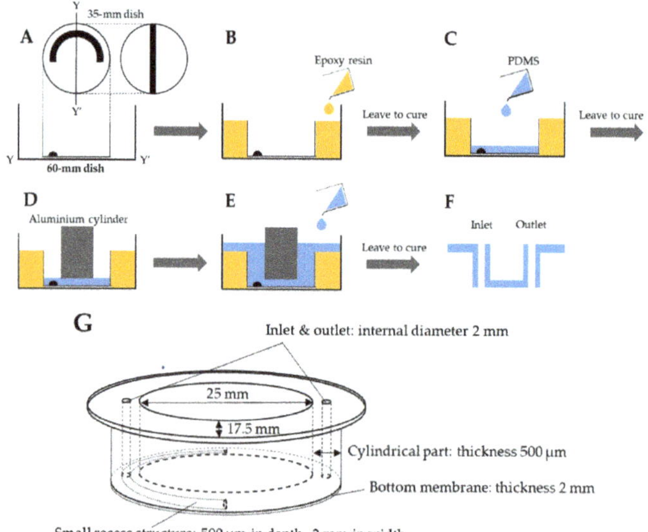

Figure 2. Manufacturing method of the culture dish adapter (CD-adapter). (**A**) Cut out a circular arc of 2.0 mm of width and 32 mm of outer diameter from a polyethylene sheet with a thickness of 500 μm and paste it on the bottom of a 35-mm culture dish from Nunc and 2.0 mm from the edge. Place the 35-mm culture dish to which the circular arc was attached in the center of a 60-mm culture dish using double-sided tape. (**B**) Pour epoxy resin into the 60-mm culture dish and leave at 25 °C for approximately 9 h to cure. (**C**) Pour 2 mL of polydimethylsiloxane (PDMS) into the 35-mm culture dish and let it stand at 25 °C for approximately 6 h to cure. (**D**) Place an aluminum cylinder with a diameter of 25 mm at the center of the 35-mm culture dish. (**E**) Pour more PDMS around the aluminum cylinder and leave at 25 °C for approximately 9 h to cure. (**F**) Remove the aluminum cylinder and culture dish from the PDMS and use a long-type biopsy punch with a diameter of 2 mm to make holes for injection and discharge. (**G**) CD-adapter made of PDMS compatible with a 35-mm culture dish. The adapter has an inlet and an outlet and a concave groove at the bottom for equalizing the flow in the culture dish. The depth of the groove is 500 μm, the diameter of the inlet and outlet is 2.0 mm, and the membrane thickness of the bottom surface of the adapter is 2.0 mm.

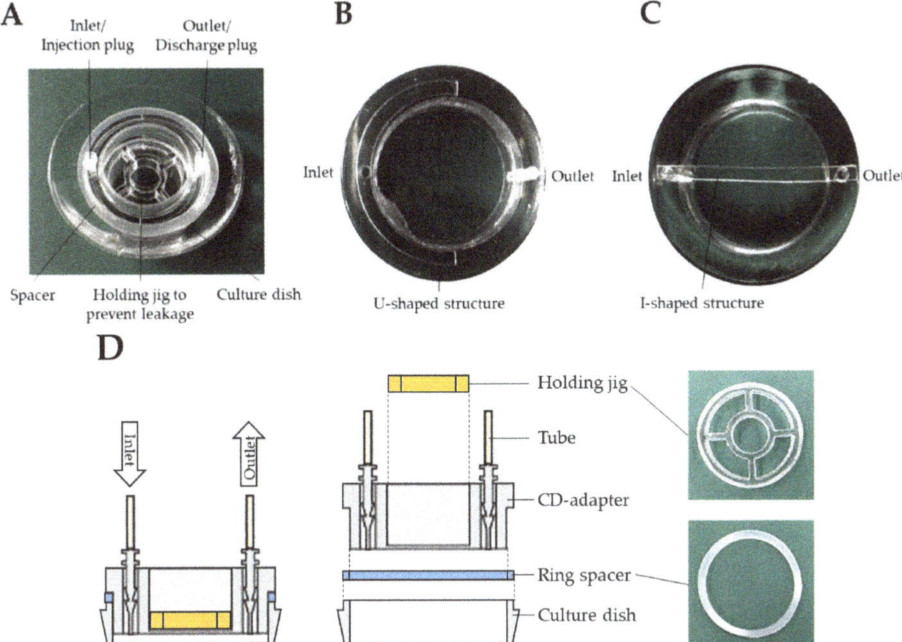

Figure 3. The fabricated CD-adapter with its small recess structure. (**A**) A CD-adapter compatible with 35-mm culture dishes. Because it is fabricated with polydimethylsiloxane (PDMS), it is highly transparent and enables cell observation by microscope. In addition, owing to good air permeability, it can supply enough oxygen to cells even in a sealed state. A plug was inserted into both inlet and outlet to connect to the tube, and the amount of medium in the culture dish can be adjusted by raising the adapter with a spacer. Furthermore, a holding jig to prevent liquid leakage was fitted in the center of the adapter, and pressure is applied from the inside toward the culture dish to increase adhesion between the adapter and the culture dish. (**B**) Small recess structure created on the bottom of the CD-adapter used in this work. A horseshoe-like small recess structure spreads from the inlet side to the equatorial line in the bottom. The width of the groove was 2.0 mm and its depth was 500 μm. (**C**) Small recess structure created on the bottom of the CD-adapter used in this work. A linear small recess structure extends from the inlet side to the outlet side. The width of the groove was 2.0 mm and its depth was 500 μm. (**D**) Procedure for setting the CD-adapter to the culture dish. A ring spacer of 2 mm thickness made of silicon was placed between the culture dish and the CD-adapter to raise the adapter and make cell culture space. A holding jig was fitted in the center of the adapter to prevent liquid leakage.

3. Experiment

3.1. Fluid Simulation of Flow Control with the CD-Adapter

Perfusion culture was performed using a CD-adapter with two types of small recess structure, and flow control was measured in each cell adhesion area. Before performing culture experiments, fluid simulation of the flow control of the small recess structure using commercial software of finite element method (COMSOL, KESCO) was performed. Figure 4A shows the flow velocity distribution in an adapter with no small recess structure when the flow rate was 1.4 μL/min, and Figure 4D shows the flow velocity distribution in the Z-Z' cross section. Since there was a strong gradient in the flow rate in the culture dish unless a small recess structure was provided in the adapter (Figure 4D), we designed a U-shaped structure that can maintain an almost uniform flow in the culture dish. And to more clearly demonstrate that flow control is possible, we also designed an I-shaped structure with a more pronounced flow bias compared to the U-shaped structure. Similarly, Figure 4B,C show flow velocity distributions in the U-shaped and I-shaped structures. Figure 4E,F show flow velocity distributions in the Z-Z' sections in Figure 4B,C. In Figure 4F, the flow of liquid increased markedly in the I-shaped structure compared to that shown in Figure 4E. From these results, we conclude that the flow velocity distribution in the culture dish can be controlled by the shape and position of the small recess structure.

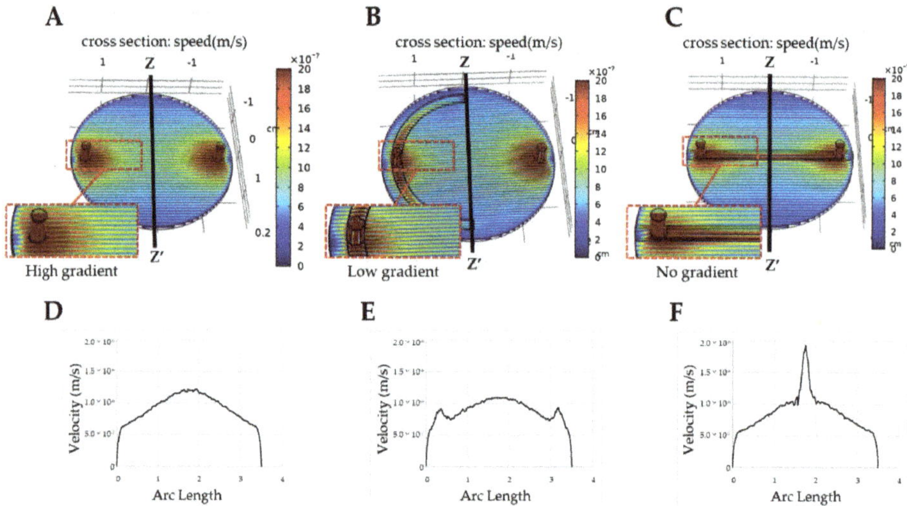

Figure 4. Simulation analysis of liquid flow in a perfusion culture dish with a CD-adapter. Liquid flow in the culture dish at 1.4 µL/min through the inlet was simulated and analyzed using COMSOL. (**A**), adapter without the small recess structure; (**B**), U-shaped structure adapter; (**C**), I-shaped structure adapter. Flow velocity on the Z-Z' line in (**A–C**) are plotted in (**D–F**), respectively. In (**F**), the liquid flow remarkably increased under the I-shaped groove compared to (**E**).

3.2. Cell Culture Test for Flow Controllability of the Small Recess Structure

Based on the results of the simulation, three adapters were prepared for each U-shaped and I-shaped structure. Cells from the 293 T cell line, which are kidney cells derived from a human fetus, were seeded in 35-mm culture dishes (IWAKI) at a seeding density of 1×10^5 cells/mL and cultured for 1 day in a 37 °C—5% CO_2 incubator. The culture medium was prepared by adding 10% fetal bovine serum (FBS) to Dulbecco's modified Eagle's medium (DMEM), and 2 mL was added to the culture dish.

The culture medium of cells cultured for 1 day was completely removed and replaced with serum-free DMEM supplemented with no FBS, the lid of the culture dish was removed, and a CD-adapter, which was autoclaved and dried, was attached to the dish. Next, the microtube pump system (Icomes Lab, Iwate, Japan), the reservoir, and the waste reservoir were connected to the clean bench as shown in Figure 5. DMEM supplemented with 10% FBS was perfused at a flow rate of 17 µL/min, which did not influence the shear stress, for 40 min by operating the pump system [14], and then the pump was stopped and the cells were cultured for 1 day at 37 °C—5% CO_2. The cells were then imaged at a magnification of 40× with a microscope (inverted microscope IX71, OLYMPUS, Tokyo, Japan) at points i, ii, and iii in Figure 6A. Images were processed using ImageJ, and the area of adhered cells was plotted (Figure 5).

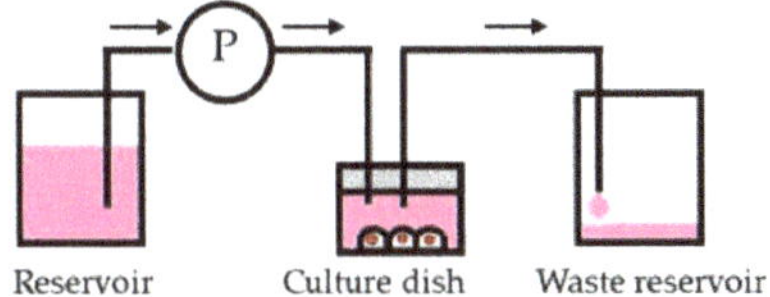

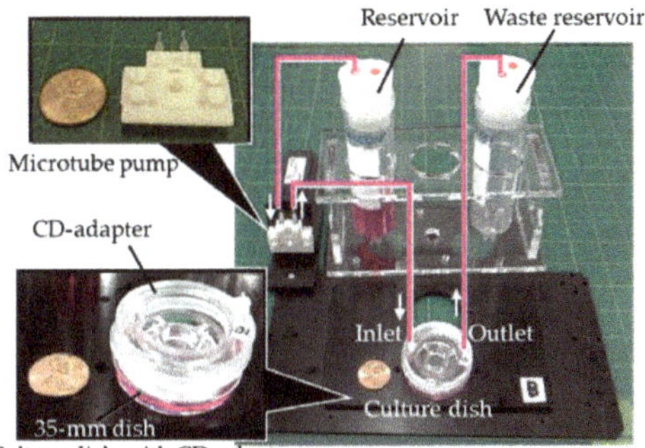

Figure 5. Schematic of the culturing system and photo of the experimental setup. A CD-adapter was attached to a 35-mm culture dish on which cells had been seeded, and connected to a micro tube pump system, reservoir, and waste reservoir. The size of the experimental setup is $200 \times 150 \times 150$ mm, which can be installed on a stage of the microscope and also can be installed in the incubator.

4. Results and Discussion

Culture experiments were carried out three times (N = 3), and the adhesion area of cells after 3 days was averaged for each observation point and plotted. Results in i and iii are normalized by value of cell adhesion area in ii in Figure 6A, and the error bars correspond to standard deviation. Adhesion areas of cells in i, ii, and iii was 0.92, 1.0, and 0.92, respectively, in the U-shaped structure adapter, and 0.65, 1.0, and 0.73, respectively, in the I-shaped structure adapter.

It is usually necessary to add serum to the culture medium of 293T cells, which otherwise do not grow normally and die. Therefore, cells over which serum-containing medium flowed should have proliferated more than those at sites with no flow. As can be seen from the simulation results in Figure 4B, culture medium flowed evenly through the culture dish from the inlet to the outlet in the U-shaped structure. Therefore, it can be assumed that, by delivering culture medium with serum, the cells will grow uniformly throughout the culture dish. Indeed, as a result of delivering the serum-containing culture medium, the cells showed the same level of proliferation in all areas of the culture dish (Figure 6B). In contrast, the simulation results in Figure 4C show that the flow velocity of the culture fluid became faster on the straight line under the small recess structure of the I-shaped structure. Accordingly, the cell culture results show that the cell proliferation rate was higher in ii, immediately below the small recess structure, compared to other areas (Figure 6B). The culture and simulation results are consistent, and we conclude that flow control can be exerted with cell culture evaluation using the small recess structure provided in the CD-adapter.

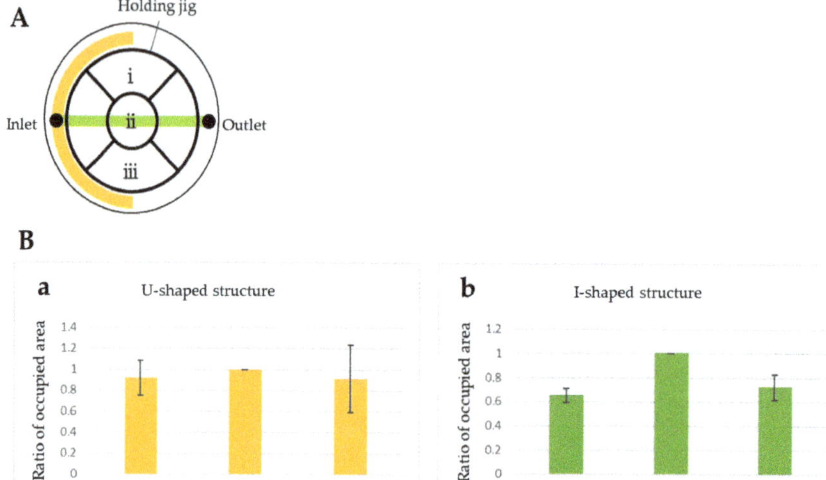

Figure 6. Culture of 293T cells using the CD-adapter. (**A**) Cells were observed at points i, ii, and iii on the vertical line intersecting the center of the straight line that connects the inlet and the outlet. DMEM culture medium supplemented with 10% fetal bovine serum was perfused at a flow rate of 17 μL/min through the inlet for 40 min, and the cells were then cultured for 1 day. (**B**) Cultured cells were imaged at all points, and the area occupied by cells was quantified using ImageJ software. The area occupied at ii was set as 1, and then the relative areas at i and iii were obtained. With the adapter with the I-shaped structure (**b**), the rate of cell proliferation under the small recess structure was significantly higher than that in other places, and no difference in proliferation rates was observed with the adapter with the U-shaped structure (**a**) at points i–iii, and it was possible to culture cells uniformly.

5. Conclusions

In this study, we developed a CD-adapter with recess structure to exert flow control in a closed environment using commercially available culture dishes. We successfully fabricated the CD-adapter using inexpensive PDMS by an easy process without any special equipment for microfabrication. The flow controllability of the recess structure was estimated using the finite element method and demonstrated by cell culturing in the closed chamber to confirm that flow control has an influence on cultured cells. Although precise control of microscale spatial resolution was not obtained, liquid flow was modulated using the developed CD-adapter with a small recess structure, and cell culture was controlled by modulating the flow of culture medium.

A more delicate control of the flow may be obtained by further elaborating the shape of the small recess structure of the adapter, and adapters for various culture dishes can be fabricated cheaply and freely. In the future, we will evaluate the flow in different shapes of the small recess structure in various adapters and their influence on cells. As this research progresses, researchers and research institutes conducting cell culture experiments will be able to perform more sophisticated experiments without greatly changing their current experimental setup, and it may become possible to conduct complex research that, until now, could only be performed using microfluidic devices in a macro environment.

Author Contributions: Conceptualization, R.Y., K.M., and S.A.; formal analysis, R.Y. and K.M.; funding acquisition, O.A., S.A., K.K., and T.N.; investigation, R.Y.; methodology, R.Y. and K.M.; project administration, R.Y.; resources, Y.A. and T.K.; supervision, N.H. and T.N.; validation, R.Y.; visualization, R.Y.; writing—original draft preparation, R.Y.; writing—review and editing, M.K., S.A., N.H., and T.N.

Funding: Part of this perfusion system was developed with financial support from the Grant-in-Aid for Sample creation support project No. 408005 of the New Energy and Industrial Technology Development Organization (NEDO) of Japan and the Adaptable and Seamless Technology Transfer Program through Target-driven R&D

(A-STEP) No. AS2815006U from the Japan Science and Technology Agency (JST). This work was partially supported by Grant-in-Aid for young scientists (B) (No. 17K17715) and Leading Initiative for Excellent Young Researchers (LEADER) in FY 2016 of the Ministry of Education, Culture, Sports, Science and Technology, Japan.

Acknowledgments: The authors would like to thank H. Kotera and H. Shintaku (Institute of Physical and Chemical Research) for assistance with the numerical simulations.

Conflicts of Interest: The authors declare no conflict of interest.

References

1. Ouyang, A.; Ng, R.; Yang, S.T. Long-term culturing of undifferentiated embryonic stem cells in conditioned media and three-dimensional fibrous matrices without extracellular matrix coating. *Stem Cells* **2007**, *25*, 447–454. [CrossRef] [PubMed]

2. Yoon, S.K.; Kim, S.H.; Lee, G.M. Effect of low culture temperature on specific productivity and transcription level of anti-4-1bb antibody in recombinant chinese hamster ovary cells. *Biotechnol. Prog.* **2003**, *19*, 1383–1386. [CrossRef] [PubMed]

3. Rashidi, H.; Alhaque, S.; Szkolnicka, D.; Flint, O.; Hay, D.C. Fluid shear stress modulation of hepatocyte-like cell function. *Arch. Toxicol.* **2016**, *90*, 1757–1761. [CrossRef]

4. Close, D.A.; Camarco, D.P.; Shan, F.; Kochanek, S.J.; Johnston, P.A. The generation of three-dimensional head and neck cancer models for drug discovery in 384-well ultra-low attachment microplates. *Methods Mol. Biol.* **2018**, *1683*, 355–369. [PubMed]

5. Seah, Y.F.S.; Hu, H.; Merten, C.A. Microfluidic single-cell technology in immunology and antibody screening. *Mol. Asp. Med.* **2018**, *59*, 47–61. [CrossRef] [PubMed]

6. Fang, Y.E.; Richard, M. Three-dimensional cell cultures in drug discovery and development. *SLAS Discov. Adv. Life Sci. RD* **2017**, *22*, 456–472.

7. Kano, M.; Suga, H.; Kasai, T.; Ozone, C.; Arima, H. Functional pituitary tissue generation from human embryonic stem cells. *Curr. Protoc. Neurosci.* **2018**, *83*, e48. [CrossRef] [PubMed]

8. Gosavi, R.A.; Salwe, S.; Mukherjee, S.; Dahake, R.; Kothari, S.; Patel, V.; Chowdhary, A.; Deshmukh, R.A. Optimization of ex vivo murine bone marrow derived immature dendritic cells: A comparative analysis of flask culture method and mouse cd11c positive selection kit method. *Bone Marrow Res.* **2018**, *2018*, 3495086. [CrossRef]

9. Kondo, E.; Wada, K.; Hosokawa, K.; Maeda, M. Microfluidic perfusion cell culture system confined in 35 mm culture dish for standard biological laboratories. *J. Biosci. Bioeng.* **2014**, *118*, 356–358. [CrossRef]

10. Yuan, D.; Tan, S.H.; Sluyter, R.; Zhao, Q.; Yan, S.; Nguyen, N.T.; Guo, J.; Zhang, J.; Li, W. On-chip microparticle and cell washing using coflow of viscoelastic fluid and newtonian fluid. *Anal. Chem.* **2017**, *89*, 9574–9582. [CrossRef]

11. Tang, S.Y.; Yi, P.; Soffe, R.; Nahavandi, S.; Shukla, R.; Khoshmanesh, K. Using dielectrophoresis to study the dynamic response of single budding yeast cells to lyticase. *Anal. Bioanal. Chem.* **2015**, *407*, 3437–3448. [CrossRef] [PubMed]

12. Mogi, K.; Fujii, T. A novel assembly technique with semi-automatic alignment for pdms substrates. *Lab Chip* **2013**, *13*, 1044–1047. [CrossRef] [PubMed]

13. Atencia, J.; Beebe, D.J. Controlled microfluidic interfaces. *Nature* **2005**, *437*, 648–655. [CrossRef] [PubMed]

14. Kim, L.; Toh, Y.C.; Voldman, J.; Yu, H. A practical guide to microfluidic perfusion culture of adherent mammalian cells. *Lab Chip* **2007**, *7*, 681–694. [CrossRef] [PubMed]

15. Toh, Y.C.; Voldman, J. Fluid shear stress primes mouse embryonic stem cells for differentiation in a self-renewing environment via heparan sulfate proteoglycans transduction. *FASEB J.* **2011**, *25*, 1208–1217. [CrossRef] [PubMed]

16. Burdon, T.; Smith, A.; Savatier, P. Signalling, cell cycle and pluripotency in embryonic stem cells. *Trends Cell Biol.* **2002**, *12*, 432–438. [CrossRef]

17. Corrigan, M.A.; Johnson, G.P.; Stavenschi, E.; Riffault, M.; Labour, M.N.; Hoey, D.A. Trpv4-mediates oscillatory fluid shear mechanotransduction in mesenchymal stem cells in part via the primary cilium. *Sci. Rep.* **2018**, *8*, 3824. [CrossRef] [PubMed]

18. Mehling, M.; Tay, S. Microfluidic cell culture. *Curr. Opin. Biotechnol.* **2014**, *25*, 95–102. [CrossRef]

19. Lee, S.Y.; Yang, S. A microfluidic-based lid device for conventional cell culture dishes to automatically control oxygen level. *Biotechniques* **2018**, *64*, 231–234. [CrossRef]
20. Chung, I.S.; Cho, H.S.; Kim, J.A.; Lee, K.H. The flow rate of the elastomeric balloon infusor is influenced by the internal pressure of the infusor. *J. Korean Med. Sci.* **2001**, *16*, 702–706. [CrossRef]
21. DasGupta, S.S.; Jeffrey, A.; Kim, I.Y.; Wayner, P.C., Jr. Use of the augmented young-laplace equation to model equilibrium and evaporating extended menisci. *J. Colloid Interface Sci.* **1993**, *157*, 332–342. [CrossRef]
22. Yuan, D.; Zhang, J.; Yan, S.; Peng, G.; Zhao, Q.; Alici, G.; Du, H.; Li, W. Investigation of particle lateral migration in sample-sheath flow of viscoelastic fluid and newtonian fluid. *Electrophoresis* **2016**, *37*, 2147–2155. [CrossRef] [PubMed]

 *applied sciences*

Article

Improving Surface Roughness of Additively Manufactured Parts Using a Photopolymerization Model and Multi-Objective Particle Swarm Optimization

Namjung Kim [1,*,†]**, Ishan Bhalerao** [2,†]**, Daehoon Han** [2]**, Chen Yang** [2] **and Howon Lee** [2,*]

1 Department of Mechanical Science and Engineering, University of Illinois at Urbana-Champaign, 1206 W. Green ST., Urbana, IL 61801, USA

2 Department of Mechanical and Aerospace Engineering, Rutgers University—New Brunswick, 98 Brett Rd., Piscataway, NJ 08854, USA; imb37@scarletmail.rutgers.edu (I.B.); daehoon.han@rutgers.edu (D.H.); chen.yang@rutgers.edu (C.Y.)

* Correspondence: kim847@illinois.edu (N.K.); howon.lee@rutgers.edu (H.L.); Tel.: +1-848-445-2213 (H.L.)

† These authors contributed equally.

Received: 30 November 2018; Accepted: 21 December 2018; Published: 3 January 2019

Featured Application: The optimization framework using a computational process modeling and a multi-objective optimization technique presented here will help to determine process parameters to produce high-quality additively manufactured parts while minimizing printing time. This framework can also be applied to other additive manufacturing techniques.

Abstract: Although additive manufacturing (AM) offers great potential to revolutionize modern manufacturing, its layer-by-layer process results in a staircase-like rough surface profile of the printed part, which degrades dimensional accuracy and often leads to a significant reduction in mechanical performance. In this paper, we present a systematic approach to improve the surface profile of AM parts using a computational model and a multi-objective optimization technique. A photopolymerization model for a micro 3D printing process, projection micro-stereolithography (PµSL), is implemented by using a commercial finite element solver (COMSOL Multiphysics software). First, the effect of various process parameters on the surface roughness of the printed part is analyzed using Taguchi's method. Second, a metaheuristic optimization algorithm, called multi-objective particle swarm optimization, is employed to suggest the optimal PµSL process parameters (photo-initiator and photo-absorber concentrations, layer thickness, and curing time) that minimize two objectives; printing time and surface roughness. The result shows that the proposed optimization framework increases 18% of surface quality of the angled strut even at the fastest printing speed, and also reduces 50% of printing time while keeping the surface quality equal for the vertical strut, compared to the samples produced with non-optimized parameters. The systematic approach developed in this study significantly increase the efficiency of optimizing the printing parameters compared to the heuristic approach. It also helps to achieve 3D printed parts with high surface quality in various printing angles while minimizing printing time.

Keywords: micro 3D printing; micro stereolithography; process parameter optimization; Taguchi's method; multi-objective particle swarm optimization

---

1. Introduction

Additive manufacturing (AM) is a set of manufacturing processes that produce three-dimensional (3D) physical objects by adding materials in a layer-by-layer fashion. The use of AM has been gradually changing from prototyping to manufacturing of end products, replacing traditional manufacturing processes [1–3]. Furthermore, AM enables manufacturing of complex geometries that are impossible to produce with traditional subtractive manufacturing techniques [4–7]. In addition, there exists a distinctive advantage in manufacturing time as well. The manufacturing time of a subtractive process is highly dependent on the geometrical complexity of parts, while process time and cost in AM are relatively less dependent of part geometry. Given these advantages, AM has been creating new opportunities in various areas; personalized healthcare products [8], reducing environmental impact for sustainability by saving raw materials, simplification of supply chain and responsiveness in demand fulfillment [9]. However, achieving dimensional accuracy in AM parts is still challenging due to surface roughness caused by the inherent layer-wise process of AM.

Surface roughness is one of the universal defects that AM products have due to a staircase effect caused by the inherent nature of layer-by-layer AM processes. Not only does rough surface profile induce dimensional inaccuracy, it also results in a significant reduction in mechanical performance of AM parts [10]. Various approaches have been proposed to address the surface roughness issue in AM parts. A predictive model for surface roughness of AM parts using an interpolation equation was introduced by Ahn et al. [11]. An interesting study done by Sager et al. [12] used a parameter estimation (PE) method to improve surface quality in stereolithography. Recently, it is also found that the stiffness variations of AM parts is induced by the geometrical differences between computer-aided design (CAD) models and the printed parts [13]. The effect of build direction that controls the directional surface roughness on tensile strength and stiffness of additively manufactured parts was also researched by Quintana et al. [14]. The numerical method was also used to optimize printing orientation in order to minimize the effect of surface roughness on mechanical properties [15]. Chockalingam et al. reported the close correlation between the mechanical properties of stereolithography components and the layer thickness and surface roughness [16]. However, a systematic approach to understand and control surface profiles of a 3D printed part while accounting for process throughput has not been reported.

This study presents a systematic approach to identify optimal printing process parameters using a computational model for projection micro-stereolithography (PμSL), a digital light processing (DLP) based AM technique shown in Figure 1 [17]. Figure 1a shows a schematic diagram of PμSL and a 3D printed part with surface roughness. CAD file is sliced in the printing software and the UV light corresponding to each cross-sectional image is projected on top of the resin vat using DMD. The liner stage moves vertically to successively build polymerized layers. Figure 1b shows microscope images of 3D printed struts with surface roughness. The polymer used is HDDA and the strut diameter is 200 μm. The three printing angles studied are 90° (vertical) and 60°. The thickness of each layer is 80 μm and a curing time per layer of 3 s was used. The surface profile of printed struts clearly demonstrates the geometrical deviations between CAD and the actual printed shape. In addition, the strut with an inclined printing angle have different level of surface roughness on each side, which will be further discussed in detail later. In our systematic approach, a mathematical model is developed based on the photopolymerization process [18,19]. Then, a computational model is implemented by commercial finite element software and validated using experimental data obtained from a custom-built PμSL apparatus. Taguchi method is used to understand the effect of printing parameters on surface roughness. A meta-heuristic optimization algorithm, called multi-objective particle swarm optimization (MOPSO), is performed to find optimal printing parameters that minimize surface roughness as well as printing time. The optimized printing parameters for micro-struts in different printing angles are also suggested. Definitions of all the acronyms used in this study are listed in Appendix E.

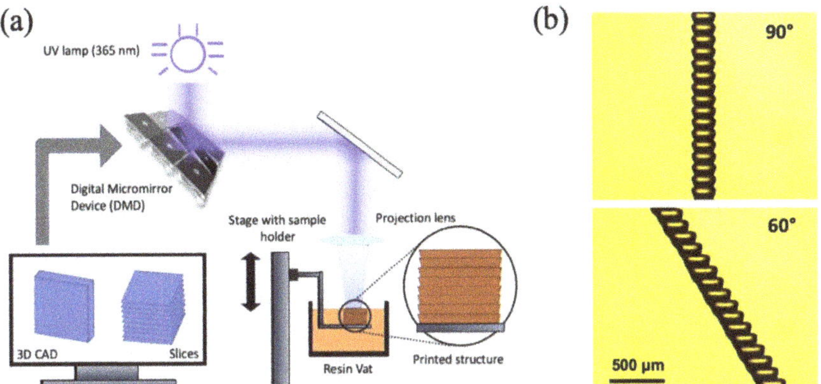

Figure 1. (**a**) Schematic diagram of PµSL and resulting surface roughness on a printed part. (**b**) Microscope images of 3D printed struts in different printing angles (90° and 60°). The polymer used is HDDA and curing time per 80 µm thick layer is 3 s. These side profiles clearly display the surface roughness caused by the layer-wise process.

2. Materials and Methods

2.1. Projection Micro-Stereolithograph (PµSL)

2.1.1. Materials for 3D Printing

In this work, the monomer was 1,6-Hexanediol diacrylate (HDDA), technical grade 80% (Sigma-Aldrich, St. Louis, MO, USA). Phenylbis (2,4,6-trimethylbenzoyl) phosphine oxide (Sigma-Aldrich, St. Louis, MO, USA), also known as its commercial name Irgacure 819, was used as a photoinitiator (PI), and 1-Phenylazo-2-napththol, also called SUDAN-1 (Sigma-Aldrich, St. Louis, MO, USA) was used as a photo-absorber (PA). Ethanol was used to wash away the excess resin after 3D printing.

2.1.2. Projection Micro-Stereolithography (PµSL) Experimental Set-Up

The AM system used in this work is a custom-built PµSL system capable of manufacturing micro-scale features. It consists of a linear stage (Thorlabs), on which the sample holder is attached. A projection lens (Thorlabs) is used to achieve a lateral resolution of 12 µm. A digital micromirror device (DMD) (Texas Instruments, Dallas, TX, USA) is used for generating projection patterns according to cross-sectional digital images of a 3D model. UV LED (365 nm, Hamamatsu, Hamamatsu City, Japan) is used as a UV illumination source. The UV light reflected from the DMD is projected on the surface of the resin inside the vat. Once a layer is formed, the sample holder is lowered by the layer thickness, and the next image is projected to cure a new layer on top of the previous one. This process repeats until all layers are completed. The actual system used in this study is shown in Figure 2. The setup is kept in a printing chamber where environmental factors such as external light and oxygen concentration are controlled.

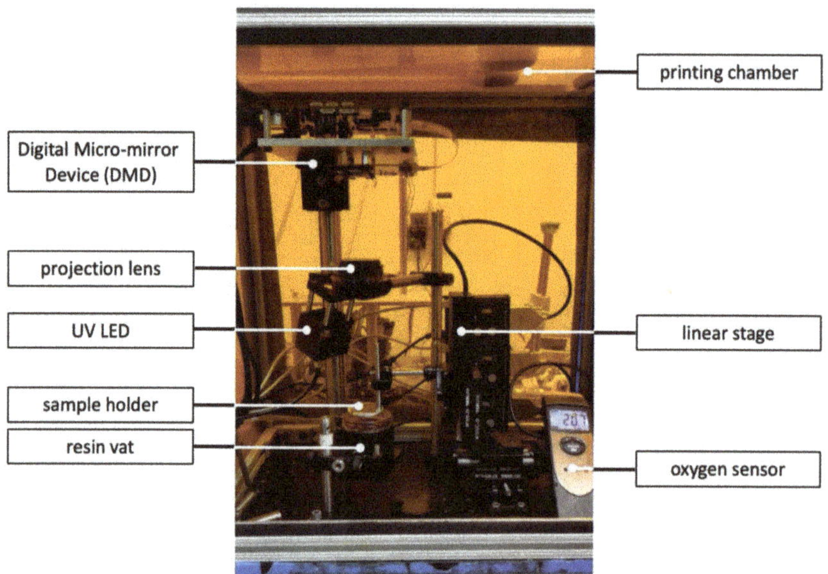

Figure 2. PμSL experimental setup.

2.2. Computational Model for Photopolymerization Process

2.2.1. Photopolymerization Model

Photopolymerization plays a central role in stereolithography AM processes including PμSL. Photopolymerization occurs in three steps; initiation, propagation, and termination [20]. When UV light is projected on a photocurable resin, free radicals are generated from photoinitiator (initiation). Free radicals readily react with monomer molecules, connecting monomers to form long polymer chains (propagation). Propagation continues until two large chains of polymer cross-link with each other (termination). The liquid resin is converted into a solid when cross-linked polymer network is formed. Environmental conditions are also an important factor that influences the reaction kinetics. Oxygen acts as an inhibitive agent because free radicals react not only with monomer molecules, but also with oxygen molecules when present, forming peroxides. These peroxides do not take part in the polymerization process, thereby inhibiting the overall conversion of liquid resin to solid. This photopolymerization process can be modeled as follows.

Light irradiation intensity (I) which decays as light travels through the resin can be modeled as

$$\frac{dI}{dz} = -(\alpha[\text{PI}] + \alpha_a[\text{PA}])I, \tag{1}$$

where [PI] and [PA] represent photoinitiator and photo absorber concentrations, α is molar absorptivity of photoinitiator, and α_a is the molar absorptivity of photo-absorber [21]. The term $(\alpha[\text{PI}] + \alpha_a[\text{PA}])$ represents overall absorption coefficient of the resin which follows the Beer–Lambert law [22].

The light intensity of the projected beam is modeled as a convolution of unit light intensity profile which is modeled as a Gaussian function [17]. In Equation (2), w is the beam width of the Gaussian function, I_0 is the peak light intensity, and r is the radial position. For n number of activated pixels on DMD, the light intensity distribution on the surface of resin is therefore given by Equation (3).

$$I = I_0 \times e^{\frac{-2(r)^2}{w^2}}, \tag{2}$$

$$I = I_0 \times \sum_0^n e^{\frac{-2(r-w\times n)^2}{w^2}}, \tag{3}$$

As described in the photopolymerization principle, initiation process consumes photoinitiator molecules which split into free radicals upon irradiation. Accounting for diffusion flux, photoinitiator concentration can be written as [21]

$$\frac{\partial[\text{PI}]}{\partial t} = \nabla(D_{\text{PI}}\nabla[\text{PI}]) - \frac{1}{2}\,\varphi\alpha\beta[\text{PI}]I, \tag{4}$$

where D_{PI} is diffusivity of photoinitiator, φ is quantum yield of free radicals, and $\beta = 1/3.27 \times 10^5$ mol/J is the amount of energy contained in one photon [23].

Upon UV exposure, photoinitiator molecules split to generate free radicals which react with monomers and activate their functional groups. These active monomers react with other monomers and begin a propagation reaction [24], given by

$$\frac{\partial[\text{M}]}{\partial t} = -k_p[\text{M}][\text{R}], \tag{5}$$

where [M] is monomer concentration, k_p is propagation rate constant and [R] is radical concentration. A negative sign indicates that monomer concentration decreases as they are converted into polymer. The radical concentration is given by [25]

$$\frac{\partial R}{\partial t} = \nabla(D_r\nabla[\text{R}]) + R_g - R_c, \tag{6}$$

where the first term represents radical diffusion with diffusion coefficient D_r, and R_g and R_c account for generation and consumption of free radicals, respectively.

$$R_g = \varphi\alpha\beta[\text{PI}]I_0\exp(-\alpha[\text{PI}]z), \tag{7}$$

$$R_c = k_t[\text{R}]^2 + k_o[\text{O}][\text{R}], \tag{8}$$

[O] is oxygen concentration, k_t is termination rate constant, and k_o is oxygen rate constant. The consumption term R_c has two parts: (1) reaction between radicals and (2) reaction with oxygen (oxygen inhibition).

The oxygen concentration is given by [26]

$$\frac{\partial O}{\partial t} = \nabla(D_o\nabla[\text{O}]) - k_o[\text{O}][\text{R}], \tag{9}$$

where D_o is oxygen diffusion constant.

The conversion ratio C can be determined from monomer concentration [21].

$$c = 1 - \sqrt{\frac{[\text{M}]}{[\text{M}]_0}}. \tag{10}$$

2.2.2. Modeling of Photopolymerization in COMSOL Multiphysics

The photopolymerization model is solved by commercial finite element analysis (FEA) software, COMSOL Multiphysics. We created a 2D axisymmetric domain representing a cross-section of cylindrical volume of resin in the vat of the PµSL system, as shown in Figure 3a.

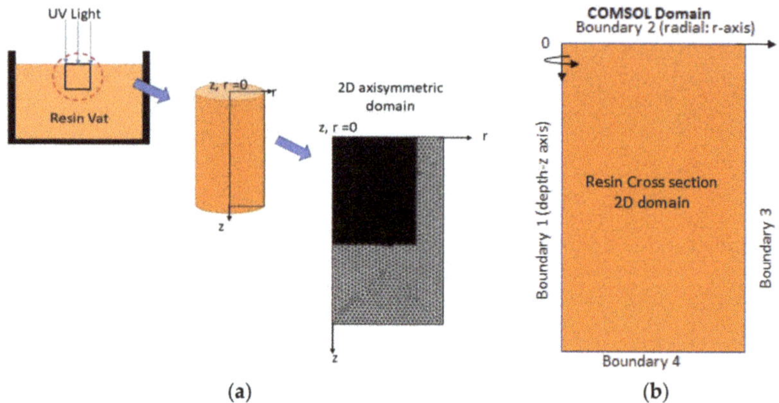

Figure 3. (**a**) 2D axisymmetric computational domain representation of UV exposed resin region in the vat. 2D axisymmetric domain consists of two domains that have different mesh sizes due to computational efficiency. (**b**) Computational domain with boundary conditions.

Here z represents the depth direction along which UV light travels and r is the radial coordinate. The domain is selected in such a way that the light incident on the surface will allow enough area for resin components to diffuse, i.e., the equations for parameter concentrations will be able to converge within the refined mesh area of the domain. The computational domain is divided into two sub regions (higher and lower mesh densities) for computational efficiency. Adaptive time step is used for stability and consistency of the algorithm.

Figure 3b represents the 2D axisymmetric domain and boundary conditions. As the partial differential equations (PDEs) of the photopolymerization process are strongly coupled, COMSOL Multiphysics solves the PDEs iteratively to obtain converged solutions in each time-step. Table 1 lists the initial and boundary conditions.

Table 1. Initial and boundary conditions.

Equation	Initial Condition	Boundary Condition
Light intensity I	-	$I(r,\ z\ =\ 0)\ =\ I_0 \times \sum_{0}^{n} e^{\frac{-2(r-w\times n)^2}{w^2}}$
Photoinitiator [PI]	$[PI]\ (t = 0, r, z) = [PI]_0$	-
Free radical [R]	$[R]\ (t = 0, r, z) = 0$	$[R]\ (r, z = 0) = 0$
Oxygen [O]	$[O]\ (t = 0, r, z) = [O]_1$	$[O]\ (r, z = 0) = [O]_0$
Monomer [M]	$[M]\ (t = 0, r, z) = [M]_0$	-

r = radial direction, z = depth direction, and t = time.

Following assumptions are made while setting up the simulation.

- Thermal properties are considered to be constant during polymerization reactions. This assumption is made based on the fact that when polymerization on micro scale is limited on a small area, surrounding resin acts a heat sink.
- The rate constants k_t and k_p are kept constant for simplification of system. Tryson et al. [24] explained more detailed information regarding the change of k_t and k_p with respect to monomer conversion C.
- Optical effects such as refraction or reflection are not considered.

Table 2 lists constants and their values. When computation is complete, it gives a continuous field of conversion ratio, defined as Equation (10). Interface between cured solid and remaining uncured liquid resin is determined by a cut-off conversion ratio contour, from which curing depth, curing width, and surface profile can be extracted. Detailed method is given in Appendices C and D.

Table 2. List of PµSL process parameters. * is from experiments.

Symbol	Value	Description
w	12 µm	Gaussian radius *
$[M]_0$	$4.46 \times 10^3 \, mol/m^3$	Monomer initial concentration *
$(PI)_0$	$48.27 \, mol/m^3$	Photoinitiator initial concentration *
$(PA)_0$	$4.06 \, mol/m^3$	Stabilizer concentration *
α	$11.9 \, m^2/mol$	Molar absorptivity of photoinitiator [21]
α_a	$4600 \, m^2/mol$	Molar absorptivity of stabilizer [21]
Φ	0.59	Quantum yield for initiator [27]
T_0	303 K	Environmental temperature *
D_r	$3.0 \times 10^{-10} \, m^2/s$	Radical diffusion coefficient [28]
D_{PI}	$3.0 \times 10^{-10} \, m^2/s$	Initiator diffusion coefficient [28]
I_0	$24.5 \, mW/cm^2$	Incident light intensity *
k_{p0}	$25 \, m^3/mol/s$	Propagation rate constant [29]
k_{t0}	$2520 \, m^3/mol/s$	Termination rate constant [29]
β	$1/(3.27 \times 10^5) \, mol/J$	Amount of energy contained in one photon [23]
t_0	5 s	Time for which light is incident on resin surface *
T_{last}	0.2 s	Decay time for the turning the illumination off *
k_o	$15 \, m^3/mol/s$	Oxygen diffusion constant [25,26]
$(O)_0$	$0.9 \, mol/m^3$	Initial oxygen concentration in resin [26]
$(O)_1$	$8.69 \, mol/m^3$	Environmental oxygen concentration *

2.3. Meta-Heuristic Optimization Technique for Multiple Objectives: Multi-Objective Particle Swarm Optimization (MOPSO)

Improving surface roughness of additively manufactured parts while keeping printing time as small as possible is considered as finding sub-optimal solutions in a parameter search space that satisfies multiple objectives. Because the photopolymerization process involves many parameters and printing variables and a system of PDEs, this problem has highly nonlinear and multivariate response surface. In addition, it is often under various complex constraints. Traditional gradient-based optimization algorithms often fail to find an optimizer of this kind of problems or may be computationally too expensive to calculate derivatives. Recently, various population-based optimization algorithms such as genetic algorithm [30] and ant colony optimization [31] have gained attention because of their derivative-free characteristics and efficiency.

Particle swarm optimization (PSO) is a meta-heuristic optimization algorithm that mimics the social behavior of birds or fish [32]. Each agent evolves and iteratively searches the global optimizer based on its path as well as its neighbors' paths. Each agent represents a solution of the problem and it will converge to the global optimizer when the iteration ends. PSO has become a useful tool for multiple reasons [33]: (1) since it is not problem specific, it offers a general framework that can be applied to all optimization problems; (2) the algorithm is straightforward and relatively simple to be implemented; (3) it is computationally efficient to find the global optimizer.

Multi-objective particle swarm optimization (MOPSO) is a PSO for multiple objective functions [34]. In MOPSO, the result is a set of different solutions instead of a single global optimizer. It is called a Pareto optimal set. There are three main issues that need to be addressed when PSO extends to MOPSO.

- Selecting a leader for evolving agents
- Retaining the non-dominant solutions in each iteration
- Maintaining diversity of agents during iterations

These issues are addressed in detail when the pseudo-code for MOPSO in Figure 4 is explained.

BEGIN
Initialize agents' position
Initialize agents' velocity
Evaluate each of agents in population (POP) using objective functions
Store the non-dominated agents in the repository (REP)
Generate hypercubes of the search space (grids) and locate the agents where the
coordinates are its values from the objective functions.
Initialize agent's history in PBEST, which keeps the best position of the agent
Iteration = 0
WHILE iteration < threshold
FOR each agent
Select leader, which is used to guide the agent towards the better position
Compute velocity of agent
Update agent's position
Maintain agent within the upper and lower bounds
Mutation for diversity
Evaluate using objective functions
Update PBEST
END
Update leaders in REP
Quality measure for leaders using objective functions
Update grid
Iteration = Iteration + 1
END
Report results in REP
END

Figure 4. Pseudo-code for MOPSO.

First, the algorithm initializes agents' position and velocity. Population (POP) is the memory where all agents' information in certain iteration is stored. For the initial set, all agents are evaluated based on its randomly distributed positions. Repository (REP) is the external repository where the non-dominated agents in POP are stored. The dominance of each agent is determined by the objective functions. The MOPSO algorithm generates grids that cover all the search space and locate all agents in the grids where the coordinates are its objection function. This controls the density of local agents by determining the local optimizer (leader) that leads neighboring agents. Personal (agent) best position, called PBEST, in the search space keeps the best local optimizer that each agent experiences when it travels through the search space.

When iteration starts, the algorithm updates the velocity of each agent based on Equation (11) [35].

$$\underline{v}_i(t) = k\underline{v}_i(t-1) + C_1 r_1 \left(\underline{x}_{PBEST,i} - \underline{x}_i(t) \right) + C_2 r_2 (\underline{x}_{leader} - \underline{x}_i(t)), \tag{11}$$

where k is inertia weight, $\underline{v}_i(t)$ and $\underline{x}_i(t)$ denote velocity and position of an agent i, at time t, respectively. The underscore indicates that the velocity and position are n-dimensional vectors where n is the number of optimizing parameters. r_1 and r_2 are random values, $r_1, r_2 \in [0,1]$. C_1 and C_2 are constants, called cognitive and social learning factors, respectively. These constants define the amount of attraction toward the agent's own best experience or that of its neighbors. $x_{PBEST,i}$ is the best position that the agent i experienced and x_{leader} is the best position that its neighbor experienced. Each grid has its own x_{leader} in it, so that it helps to explore the entire search space. REP is taken from repository which attracts the agent to the global optimizer in each iteration. By balancing the terms related to C_1 and C_2, the algorithm enhances the ability to search for the global optimum in the search space. Then, each agent location is updated by Equation (12) [35]

$$\underline{x}_i(t) = \underline{x}_i(t-1) + \underline{v}_i(t), \tag{12}$$

If the position of the agent is located out of the region between the lower and upper bounds that are given by the user, it is forced to be moved inside the valid search space. Then, each agent in POP is evaluated and updated the contents of REP by placing the non-dominated agents within the hypercube. Any dominated agents in REP are removed in this step. When the current status of agent is better than its PBEST, PBEST is updated by the current status. The mutation step is also included to give the diversity of searching ability and increase the opportunity to search the global optimizer. In this study, we used the source code available in the public domain [36] and modify it to match our purpose. The parameters in the algorithm are followed by the basic settings in the source, but the number of agents is reduced to 40 for computational efficiency. The detailed guideline for deciding parameters in PSO and MOPSO algorithm is explained in [33,34].

2.4. MOPSO with COMSOL Multiphysics-MATLAB LiveLink

In order to use the powerful optimization toolboxes and rich built-in libraries in MATLAB, we connected COMSOL Multiphysics with MATLAB via MATLAB LiveLink for COMSOL. Figure 5 shows a schematic description and data flow between MATLAB and COMSOL Multiphysics during our optimization process. The whole process consists of two sub-steps: (1) model calibration and (2) process parameter optimization. First, printing parameters to be calibrated are sampled from the upper and lower bounds of each parameter. Selected printing parameters are sent to the photopolymerization model in COMSOL Multiphysics. COSMOL performs photopolymerization process simulation with given parameters and passes the result (curing depth, in this case) back to MATLAB. This process iteratively finds the calibrated parameters that minimize the deviation between curing depth in experiment and simulation. The calibrated parameters are updated in photopolymerization model in COMSOL Multiphysics.

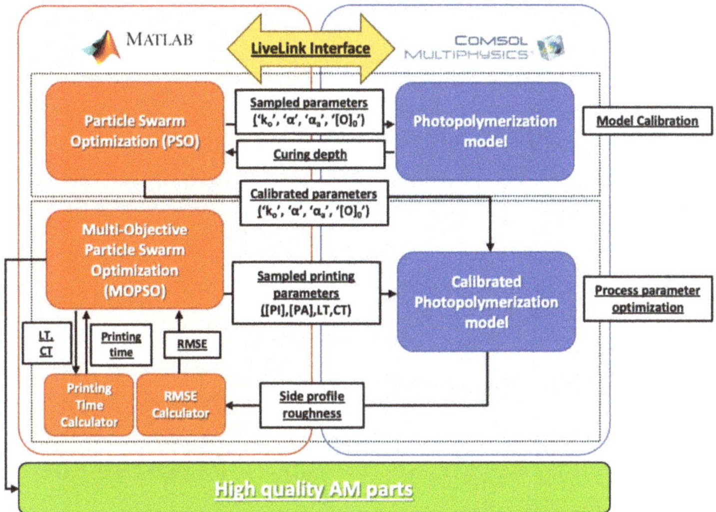

Figure 5. Flow chart of MATLAB LiveLink interface with COMSOL Multiphysics. Proposed modeling framework consists of two sub-steps: model calibration and process parameter optimization. The flow of data is shown as arrows and labels.

Second, MOPSO samples printing parameters in the parameter search space and passes it to the updated model in COMSOL Multiphysics to obtain surface profile of AM part of given inputs. The surface profile is extracted from photopolymerization simulation and returned to MATLAB. The custom-built MATLAB script calculates root mean squared error (RMSE) value of the given surface profile and uses it as an objective function in the optimization process. The method to calculate RMS for a given surface profile is explained in detail in Appendix D. Another objective function is printing time, which can be calculated by using a custom-built MATLAB script. The detailed method for obtaining printing time is in Section 3.3. This loop continues until it reaches the iteration limit.

3. Results and Discussions

3.1. Model Validation

When a UV light is projected on the surface of a photocurable resin, the liquid resin is converted into solid from the surface to a certain depth. This depth is called curing depth or C_d. The curing depth can be expressed as [20]

$$C_d = D_p \ln(\frac{E}{E_c}), \tag{13}$$

where D_p is penetration depth, E is given light energy, and E_c is critical light energy. It is seen that curing depth is proportional to natural logarithm of given light energy. This is known as a working curve for a given resin. D_p and E_c are resin specific characteristic parameters. In this study, we use a set of working curves obtained from experiment to validate our simulation model. The details for obtaining a working curve experimentally is given in Appendix B. In simulation, the conversion ratio C given in Equation (10) is obtained as a continuous field from the initial monomer concentration ($[M]_0$) and the monomer concentration ($[M]$) remaining after the applied energy dose. The value of C varies between 0 and 1 with 0 being uncured liquid and 1 being fully cured solid. Since a conversion ratio corresponding to the gel point is not readily available from experimental measurement, the conversion ratio value that defines the interface between liquid and cured solid polymer, or 'cut-off' conversion ratio, should be chosen to obtain curing depth from simulation. A various range of cut-off conversion ratios have been reported in literature [21], and we used 5% as a conversion cut-off ratio in this study.

In order to validate our computational model, we first experimentally obtained working curves of resins having different PI and PA concentrations. The details regarding sample preparation and post-processing procedure are described in Appendix A. PSO was performed on several simulation constants of 'k_o', 'α', 'α_a', '$[O]_0$', and 'C' to calibrate the computational model. The values of these parameters are reported at a varied range in literature as listed in Table 3. Using RMSE as an objective function, optimization algorithm POS was performed. Table 3 also lists calibrated parameter values after performing PSO. With the calibrated parameters, we performed numerical simulations to obtain working curves for the same resins used in the experiment. Working curves from simulation and experiment are shown in Figure 6 and they show a good agreement. Based on this result, we confirm the validity of our photopolymerization computational model.

Table 3. List of calibrating parameters.

Parameter		Lower/Upper Bounds	Calibrated Values
Symbol	Name		
k_o	Oxygen inhibition constant	5–10×10^5 m^3/mol/s [25,26]	10.02 m^3/mol/s
α	PI molar absorptivity	4.6–20 m^2/mol [21]	20 m^2/mol
α_a	PA molar absorptivity	3680–5520 m^2/mol [21]	4591.7 m^2/mol
$[O]_0$	Initial oxygen concentration	0.8–1.2 mol/m^3 [26]	1.03 mol/m^3

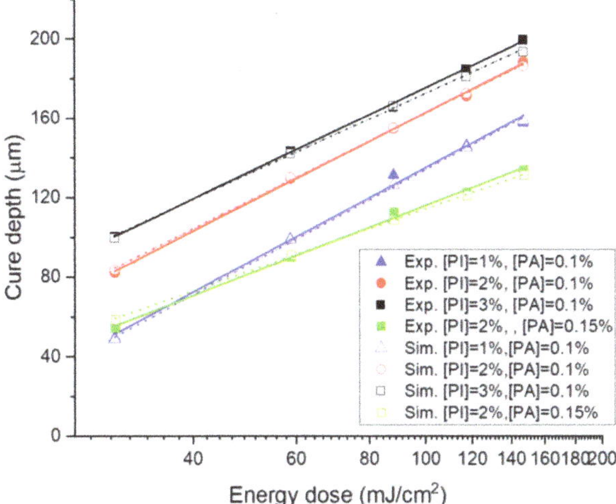

Figure 6. Comparison of curing depth between experiment and simulation. *x*-axis represents energy dosage in log scale and *y*-axis is curing depth in linear scale.

3.2. Effect of Process Parameters on Curing Depth and Width

Since each PμSL process parameter has critical effect on the printing quality, i.e., curing depth and width, evaluation of the effect of each parameter is necessary. Through multiple simulations using the computational model we developed, we studied the impact of each parameter on printing quality of AM parts. Table 4 lists the range of values for each parameter we used for this study. Note that the largest oxygen concentration used in our simulation was 21% as shown in the bold in Table 4 because it is the actual oxygen concentration in the air. Therefore, the corresponding normalized parameter value for this data point is 2.1.

Table 4. Printing parameter evaluation using normalized value concept.

Parameter	Values				
	A: 0.5	B: 1	C: 1.5	D: 2	E: 2.5
[PI] %	1	2	3	4	5
[PA] %	0.05	0.1	0.15	0.2	0.25
I_0 (mW/cm²)	10	20	30	40	50
[O] %	5	10	15	20	21 (**E: 2.1**)
Time (s)	1	2	3	4	5

First, reference curing depth and width were determined from the results produced with the set of parameters in column B. Then, simulations were performed while one parameter was varied with all other parameters being kept constant. Resulting curing depth and width were normalized by the reference value to evaluate the effect of the parameter studied.

Figure 7a shows the effect of the parameters on curing depth. Curing depth increases as (PI) increases because PI increases reactivity of the resin. When (PA) concentration increases, the cure depth reduces because light penetration depth decreases with PA. Cure depth is relatively insensitive to environmental (O). When light intensity and exposure time increase, cure depth increases because light energy is product of light intensity and exposure time. It is observed that (PA) influences cure depth the most, which is also found from results in Figure 6 where D_p and C_d both decrease as (PA) increases.

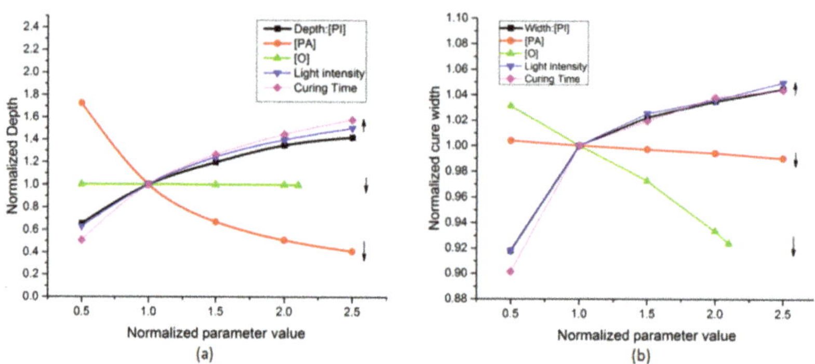

Figure 7. Effect of printing parameters on (**a**) curing depth and (**b**) curing width.

Similar analysis was performed to study the effect of the parameters on curing width. As shown in Figure 7b, it is interesting that environmental oxygen concentration has the highest effect on curing width. As environmental oxygen concentration increases, oxygen inhibition becomes more prominent at the surface of the resin, which results in decrease in curing width. As expected, when (PI), light intensity, and exposure time increase, curing width increases due to increased light energy or reactivity of resin. In contrast to curing depth case, (PA) has least effect on curing width because its role is primarily to control the penetration of light in depth direction.

3.3. Parameter Sensitivity Analysis for Surface Roughness Using Taguchi Orthogonal Array

In PµSL process, a 3D part is built in a layer-by-layer fashion. Since each layer has its own characteristic side profile as a result of photopolymerization reaction, when they are stacked together repeatedly, a distinctive surface roughness arises. This is called the staircase effect. We performed a systematic analysis to study effect of PµSL process parameters on the surface roughness of a 3D printed structure. Since there are many parameters involved, we employed design of experiment (DOE) method proposed by Taguchi to reduce the number of experiments to be performed. This method is known as Taguchi method of orthogonal arrays (OA) [37–39]. Figure 8 shows the steps involved in DOE for this study.

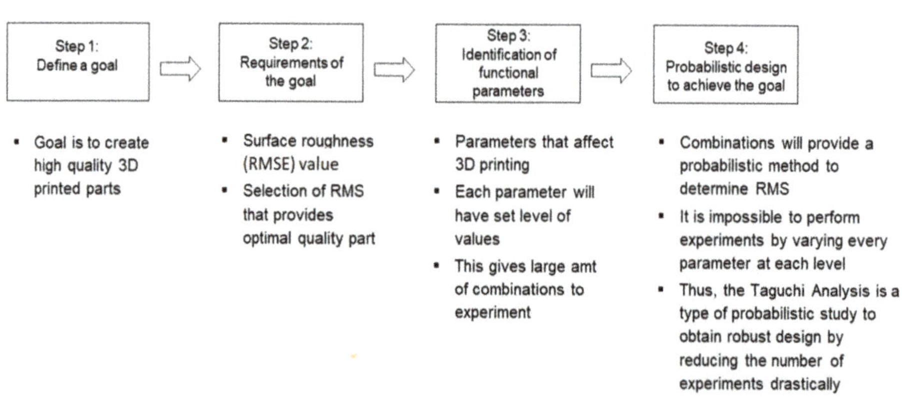

Figure 8. Steps for developing a robust DOE.

Based on the result obtained in the previous section, we selected (PI), (PA), (O), layer thickness (LT), and curing time (CT) to control surface roughness. We set three levels for each parameter, or 'factor'. Therefore, the total number of possible combinations of parameters obtainable is $3^5 = 243$. From standard Taguchi OA tables, $L_{27}(3^{13})$ meets the criteria of our analysis where there are five factors with three levels. This dramatically reduces the number of experiment necessary to only 27. Following this, we performed 27 simulations and numerically extracted surface roughness from each simulation. Detailed method to extract surface profile from photopolymerization simulation is described in Appendix D. The factors and their levels, along with surface roughness as response, are shown in Table 5. Surface roughness is measured by RMSE. After obtaining RMSE from each simulation, sensitivity analysis was performed to find the optimal levels for each factor. Signal-to-noise (SN) ratios measure how the response (RMSE value in this study) varies relative to a target value (the expectation set to find optimal factor levels). Since minimum surface roughness would give high quality AM parts, 'smaller the better' criterion was chosen to evaluate the factor levels. The formula to calculate SN ratio for 'smaller the better' criterion is given by,

$$\text{SN ratio} \ = \ 10 * \log_{10}\left(\frac{\sum Y^2}{n}\right), \tag{14}$$

where Y is the response of given factor level combination—i.e., RMSE for this case—and n is the number of responses in given factor level combination. Using equation (14), SN ratios for each experiment were obtained and listed in Table 5.

Table 5. Factor and levels for Taguchi OA.

Simulation No.	Factors					RMSE (μm)	SN Ratio
	(PI) %	(PA) %	(O) %	Layer Thickness (μm)	Curing Time (s)		
1	1	0.05	10	100	2	1	0
2	1	0.05	10	100	3	0.9	0.915
3	1	0.05	10	100	4	0.9	0.915
4	1	0.1	15	80	2	1.7	−4.609
5	1	0.1	15	80	3	1.4	−2.922
6	1	0.1	15	80	4	1.3	−2.279
7	1	0.15	21	50	2	1.3	−2.279
8	1	0.15	21	50	3	1.1	−0.828
9	1	0.15	21	50	4	1	0
10	2	0.05	15	50	2	0.4	7.959
11	2	0.05	15	50	3	0.3	10.458
12	2	0.05	15	50	4	0.3	10.458
13	2	0.1	21	100	2	2	−6.021
14	2	0.1	21	100	3	1.7	−4.609
15	2	0.1	21	100	4	1.7	−4.609
16	2	0.15	10	80	2	2.5	−7.959
17	2	0.15	10	80	3	2	−6.021
18	2	0.15	10	80	4	1.8	−5.105
19	3	0.05	21	80	2	0.7	3.098
20	3	0.05	21	80	3	0.6	4.437
21	3	0.05	21	80	4	0.6	4.437
22	3	0.1	10	50	2	0.7	3.098
23	3	0.1	10	50	3	0.6	4.437
24	3	0.1	10	50	4	0.5	6.021
25	3	0.15	15	100	2	3.4	−10.630
26	3	0.15	15	100	3	2.8	−8.943
27	3	0.15	15	100	4	2.6	−8.299

Based on this result, main effects plots for SN ratios in Figure 9 were generated. Also, a response table (Table 6) was generated, from which the parameters that have the largest effect on the response can be identified. In Figure 9, the average SN ratio of the response is presented by the dotted line. Since the goal here is to find values for each parameter that maximize the SN ratio, the optimal values for each parameter can be determined to achieve the minimum surface roughness. The rank represents which factor affects surface roughness the most. Rank is based on the delta value, which is the difference between the highest and the lowest average SN ratio value for each factor. From the response table, the parameter that affects surface roughness the most is PA concentration, and oxygen concentration has the least impact on surface roughness.

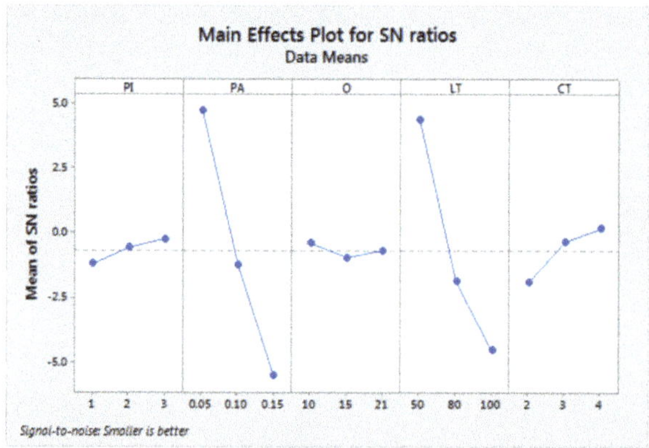

Figure 9. SN ratio analysis based on RMSE response. The level corresponding to the maximum value of mean of SN ratios is selected for each factor.

Table 6. Response table for signal-to-noise ratio: smaller the better.

Level	PI	PA	O	LT	CT
1	−1.2319	4.7418	−0.4110	4.3692	−1.9269
2	−0.6055	−1.2770	−0.9787	−1.8804	−0.3418
3	−0.2605	−5.5626	−0.7081	−4.5867	0.1708
Delta	0.9714	10.3044	0.5677	8.9559	2.0977
Rank	4	1	5	3	2

3.4. Optimizing Printing Parameters with MOPSO

Based on the rank from the sensitivity analysis, we selected four highly sensitive parameters that affect surface roughness the most: PI and PA concentrations, layer thickness (LT), curing time (CT). These parameters are used as control parameters in MOPSO algorithm. Since our goal is to determine process parameters to produce high surface quality part as fast as possible, the objectives for MOPSO are RMSE of surface profile of a printed part and printing time. Time required to print a structure (t_{total}) is given by

$$t_{total} = t_{layer}*\text{number of layers}, \tag{15}$$

$$t_{layer} = t_{exposure} + t_{stage}, \tag{16}$$

where t_{layer} is the time required to complete one cycle for a layer and $t_{exposure}$ is the curing time that the UV light is exposed on resin surface. t_{stage} is the time required for the linear stage to move sample holder in each process cycle for a layer (measured to be 5 s in experiment). The number of layers is determined by the overall height of structure divided by the layer thickness. In this study, the height that we want to print was set to be 1 mm.

3.4.1. Optimized Printing Parameters for a Vertical Strut

In this section, we apply MOPSO to determine optimal printing parameters that minimizes two objective functions: surface roughness of a printed strut and printing time when the angle of the strut is 90° from the horizontal plane (vertical strut). Table 7 shows the upper and lower bounds of the parameters considered.

Table 7. List of optimizing parameters.

Parameter		Upper/Lower Bounds	Group A	Group B
Symbol	Name			
[PI]	Photoinitiator concentration	24.14–72.4 mol/m^3	72.4 mol/m^3	72.4 mol/m^3
[PA]	Photo-absorber concentration	2.03–6.1 mol/m^3	2.03 mol/m^3	2.03 mol/m^3
LT	Layer thickness	20–120 μm	120 μm	102 μm
CT	Curing time	1–4 s	1.0 s	1.0 s

After performing MOPSO, the graphical representation of Pareto set shown in Figure 10 was generated. x- and y-axis are surface roughness and printing time, respectively. All the agents produce Pareto front, but the shape of Pareto front consists of two major groups, as listed in Table 7. The corresponding objective values are (RMSE = 1.45 μm, printing time = 54.79 s), (RMSE = 1.5 μm, printing time = 48.33 s) for group A and B, respectively. For a vertical strut, higher PI and lower PA improve surface roughness, so that it reduces the surface profile RMSE while giving the small effect on the printing time. In addition, since CT does not change the roughness significantly, it stays at the lower bound to minimize the total printing time, as expected. Table 8 shows the comparison between the several printing times and surface roughness of non-optimized printing examples and optimized examples of Group A and B, shown in the bold in Table 8. The parameter sets for non-optimized printing conditions 1, 2, and 3 are [PI = 65.44 mol/m^3, PA = 4.74 mol/m^3, LT = 53 μm, CT = 1.00 s, PI = 64.12 mol/m^3, PA = 3.04 mol/m^3, LT = 80 μm, CT = 2.00 s, PI = 54.09 mol/m^3, PA = 2.66 mol/m^3, LT = 120 μm, CT = 4.00 s], respectively. When the Group A and the result from non-optimized parameter 1 are compared to each other, it is realized that the proposed optimization framework can reduce the printing time 50% while keeping the surface quality. In addition, the comparison between the Group B and the result from non-optimized parameter 3 shows that the proposed optimization framework reduces 38% of surface roughness when the printing process is at the minimum printing time. Lastly, the result from non-optimized parameters 2 shows the higher printing time and the surface roughness, which often happens when the printing parameters are not optimized. Based on this result, it is clearly seen that the optimized parameters provide better or at least similar surface roughness while significantly reducing printing time.

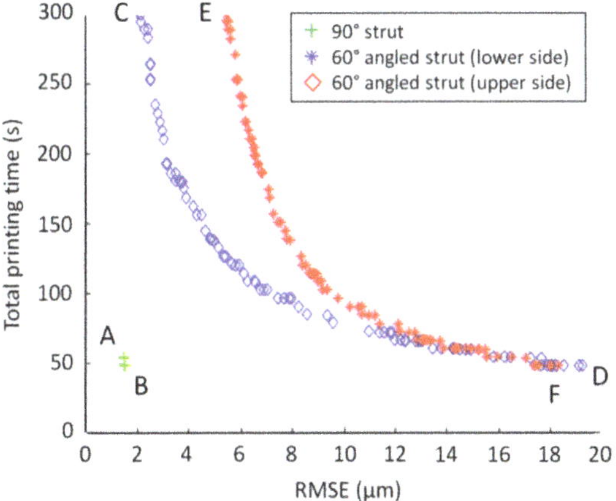

Figure 10. Pareto optimal sets for a vertical strut (green), optimizing upper side of 60° angled strut (red) in Figure 11, and optimizing lower side of 60° angled strut (blue) in Figure 11. Point C and D are two extremes of optimizing lower side of the 60°angled strut. Point E and F are two extremes of optimizing upper side of the 60° angled strut.

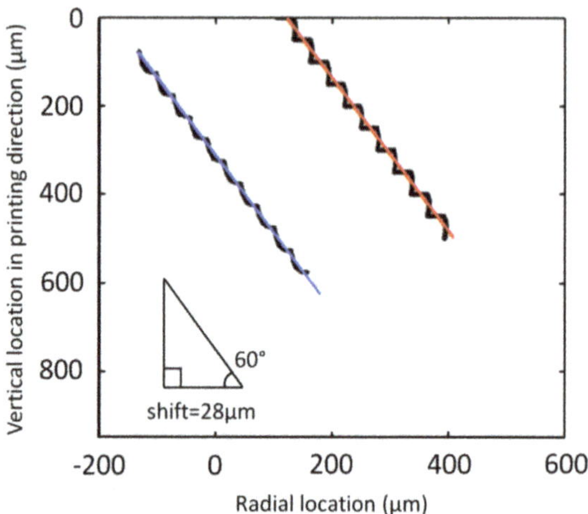

Figure 11. 60° angled strut and its mean profiles for upper (red) and lower (blue) surface profile.

Table 8. Objectives comparison between optimized and non-optimized printing conditions.

Objectives	Optimized Parameters of Group A	Optimized Parameters of Group B	Non-Optimized Parameters 1	Non-Optimized Parameters 2	Non-Optimized Parameters 3
Total printing time (s)	**54.79**	**48.33**	108.85	72.49	48.33
Surface roughness (μm)	**1.45**	**1.5**	1.45	1.94	2.39

3.4.2. Optimized Printing Parameters for an Angled Strut

To extend our study, we apply the same approach to a 60° angled strut. The objective functions and the considering printing parameters (including the upper and lower bounds) are the same as the vertical strut, but the difference is that surface roughness is different in upper and lower sides. When the angle of the strut is 60° as shown in Figure 11, the result from the upper side of the strut is expected to have no major difference from the vertical strut. However, on the lower side, there exists an additional effect on surface roughness since the light penetration from the upper layer may polymerize resin deeper to the layer below. To visualize this difference, we performed two MOPSO considering surface roughness of the lower and the upper side of a strut and compared the result in Figure 10. The Pareto optimal sets in the figure clearly display the trade-off between surface roughness and printing time; minimizing total printing time increases RMSE, and vice versa. Point C and D are two extremes of optimizing lower side of the 60° angled strut. The optimized parameters of point C and D are (PI = 53.89 mol/m^3, PA = 6.10 mol/m^3, LT = 20.00 μm, CT = 1.00 s, RMSE = 2.13 μm, printing time = 300.00 s) and (PI = 72.4 mol/m^3, PA = 2.03 mol/m^3, LT = 120.00 μm, CT = 4.00 s, RMSE = 19.25 μm, printing time = 48.33 s), respectively. It is interesting to see that when the printing time is maximized and RMSE is minimized, PA is the maximum value at the upper bound and LT is minimum value at the lower bound. On the contrary, when the printing time is minimized, LT is maximized and PA is minimized. When the upper side of surface profile is considered, the two extremes E and F are (PI = 60.56 mol/m^3, PA = 6.10 mol/m^3, LT = 20.00 μm, CT = 1.00 s, RMSE = 5.48 μm, Printing time = 300.00 s) and (PI = 64.30 mol/m^3, PA = 3.91 mol/m^3, LT = 120.00 μm, CT = 1.00 s, RMSE = 18.28 μm, Printing time = 48.33 s), respectively. The details of all four extremes are displayed in Table 9. This result also supports the high correlation between the RMSE and (PA) and LT as discussed in Section 3.2.

Table 9. List of optimizing parameters.

Parameter		Upper/Lower Bounds	Point C (Lower Side)	Point D (Lower Side)	Point E (Upper Side)	Point F (Upper Side)
Symbol	Name					
[PI]	Photoinitiator concentration	24.14–72.4 mol/m^3	53.89 mol/m^3	72.4 mol/m^3	60.56 mol/m^3	64.30 mol/m^3
[PA]	Photo-absorber concentration	2.03–6.1 mol/m^3	6.10 mol/m^3	2.03 mol/m^3	6.10 mol/m^3	3.91 mol/m^3
LT	Layer thickness	20–120 μm	20 μm	120 μm	20 μm	120 μm
CT	Curing time	1–4 s	1.0 s	4.0 s	1.0 s	1.0 s

The Pareto optimal set clearly visualizes the difference produced by the light penetration. Since the printing time is dominated by LT, the extreme in the printing times between both cases remain the same. However, the surface profile RMSE in Figure 10 shows a clear difference because it is very likely that the light penetrated from the upper layer changes the surface profile. As in the previous section, we compared the optimized examples selected from the Pareto optimal set with the non-optimized examples obtained from the validation set up, as shown in Table 10. The clear difference from the optimization is displayed in the bold in Table 10. The printing parameters for the non-optimized examples are (PI = 54.18 mol/m^3, PA = 4.62 mol/m^3, LT = 20.00 μm, CT = 1.00 s), (PI = 50.12 mol/m^3, PA = 6.08 mol/m^3, LT = 28.00 μm, CT = 2.00 s), and (PI = 56.77 mol/m^3, PA = 3.13 mol/m^3, LT = 120.00 μm, CT = 1.00 s) for example 1, 2, and 3, respectively. The comparison between Point C and the non-optimized point 1 shows the 16% surface quality increase by the proposed optimization framework. In addition, Point D shows 18% surface quality increase even in the fastest printing speed. Therefore, like the previous vertical strut, the comparison between the optimized points and the non-optimized cases, it is clearly seen that the proposed optimization framework significantly increases the efficiency of the printing process.

Table 10. Objectives comparison between optimized and non-optimized printing conditions.

Objectives	Point C	Point D	Point E	Point F	Non-Opt. 1	Non-Opt. 2	Non-Opt. 3
Total printing time (s)	300	**48.33**	300	**48.33**	300	210.95	48.33
Surface profile RMSE (μm)	**2.13**	19.25	**5.48**	18.28	2.47	3.57	22.48

The clear difference between the Pareto sets of the upper and lower side profiles, shown in Figure 10, is expected for other angled struts. The degree of effect from the light penetration from the upper layer on the surface roughness of the lower layer will be different because the area affected by the light penetration is different, depending on the printing angle. The optimization method presented here can be easily applied to other 3D printed geometries that may have different side slopes.

It is very important to know the configuration of the Pareto optimal set between these two extremes. Different points on this set allow us to have different printing set-up and corresponding printing time and RMSE, but its printing time and roughness are limited by these two extremes. Therefore, this Pareto optimal set can be a guideline to make better printing conditions depending on the need. The usability of MOPSO is even more obvious when the result from MOPSO is compared with the non-optimized printing case. The approach described in this section can be extended to other printing angles or complex surface profile. It can also be possible to use different objective functions, such as the mechanical strength, total mass, or other properties of AM parts. Based on these results, we conclude that the MOPSO for PμSL helps to optimize the cost functions of interest. In addition, it offers a design guideline for the performance measure for AM parts.

4. Conclusions

We present a systematic approach that provides the optimal printing process parameters for high quality AM parts using a computational model and the particle swarm optimization algorithm. A computational model representing the photopolymerization kinetics involved in PμSL process was implemented and process constants were carefully calibrated by using experimental data obtained from a custom-built PμSL system. Taguchi's Orthogonal Array (OA) method was used to select the parameters that affect the surface quality of AM parts significantly. Four process parameters (PI and PA concentrations, layer thickness, and curing time) selected from Taguchi's OA were used as controlling parameters for optimization algorithm. A meta-heuristic population-based optimization algorithm, particle swarm algorithm for multiple objectives, called multi-objective particle swarm optimization (MOPSO), was used to determine optimal printing process parameters with given geometry of AM parts. Two struts with different printing angles (vertical and 60°) were considered and the optimized printing process parameters were determined for each case. The Pareto optimal set in each printing angle was obtained so that it can be used as a guideline when the printing process parameters need to be set. The result showed that the proposed optimization framework reduces 50% of printing time while keeping the surface quality equal for the vertical strut, and increases 18% of surface quality of the angled strut even in the fastest printing speed, compared to the samples produced by using non-optimized parameters. This framework consisting of the computational model for photopolymerization, process parameter selection by Taguchi's method, and MOPSO for optimization printing process parameters resulted in significant improvement in the quality of AM parts while keeping the printing time minimum. By changing the objective functions of MOPSO, presented approach can also be used for optimizing various quality measures, such as minimizing printing cost and maximizing mechanical strength. Our proposed optimization technique can be easily applied to other photocurable polymers by incorporating material-specific photocuring kinetics parameters for a given polymer in the photopolymerization process modeling. In addition, increasing dimensionality of search space in MOPSO is straightforward without modification of the algorithm, so multiple printing parameters as well as external inputs can be easily increased by adding an additional dimension in the search space. We believe that the process optimization method presented in this study helps to achieve high-quality 3D printed structures and that it can be easily extended to other AM techniques where trial-and-error approaches are used to determine process parameters.

Author Contributions: Conceptualization, H.L.; Data curation, N.K., I.B., and D.H.; Formal analysis, N.K., I.B., D.H., and H.L.; Funding acquisition, H.L.; Investigation, N.K., I.B., and H.L.; Methodology, N.K., I.B., D.H., C.Y., and H.L.; Project administration, H.L.; Resources, C.Y. and H.L.; Software, N.K. and I.B.; Supervision, H.L.; Validation, N.K. and I.B.; Visualization, N.K. and I.B.; Writing—original draft, N.K. and I.B.; Writing—review and editing, N.K., I.B., and H.L.

Funding: This research was funded by Rutgers University through the School of Engineering and the Department of Mechanical and Aerospace Engineering.

Conflicts of Interest: The authors declare no conflict of interest.

Appendix A. Sample Preparation for Curing Depth Study

The cure depth study was performed by printing the bridge structures shown in Figure A1. A single layer in a rectangular shape supported at its both ends was printed. Since there is no supporting layer underneath, thickness of each bridge is the curing depth of the layer. A single structure was designed in which five bridges could be printed for each different energy dose. Each structure has two columns of bridges and each structure was printed twice, thus giving four bridge samples for each set of parameters. These protruding notches were added at the center of the bridge supports to help to bring the structure to a focal plane of a microscope during thickness measurement. After each structure was printed, it was placed in ethanol for 3 s and then let dry in the air. The measurements were done as described in Appendix B. Table A1 lists the printing parameters varied.

Figure A1. Bridge structure illustration for cure depth measurement.

Table A1. Printing parameters for HDDA.

Concentration Level	Photoinitiator (PI)		Photo-Absorber (PA)		Environmental O_2 (O)	
	(mol/m^3)	%	(mol/m^3)	%	(mol/m^3)	%
Low	24.14	1	4.06	0.1	8.69	21
Medium	48.27	2	6.1	0.15	-	-
High	72.4	3	-	-	-	-

Appendix B. Curing Depth Measurement in Experiment

A printed sample was first placed under an optical microscope with a digital camera attached. A 5× lens was selected for measurement. The stage of the microscope was adjusted until the bridge layers came into focus. Digital images of the bridges were captured. The thickness of each bridge was measured using image analysis software, ImageJ. Each cure depth was measured at the center of the bridge as shown in Figure A2. The value obtained from this measurement was in pixels and the pixel to microns conversion was done based on the image size and conversion factor which was obtained from calibration.

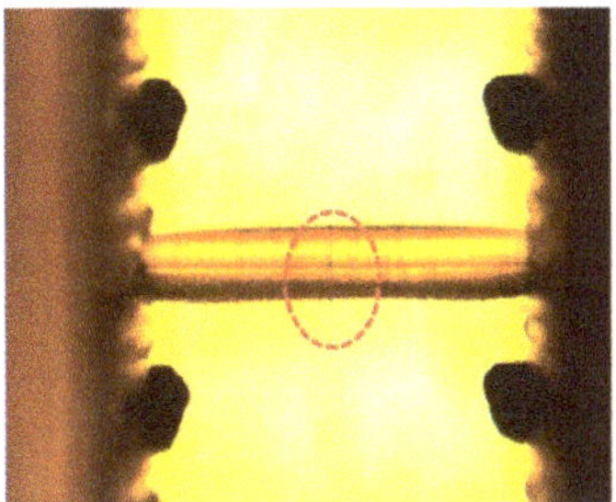

Figure A2. Use of ImageJ line tool to measure curing depth. The optical lens is focused on the notches to find optimal depth. Pixel length of line drawn at the center of the layer. Conversion is done to microns based on conversion factor.

Appendix C. Constructing Cured Profile in Simulation

Figure A3a shows a conversion ratio contour plot generated from COMSOL Multiphysics simulation. A contour line corresponding to 5% cut-off conversion ratio was extracted and mirrored to construct a full cross-section as shown in Figure A3b. Light attenuation along z-direction results in a trapezoidal shaped curing pattern. This plot was used obtain curing width and curing depth of the layer.

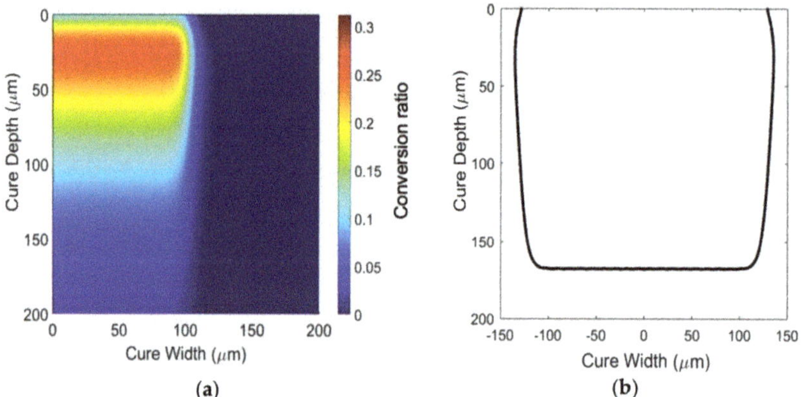

Figure A3. Conversion contour layer profile extraction: (**a**) Conversion contour for layer profile; (**b**) Extracted shape of the cured profile based on the 4% (0.04) cutoff conversion ratio contour.

Appendix D. Calculating Surface Roughness as RMSE

The RMSE was used to quantify surface roughness of AM parts. Once a layer thickness to be used in PμSL process is specified, the bottom portion of the cross-section in Figure A3b below the layer thickness was trimmed away, leaving a cross-section profile of each layer. The layer-wise PμSL process was emulated by stacking this cross-section. A 2D representation of vertically stacked layers is shown in Figure A4. The mean line of the surface profile was first obtained, from which profile deviations were calculated. Subsequently, RMSE of the surface profile was calculated. A similar approach was used for angled struts.

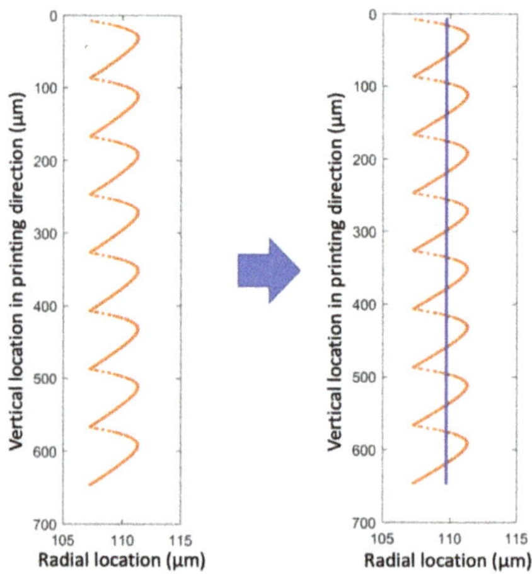

Figure A4. Mean line (blue) on top of surface profile extracted from the simulation data.

Appendix E. Table for Acronyms

Table A2. Table for acronyms.

Acronym	Definition
AM	Additive Manufacturing
PµSL	Projection Micro Stereolithograpy
CAD	Computer-Aided Design
HDDA	Hexanediol Diacrylate
PI	Photoinitiator
PA	Photo-absorber
DMD	Digital Micromirror Device
PDE	Partial Differential Equation
PSO	Particle Swarm Optimization
MOPSO	Multi-Objective Particle Swarm Optimization
RMSE	Root Mean Squared Error
DOA	Design of Experiment
OA	Orthogonal Array
LT	Layer Thickness
CT	Curing Time
DOF	Degree of Freedom

References

1. Melchels, F.P.; Feijen, J.; Grijpma, D.W. A review on stereolithography and its applications in biomedical engineering. *Biomaterials* **2010**, *31*, 6121–6130. [CrossRef]
2. Alapan, Y.; Hasan, M.N.; Shen, R.; Gurkan, U.A. Three-dimensional printing based hybrid manufacturing of microfluidic devices. *J. Nanotechnol. Eng. Med.* **2015**, *6*, 021007. [CrossRef] [PubMed]
3. Vaezi, M.; Seitz, H.; Yang, S. A review on 3D micro-additive manufacturing technologies. *Int. J. Adv. Manuf. Technol.* **2013**, *67*, 1721–1754. [CrossRef]
4. Frenzel, T.; Findeisen, C.; Kadic, M.; Gumbsch, P.; Wegener, M. Tailored Buckling Microlattices as Reusable Light-Weight Shock Absorbers. *Adv. Mater.* **2016**, *28*, 5865–5870. [CrossRef] [PubMed]
5. Maloney, K.J.; Fink, K.D.; Schaedler, T.A.; Kolodziejska, J.A.; Jacobsen, A.J.; Roper, C.S. Multifunctional heat exchangers derived from three-dimensional micro-lattice structures. *Int. J. Heat Mass Transf.* **2012**, *55*, 2486–2493. [CrossRef]
6. Warnke, P.H.; Douglas, T.; Wollny, P.; Sherry, E.; Steiner, M.; Galonska, S.; Becker, S.T.; Springer, I.N.; Wiltfang, J.; Sivananthan, S. Rapid prototyping: porous titanium alloy scaffolds produced by selective laser melting for bone tissue engineering. *Tissue Eng. Part C: Methods* **2008**, *15*, 115–124. [CrossRef] [PubMed]
7. Han, D.; Farino, C.; Yang, C.; Scott, T.; Browe, D.; Choi, W.; Freeman, J.W.; Lee, H. Soft Robotic Manipulation and Locomotion with a 3D Printed Electroactive Hydrogel. *ACS Appl. Mater. Interfaces* **2018**, *10*, 17512–17518. [CrossRef] [PubMed]
8. Hornick, J. 3D printing in Healthcare. *J. 3D Printing Med.* **2017**, *1*, 13–17. [CrossRef]
9. Campbell, T.; Williams, C.; Ivanova, O.; Garrett, B. *Could 3D Printing Change the World. Technologies, Potential, and Implications of Additive Manufacturing*; Atlantic Council: Washington, DC, USA, 2011.
10. Meza, L.R.; Greer, J.R. Mechanical characterization of hollow ceramic nanolattices. *J. Mater. Sci.* **2014**, *49*, 2496–2508. [CrossRef]
11. Ahn, D.; Kim, H.; Lee, S. Surface roughness prediction using measured data and interpolation in layered manufacturing. *J. Mater. Process. Technol.* **2009**, *209*, 664–671. [CrossRef]
12. Sager, B.; Rosen, D.W. Use of parameter estimation for stereolithography surface finish improvement. *Rapid Prototyp. J.* **2008**, *14*, 213–220. [CrossRef]
13. Hasan, R. Progressive Collapse of Titanium Alloy Micro-Lattice Structures Manufactured Using Selective Laser Melting. Ph.D. Thesis, University of Liverpool, Liverpool, UK, 2013.
14. Quintana, R.; Choi, J.-W.; Puebla, K.; Wicker, R. Effects of build orientation on tensile strength for stereolithography-manufactured ASTM D-638 type I specimens. *Int. J. Adv. Manuf. Technol.* **2010**, *46*, 201–215. [CrossRef]

15. Ancau, M.; Caizar, C. The optimization of surface quality in rapid prototyping. In Proceedings of the WSEAS International Conference on Engineering Mechanics, Structures, Engineering Geology (EMESEG '08), Crete Island, Greece, 22–24 July 2008.

16. Chockalingam, K.; Jawahar, N.; Chandrasekhar, U. Influence of layer thickness on mechanical properties in stereolithography. *Rapid Prototyp. J.* **2006**, *12*, 106–113. [CrossRef]

17. Sun, C.; Fang, N.; Wu, D.; Zhang, X. Projection micro-stereolithography using digital micro-mirror dynamic mask. *Sens. Actuators A: Phys.* **2005**, *121*, 113–120. [CrossRef]

18. Decker, C.; Jenkins, A.D. Kinetic approach of oxygen inhibition in ultraviolet-and laser-induced polymerizations. *Macromolecules* **1985**, *18*, 1241–1244. [CrossRef]

19. Jariwala, A.S.; Ding, F.; Boddapati, A.; Breedveld, V.; Grover, M.A.; Henderson, C.L.; Rosen, D.W. Modeling effects of oxygen inhibition in mask-based stereolithography. *Rapid Prototyp. J.* **2011**, *17*, 168–175. [CrossRef]

20. Odian, G. *Principles of Polymerization*; John Wiley & Sons: Hoboken, NJ, USA, 2004.

21. Fang, N.; Sun, C.; Zhang, X. Diffusion-limited photopolymerization in scanning micro-stereolithography. *Appl. Phys. A* **2004**, *79*, 1839–1842. [CrossRef]

22. Ingle, J.D., Jr.; Crouch, S.R. *Spectrochemical Analysis*; Prentice-Hall: Inglewood Cliffs, NJ, USA, 1988.

23. Rohatgi-Mukherjee, K. *Fundamentals of Photochemistry*; Wiley Eastern Ltd.: New Delhi, India, 1978.

24. Tryson, G.; Shultz, A. A calorimetric study of acrylate photopolymerization. *J. Polym. Sci.: Polym. Phys. Ed.* **1979**, *17*, 2059–2075. [CrossRef]

25. Zhao, Z.; Mu, X.; Wu, J.; Qi, H.J.; Fang, D. Effects of oxygen on interfacial strength of incremental forming of materials by photopolymerization. *Extreme Mech. Lett.* **2016**, *9*, 108–118. [CrossRef]

26. Dendukuri, D.; Panda, P.; Haghgooie, R.; Kim, J.M.; Hatton, T.A.; Doyle, P.S. Modeling of oxygen-inhibited free radical photopolymerization in a PDMS microfluidic device. *Macromolecules* **2008**, *41*, 8547–8556. [CrossRef]

27. Florio, J.J.; Miller, D.J. *Handbook of Coating Additives*; CRC Press: Boca Raton, FL, USA, 2004.

28. Brandrup, J.; Immergut, E.H.; Grulke, E.A.; Abe, A.; Bloch, D.R. *Polymer Handbook*; Wiley: New York, NY, USA, 1989; Volume 7.

29. Flach, L.; Chartoff, R.P. A process model for nonisothermal photopolymerization with a laser light source. I: Basic model development. *Polym. Eng. Sci.* **1995**, *35*, 483–492.

30. Deb, K.; Pratap, A.; Agarwal, S.; Meyarivan, T. A fast and elitist multiobjective genetic algorithm: NSGA-II. *IEEE Trans. Evol. Comput.* **2002**, *6*, 182–197. [CrossRef]

31. Dorigo, M.; Birattari, M. Ant colony optimization. In *Encyclopedia of Machine Learning*; Springer: Berlin, Germany, 2011; pp. 36–39.

32. Eberhart, R.; Kennedy, J. A new optimizer using particle swarm theory. In Proceedings of the Sixth International Symposium on Micro Machine and Human Science, MHS'95, Nagoya, Japan, 4–6 October 1995; pp. 39–43.

33. Reyes-Sierra, M.; Coello, C.C. Multi-objective particle swarm optimizers: A survey of the state-of-the-art. *Int. J. Comput. Intell. Res.* **2006**, *2*, 287–308.

34. Coello, C.A.C.; Pulido, G.T.; Lechuga, M.S. Handling multiple objectives with particle swarm optimization. *IEEE Trans. Evol. Comput.* **2004**, *8*, 256–279. [CrossRef]

35. Kennedy, J. Particle swarm optimization. In *Encyclopedia of Machine Learning*; Springer: Berlin, Germany, 2011; pp. 760–766.

36. Yarpiz. Multi-Objective Particle Swarm Optimization (MOPSO). Available online: www.yarpiz.com (accessed on 3 November 2018).

37. Taguchi, G. *Introduction to Quality Engineering: Designing Quality into Products and Processes*; Asian Productivity Organization: Tokyo, Japan, 1986.

38. Ross, P.J. *Taguchi Techniques for Quality Engineering: Loss Function, Orthogonal Experiments, Parameter and Tolerance Design*; McGraw-Hill: New York, NY, USA, 1988.

39. Peace, G.S. *Taguchi Methods: A Hands-On Approach*; Addison-Wesley: Reading, MA, USA, 1993.

Article

Manufacturing of Micro-Lens Array Using Contactless Micro-Embossing with an EDM-Mold

Kangsen Li [1], Gang Xu [1], Xinfang Huang [2], Zhiwen Xie [2] and Feng Gong [1,*]

[1]   Guangdong Provincial Key Laboratory of Micro/Nano Optomechatronics Engineering,
    College of Mechatronics and Control Engineering, Shenzhen University, Shenzhen 518060, China;
    likangsenszu@163.com (K.L.); xugang@szu.edu.cn (G.X.)

[2]   School of Mechanical Engineering and Automation, University of Science and Technology Liaoning,
    Anshan 114051, China; xfgg527@163.com (X.H.); xzwustl@126.com (Z.X.)

*   Correspondence: gongfeng186@163.com; Tel./Fax: +867-552-655-8509

Received: 30 November 2018; Accepted: 22 December 2018; Published: 26 December 2018

Abstract: Micro embossing is an effective way to fabricate a polymethyl methacrylate (PMMA) specimen into micro-scale array structures with low cost and large volume production. A new method was proposed to fabricate a micro-lens array using a micro-electrical discharge machining (micro-EDM) mold. The micro-lens array with different shapes was established by controlling the processing parameters, including embossing temperature, embossing force, and holding time. In order to obtain the friction coefficient between the PMMA and the mold, ring compression tests were conducted on the Shenzhen University's precision glass molding machine (SZU's PGMM30). It was found that the friction coefficient between the PMMA specimen and the mold had an interesting change process with increasing of temperature, which affected the final shape and stress distribution of the compressed PMMA parts. The results of micro-optical imaging of micro-lens array indicated that the radius of curvature and local length could be controlled by adjusting the processing parameters. This method provides a basis for the fabrication and application of micro-lens arrays with low-cost, high efficiency, and mass production.

Keywords: micro-EDM molds; micro-lens array; contactless embossing; friction coefficient

---

1. Introduction

The micro-lens array (MLA), called a Fly-eye lens, is widely used in optical communication, lighting displays, optical sensors, and illumination because of its unique structures and function. There are many methods that can fabricate micro-lenses, including plasma etching [1], laser ablation [2], micro-milling [3], photolithography [4], micro-injection [5], glass reflow [6] and micro hot embossing [7–12]. Wherein, micro hot embossing is regarded as one of the most promising processes for mass production.

Micro hot embossing is a flexible and low-cost technique, consisting of heating, embossing, and cooling. In recent years, many types of hot embossing processes have been reported, including PDMS (polydimethylsiloxane) soft mold imprinting [13–17], Deep reactive ion etching (DRIE) Si template duplicating [18], LIGA (Acronym of German words: Lithografie, Galvanik, Abformung) Ni mold copying [19], focused ion beam (FIB) mold embossing [20,21], and roll-to-roll embossing [22,23]. The quality of the embossed micro-lens depends on the precision of the mold insert. For morphologic accuracy of mold fabrication, the mold with the micro-hole array is an effective tool to form the micro-lens array. Schulze et al. [24,25] first proposed the contactless embossing technology to fabricate a micro-lens array (CEM). CEM is a potential method for the achievement of mass production. Xie et al. [26] fabricated a polymer refractive micro-lens array on a stainless steel through-holes template by contactless hot embossing. Moore et al. [27] investigated the effects of processing parameters in the polymer substrates contactless embossing process. The method is a reliable way to produce a micro-lens. In recent years, micro-EDM was widely investigated [28,29]. However, it was first used to fabricate the micro-hole array in this study, which was then used for the manufacturing of the micro-lens array.

The friction coefficient between polymethyl methacrylate (PMMA) and the mold interface plays an important role in the process, affecting the molding force and demolding force [30,31]. The friction at the interface of mold/PMMA affects the wear of the mold surface (the mold life and the quality of the embossed lens) [32]. Male and Depierre [33] used a ring compression test to characterize the friction behavior, indicating the universal friction calibration curves. The ring compression test is a reliable way to determine the friction behaviors between PMMA and the mold [34–36].

The fabrication of a micro-lens array was investigated using contactless micro-embossing with an EDM-mold. To verify the friction behaviors between PMMA and the mold, ring compression tests were carried out on a Shenzhen University's precision glass molding machine. Then, the micro-hole of the SKD-11 mold was fabricated by electric discharge machining. After that, we obtained a series of micro-lens arrays by contactless embossing at different processing parameters. ABAQUS software was used for the finite element simulation of the PMMA embossing process. Finally, the micro-optical performance of the micro-lens array was measured using optical experiment apparatus.

2. Materials and Methods

2.1. Materials

Table 1 shows the characteristics of PMMA and mold tools (SKD-11). SKD-11 (wt%: Cr(15%), C(1.55%), Mo(0.7%), V(1.0%), Mn(0.3%), Si(0.25%)) is a tool steel widely used in the fabrication of molds. At room temperature, PMMA is a glassy polymer, with a glass transition temperature of 105 °C and melting temperature of 220 °C.

Table 1. The physical properties of polymethyl methacrylate (PMMA) and SKD-11 mold.

Property	PMMA	SKD-11
Young's Modulus (GPa)	2.4	210
Poisson ratio	0.37	0.2
Density (kg/m^3)	1185	7700
Thermal conductivity (W/m^2·K)	0.2	200
Specific heat (J/kg·K)	1466	460
Thermal expansion (K^{-1})	4.4×10^{-4}	11×10^{-6}
Glass transition temperature (°C)	105	-
Melting temperature (°C)	220	-

2.2. Ring Compression Test

Figure 1a shows that the PMMA rings were fabricated with an internal diameter (ID) of 6 mm, an outer diameter (OD) of 12 mm, and a height (H) of 4 mm, which is a standard ring ratio of OD:ID:H = 6:3:2. Figure 1b shows that effect of friction magnitude on the PMMA deformation during the ring compression test. The high friction force leads to an inward flow of the PMMA material when a ring is compressed between flat molds. On the contrary, the inner diameter of ring increases when the friction is low. Therefore, we can obtain the friction coefficient between PMMA and mold by this relationship. The surface roughness of the PMMA ring is between Ra 5 to 7 nm (see Figure 1f). Our team conducted a series of ring compression tests on SZU's PGMM30 (see Figure 1c,d). The heating chamber provided a vacuum, or Nitrogen environment, for the embossing process. The upper mold remained stationary, while the lower mold was driven upward and downward by an AC servomotor system. The load of the system was monitored by a load cell with a resolution of 1 N. During the embossing stage, the lower mold position was recorded by a position sensor with a resolution of 1 μm. To obtain the relationship between friction coefficient and temperature, a series of rings were compressed under different temperatures (90, 105, 120, 135, and 150 °C). Figure 1e shows the history of temperature, force, and position time. Figure 1f shows the compressed PMMA ring under different temperatures.

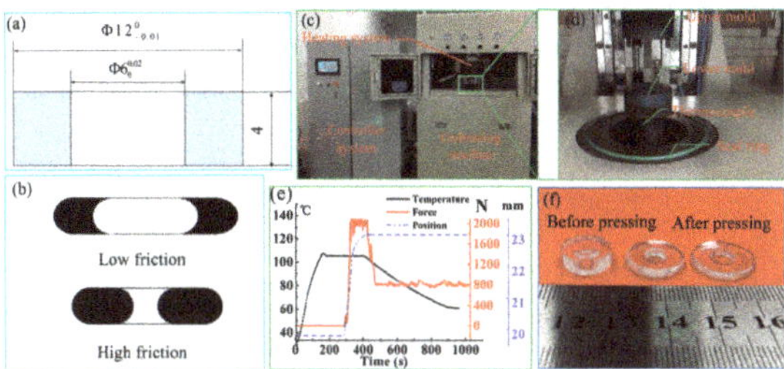

Figure 1. (**a**) Dimension of the polymethyl methacrylate (PMMA) rings; (**b**) effect of friction magnitude with PMMA deformation during ring compression test; (**c**) Shenzhen University's precision glass molding machine (SZU's PGMM30); (**d**) mold assembly; (**e**) time history of the force, temperature, and position; (**f**) before and after pressing of the PMMA ring.

2.3. The Manufacture of the Micro-EDM Mold Insert

The copper micro-electrode, of which diameter was 250 µm, was used in micro-EDM to process micro-hole arrays. Figure 2a shows the schematic of the micro-EDM. Figure 2b shows the image of the electrical discharge machine (Sodick, Japan). The processing parameters of micro-EDM consist of voltage, pulse frequency, pulse width, and pulse interval [37,38]. Through adjusting the processing parameters, micro-EDM can effectively fabricate micro-hole arrays. In the micro-EDM process, the voltage was set as 100 V, applied to the material. The workpiece material was a SKD-11 mold tool. The pulse frequency was set to 0.2 MHz. The pulse width and pulse internal was set to 500 and 400 nanoseconds, respectively. Figure 2c,d show the experimental results. From the local SEM image of micro-hole array, the diameter of micro-holes is consistent. It is noteworthy that the side wall of micro-holes is not smooth. Therefore, the micro-hole array needs to be polished being applying in embossing. The micro-electrodes always have wear during micro-EDM process, which results in the inconsistent depth of micro-holes [37]. In the study, the fabrication micro-hole was blind. The wear of micro-electrodes affected the depth accuracy of micro-holes, which should be avoided. Thus, the depth of micro-holes was ensured by position compensation. The diameter and depth of each hole was around 320 and 300 µm, respectively.

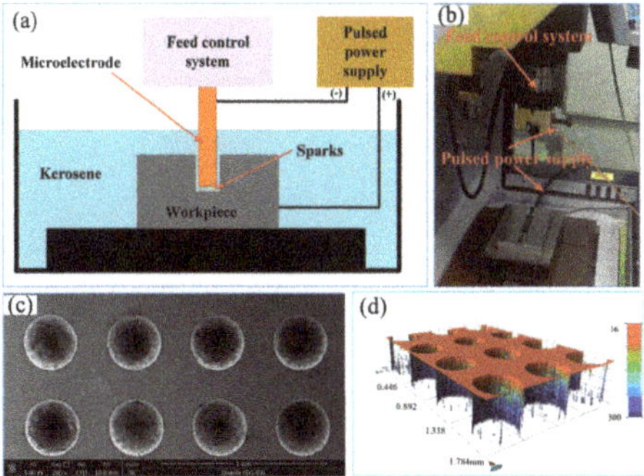

Figure 2. (**a**) The schematic of the micro-electric discharge machining (EDM); (**b**) picture of the experiment apparatus (AP1L); (**c**) the SEM diagram of the micro-hole array of mold; (**d**) the dimensions and layout of the micro-holes array.

2.4. Fabrication of Micro-Lens Array

The micro-embossing process includes presetting, heating, embossing, cooling, and demolding (see Figure 3). The processing parameters, such as embossing temperature, embossing pressure, keeping pressure, embossing time, and cooling velocity, have great influence on the friction characteristic, deformation behavior, and replication quality of hot embossing micro structures.

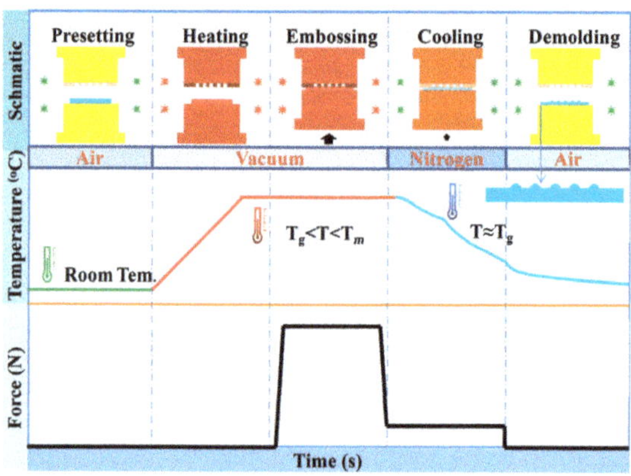

Figure 3. Schematic of micro-embossing process using electrical discharge machining (EDM) mold.

At the beginning of embossing, the chamber of the embossing apparatus was vacuumed using a vacuum pump. Then, the mold and the PMMA material were heated to a temperature between glass transition and melting point, and soaked for about 100 s to make temperature uniformly distributed. Furthermore, the lower mold was moved by the servomotor drive, controlled by load cell and position sensor. The embossing pressure was in the range of 200–800 N. Finally, nitrogen flowed through the embossing chamber to cool the molds and substrate to a certain temperature. When the mold cooled to room temperature, the PMMA substrates were demolding.

The main influencing factors, including embossing temperature, embossing pressure, and holding time, were performed with sensitivity analysis by investigating orthogonal experiment. In our case, the orthogonal experiment was composed of three factors with three levels, namely embossing temperature, embossing force, and holding time (see Table 2).

Table 2. List of factors, levels for orthogonal design.

Levels	Factors		
	Embossing Temperature (°C)	Embossing Force (N)	Holding Time (s)
1	105	200	120
2	120	400	180
3	135	800	240

3. Finite Element Model

Between glass transition and melting temperature, PMMA shows as viscoelastic behaviors. In this study, the viscoelastic material was treated as a general Maxwell model. The temperature dependence of PMMA material can be treated as a Thermally Rhetorically Simplicity (TRS) model by the Williams-Landel-Ferry (WLF) equation [39]. C1 and C2 are the WLF equation constants that can be fitted.

The response of PMMA material to the continuous stress history can be expressed by the general Maxwell model

$$\sigma(t) = \int_0^t G(t-\tau)\frac{d\varepsilon(t)}{d\tau}d\tau + G(t)\varepsilon(0) \tag{1}$$

$$G(t) = G_i \sum_{i=1}^{N} w_i \exp(-t/\tau_{vi}) + G_\infty, \tag{2}$$

where t is the current time; τ the past time; w_i are the weigh factors; and G is the modulus of material.

Stress relaxation at different temperatures can be expressed by WLF Equation [39]

$$\log a_T = -\frac{C_1(T - T_{ref})}{C_2 + (T - T_{ref})}, \tag{3}$$

where T_{ref} is the reference temperature and C_1 and C_2 are the WLF equation constants that can be fitted. As for the PMMA material, C1 and C2 can be taken as 17.4 and 51.6, respectively [40]. The mechanical properties were obtained from the literature [41,42].

Figure 4 shows the initial geometry of the model, including the PMMA, the upper mold, lower mold, and mold core. In order to reduce the computation, one fourth of the coupled thermal-displacement model was developed in the software. The PMMA was modeled as linear viscoelastic material, and the molds were treated as elastic material. Two master-slave types of contact interaction pairs were established when the interface behaviors were treated as hard contact and coulomb friction model with penalty formulation. The PMMA/mold interface friction coefficient was obtained from the ring compression tests. The surfaces of upper and lower mold were defined as the master surface. The vertical displacement of the upper mold was fixed with the horizontal displacement to keep free. The heating mechanism in the PMMA embossing refers to gap heat conductance and radiation.

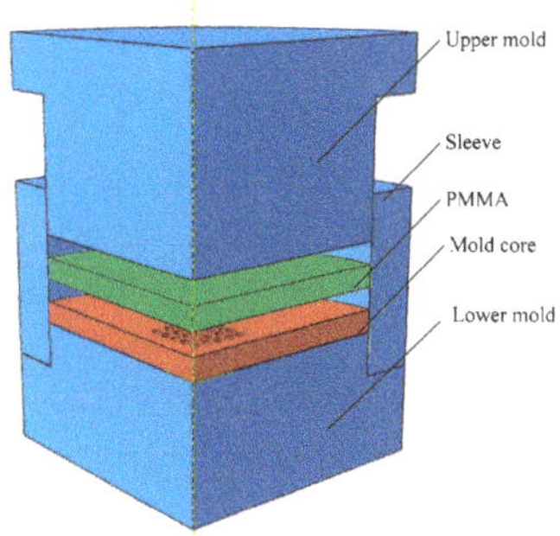

Figure 4. The initial geometry model used for simulation of the contactless micro-embossing process.

4. Results and Discussion

4.1. Friction Coefficient between PMMA and Mold

Friction at the mold/PMMA interface affects the flow ability, molding force, shape of molded PMMA, and mold life in forming process. To understand the deformation behaviors of PMMA, numerical simulation is an effective method to predict the shape of molded PMMA. Figure 5a shows the simulation results of the PMMA ring for different interface friction coefficients, assuming the PMMA ring was pressed to the same vertical displacement. If the interfacial friction was low, the deformation configuration of the PMMA material was radially outward. On the contrary, the inner diameter reduced when the external diameter increased with the increased friction coefficient. Figure 5b shows the change curve of the ring deformation during the pressing stage. Figure 5c shows the numerical friction calibration curves by extracting the dimensional changes of the PMMA ring.

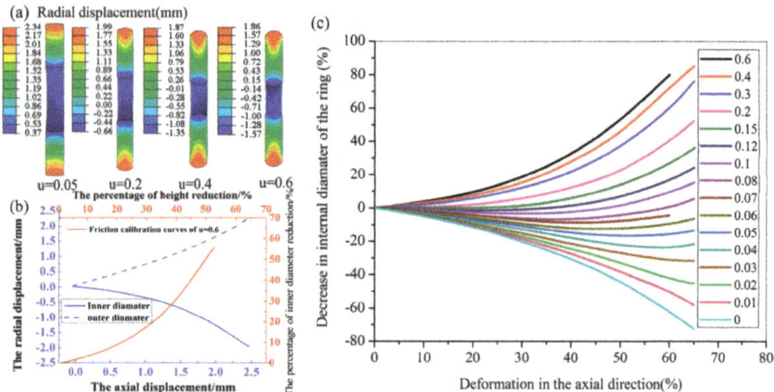

Figure 5. The simulation results. (**a**) The PMMA ring compression for different interface friction coefficient; (**b**) the displacement change of the diameters, and the friction calibration curve obtained from the simulation results; (**c**) numerical friction calibration curves under different friction coefficient.

To investigate the influence of temperature, the forming temperature of the PMMA ring was set at 90, 105, 120, 135 and 150 °C, respectively. Figure 6a shows the deformation dimension of the pressed rings at the range of 90–150 °C. Interesting changes were that the inner diameter of PMMA ring largened as the temperature rose, and the outer diameter was slightly bulged. Therefore, we obtained the friction condition by observing the dimensional changes of the inner diameter. The inner diameter and height of the ring after cooling were measured by digital Vernier caliper. Table 3 shows the experimental results of the ring compression test. To reduce the size error, each dimension was measured ten times in each different position, with the average taken. Figure 6b shows the numerical calibration curves and measured data points. The friction coefficient at the PMMA/mold was obtained by comparing the friction curves and experimental data. In the range of 90–150 °C, the friction coefficient decreased with the increasing pressing temperature. This was because the deformation resistance gradually reduced by raising the temperature.

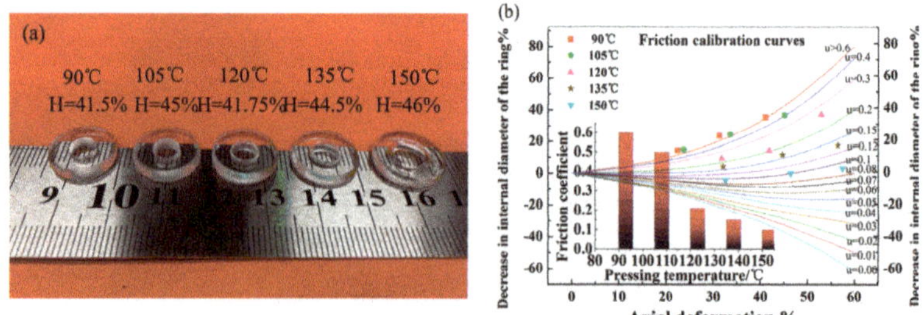

Figure 6. PMMA compression test: (**a**) The ring after compression under different temperatures; (**b**) numerical friction calibration curves and experimental data.

Table 3. The experimental results of the ring compression test.

No.	Temp./ °C	Force/N	ID/mm	OD/mm	T/mm
1	90	3000	5.21	13.26	3.10
2	90	3000	4.66	13.85	2.75
3	90	3000	2.34	14.66	3.96
4	105	2000	5.17	13.37	3.05
5	105	2000	4.61	14.08	2.66
6	105	2000	3.88	14.48	2.20
7	120	1500	5.51	14.25	2.73
8	120	1500	5.22	15.06	2.33
9	120	1500	3.87	16.09	1.89
10	135	1000	5.81	14.47	2.72
11	135	1000	5.37	15.48	2.22
12	135	1000	5.01	17.00	1.75
13	150	500	6.30	15.26	2.70
14	150	500	6.06	15.78	2.15
15	150	500	5.88	17.51	1.71

4.2. Fabrication of Micro-Lens Array

The height ratios of the micro-lens array change with different processing parameters. It is worth mentioning that the profile of micro-lens array can be replicated randomly by controlling the embossing conditions. The three processing parameters had significant effects on the deformation of the micro-lens array. The results indicated that the flow behaviors of PMMA were much better with the increase of embossing temperature. In addition, the increasing holding pressure and time can reduce the recovery deformation of micro-lens array. The height ratios of micro-lens array became larger as the embossing force rose.

Figure 7 shows the embossed PMMA microstructure under different processing parameters. The aspect ratios of the microstructure changed with different processing parameters. Much like the profile of micro-lens array, the profile of microstructure can also be replicated randomly by controlling the embossing conditions. With the increase of temperature, the depth of the replicated structures increased, and their recovery deformation decreased. The defects occurred on the edge of the microstructures, due to the mold quality, because the edge of mold machined by EDM appeared with some defects affecting the replication quality. In the future studies, the problem of fabrication quality will be solved by polishing or coating.

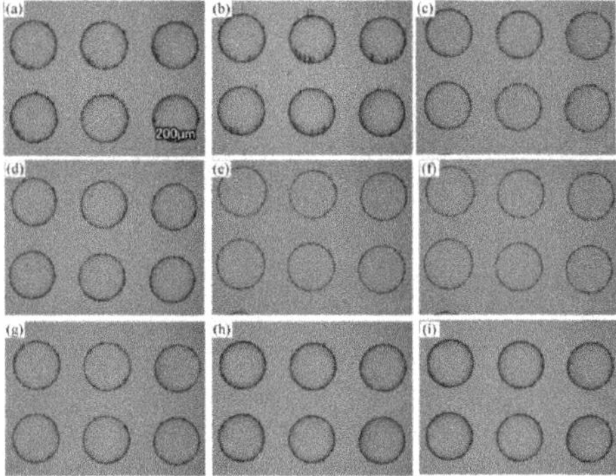

Figure 7. Micro-Embossing PMMA micro-lens array under different processing parameters. (**a**) 105 °C/200 N/120 s; (**b**) 105 °C/400 N/180 s; (**c**) 105 °C/800 N/240 s; (**d**) 120 °C/200 N/120 s; (**e**) 120 °C/400 N/240 s; (**f**) 120 °C/800 N/180 s; (**g**) 135 °C/200 N/240 s; (**h**) 135 °C/400 N/180 s; (**i**) 135 °C/800 N/120 s.

The ContourGT-X 3D Optical Profiler (Bruker, Germany) was used to observe the surface morphology and profile shape. Figure 8 shows the measurement results of different embossing processing parameters, and the 3D profile and shape of the formed micro-lens array (with diameter ≈ 300 μm, and height ≈ 10 μm) corresponding to Figure 7a. Figure 8b shows the profile shape of the micro-lens array (with diameter ≈ 300 μm, and height ≈ 33 μm) by filtering. It can be concluded that embossing temperature and force are the important factors on the contactless hot embossing process.

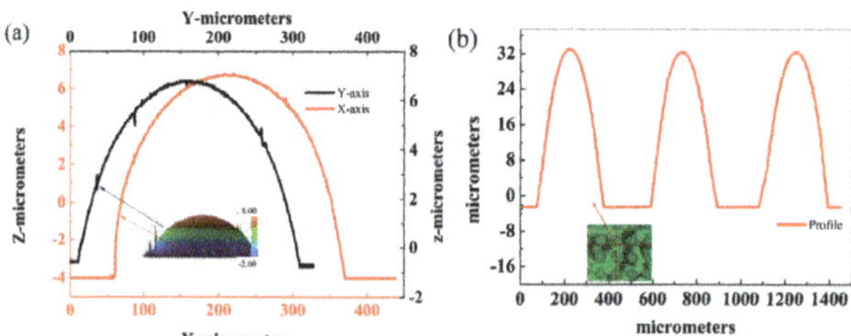

Figure 8. Profile of micro-lens measured using ContourGT-X 3D Optical Profiler. (**a**) Forming under conditions of 105 °C/200 N/120 s; (**b**) forming under conditions of 135 °C/800 N/120 s.

The stress inside the microstructure affected the optical quality of the embossed micro-lens because of temperature gradient during cooling. The simulation can be used to investigate the evolution mechanism of residual stress during hot embossing. Figure 9 shows the Von Mise stress and friction shear stress of the micro-lens array from FE simulation. The maximum stress focused on the edge of the microstructure, and the lowest stress took place around the center of microstructure. The deformation characteristic of the PMMA during hot embossing is dependent on temperature. The stress concentrated in the contact edge region of the lens, during thermal loading process, which led to large plastic deformation in the region. Stress concentration and shear flow led to the defects and strain hardening of the micro-lens, affecting the optical quality. The friction coefficient had a great effect on the shear deformation. To reduce the stress concentration and shear stress, we increased the temperature of mold in the certain range.

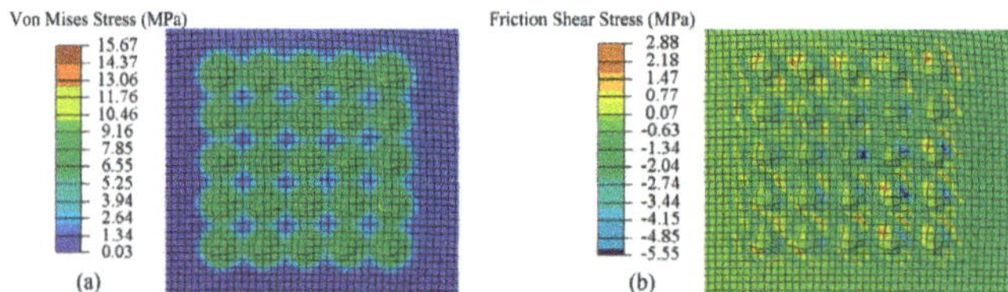

Figure 9. The stress distribution of micro-lens array. (**a**) Von Mises stress; (**b**) friction shear stress.

4.3. Micro-Optics of Micro-Lens Array

The optical quality of the micro-lens array was the most important parameter. To verify the imaging performance of the micro-lens array, it can be detected by optical measurement system. Figure 10a shows the schematic diagrams of the system. The experiment system consisted of a charge-coupled device (CCD) camera, workbench, directional light source, and computer system. The system had the advantage of convenience and rapid response. Through the projection information, we can estimate the imaging quality.

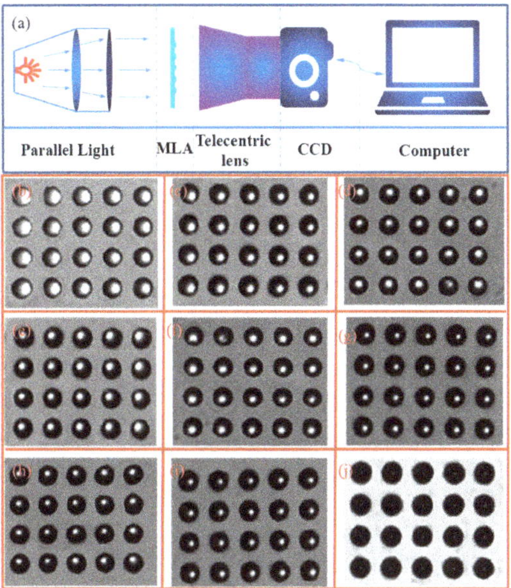

Figure 10. (**a**) Schematic diagram of the experimental system; (**b**) the light spots of the micro-lens array from Figure 7a; (**c**) The light spots of the micro-lens array from Figure 7b; (**d**) the light spots of the micro-lens array from Figure 7c; (**e**) the light spots of the micro-lens array from Figure 7d; (**f**) the light spots of the micro-lens array from Figure 7e; (**g**) the light spots of the micro-lens array from Figure 7f; (**h**) The light spots of the micro-lens array from Figure 7f; (**i**) the light spots of the micro-lens array from Figure 7h; (**j**) the light spots of the micro-lens array from Figure 7i.

Figure 10b–j shows the focused light spots of the embossed micro-lens array corresponding to micro-lens array of Figure 7. The light spots of the micro-lens array had a great difference in the same projection position. From the imaging performance of the micro-lens array, the processing parameters had a great influence on the profile of the lens array. The light spots are non-uniform due to the non-uniformity of the lens shape. In the embossing process, the deformation of the PMMA material had a recovery stage, which was related to the temperature and pressure. When the embossing temperature is 105 °C with embossing force of 200 N and holding time of 120 s, the micro-lens endured larger recovery deformation than that of other processing parameters. Large recovery led to irregular deformations if the process had no intervention. An apparent optical imaging difference can be observed among these micro-lenses under different processing parameters. The temperature and the embossing force were the two crucial parameters affecting the fabrication of micro-lens array.

5. Conclusions

This work reported the embossed micro-lens arrays based on micro-EDM micro-hole array molds. The ring compression test indicated that in the viscous region, the friction force between PMMA and mold decreased as temperature rose. The recovery occurred in the cooling stage without loading, which affected the replication precision. The embossing temperature and force affected deformation and surface morphology of the embossed micro-lens array. The results of this work indicated that the deformation behaviors can be used to fabricate different local lengths of micro-lens. Contactless embossing on the micro-EDM molds was a flexible method with low cost for fabrication of micro-structures arrays.

Author Contributions: Conceptualization, F.G., G.X. and K.L.; Methodology, F.G., K.L. and X.H.; Software, K.L.; Writing-original Draft Preparation, K.L. and Z.X.; Funding Acquisition, F.G.

Funding: The authors gratefully acknowledge the financial support of the Natural Science Foundation of Guangdong Province (2018A030313466) and the Shenzhen Science and Technology Innovation Commission (SGLH20150213170331329).

Acknowledgments: The author would like to appreciate the colleagues for their help to improve the paper.

Conflicts of Interest: The authors declare no conflict of interest.

References

1. Kuang, D.; Zhang, X.; Gui, M.; Fang, Z. Hexagonal microlens array fabricated by direct laser writing and inductively coupled plasma etching on organic light emitting devices to enhance the outcoupling efficiency. *Appl. Opt.* **2009**, *48*, 974. [CrossRef] [PubMed]

2. Lim, C.S.; Hong, M.H.; Kumar, A.S.; Rahman, M.; Liu, X.D. Fabrication of concave micro lens array using laser patterning and isotropic etching. *Int. J. Mach. Tool Manuf.* **2006**, *46*, 552–558. [CrossRef]

3. Mccall, B.; Descour, M.; Tkaczyk, T. Fabrication of plastic microlens arrays for array microscopy by diamond milling techniques. *Opt. Eng.* **2010**, *49*, 730. [CrossRef] [PubMed]

4. Wu, H.; Odom, T.W.; Whitesides, G.M. Reduction photolithography using microlens arrays: Applications in gray scale photolithography. *Anul. Chem.* **2002**, *74*, 3267–3273. [CrossRef]

5. Luo, C.W.; Chiang, Y.C.; Cheng, H.C.; Wu, C.Z.; Huang, C.F.; Wu, C.W.; Shen, Y.K.; Lin, Y. A novel and rapid fabrication for microlens arrays using microinjection molding. *Poly. Eng. Sci.* **2011**, *51*, 391–402. [CrossRef]

6. Chang, C.Y.; Yang, S.Y.; Huang, L.S.; Chang, J.H. Fabrication of plastic microlens array using gas-assisted micro-hot-embossing with a silicon mold. *Infrared Phys. Technol.* **2006**, *48*, 163–173. [CrossRef]

7. Albero, J.; Perrin, S.; Bargiel, S.; Passilly, N.; Baranski, M.; Gauthier-Manuel, L.; Bernard, F.; Lullin, J.; Froehly, L.; Krauter, J.; et al. Dense arrays of millimeter-sized glass lenses fabricated at wafer-level. *Opt. Express* **2015**, *23*, 11702–11712. [CrossRef] [PubMed]

8. Ong, N.S.; Koh, Y.H.; Fu, Y.Q. Microlens array produced using hot embossing process. *Microelectron. Eng.* **2002**, *60*, 365–379. [CrossRef]

9. Chang, C.Y.; Yang, S.Y.; Chu, M.H. Rapid fabrication of ultraviolet-cured polymer microlens arrays by soft roller stamping process. *Microelectron. Eng.* **2007**, *84*, 355–361. [CrossRef]

10. Kunnavakkam, M.V.; Houlihan, F.M.; Schlax, M.; Liddle, J.A.; Kolodner, P.; Nalamasu, O.; Rogers, J.A. Low-cost, low-loss microlens arrays fabricated by soft-lithography replication process. *Appl. Phys. Lett.* **2003**, *82*, 1152–1154. [CrossRef]

11. Lin, C.R.; Chen, R.H.; Hung, C. Preventing non-uniform shrinkage in open-die hot embossing of PMMA microstructures. *J. Mater. Process. Technol.* **2003**, *140*, 173–178. [CrossRef]

12. Chang, C.Y.; Yu, C.H. A basic experimental study of ultrasonic assisted hot embossing process for rapid fabrication of microlens arrays. *J. Micromech. Microeng.* **2015**, *25*. [CrossRef]

13. Liu, Y.; Zhang, P.; Deng, Y.; Hao, P.; Fan, J.; Chi, M.; Wu, Y. Polymeric microlens array fabricated with PDMS mold-based hot embossing. *J. Micromech. Microeng.* **2014**, *24*, 95028. [CrossRef]

14. Chang, C.Y.; Yang, S.Y.; Huang, L.S.; Hsieh, K.H. Fabrication of polymer microlens arrays using capillary forming with a soft mold of micro-holes array and UV-curable polymer. *Opt. Express* **2006**, *14*, 6253–6258. [CrossRef] [PubMed]

15. Hu, C.N.; Hsieh, H.T.; Su, G.D.J. Fabrication of microlens arrays by a rolling process with soft polydimethylsiloxane molds. *J. Micromech. Microeng.* **2011**, *21*, 65013. [CrossRef]

16. Kim, M.; Moon, B.U.; Hidrovo, C.H. Enhancement of the thermo-mechanical properties of PDMS molds for the hot embossing of PMMA microfluidic devices. *J. Micromech. Microeng.* **2013**, *23*, 95024. [CrossRef]

17. Goral, V.N.; Hsieh, Y.C.; Petzold, O.N.; Faris, R.A.; Yuen, P.K. Hot embossing of plastic microfluidic devices using poly(dimethylsiloxane) molds. *J. Micromech. Microeng.* **2010**, *21*, 17002–17009. [CrossRef]

18. Pan, L.W.; Shen, X.; Lin, L. Microplastic lens array fabricated by a hot intrusion process. *J. Micoelectromech. Syst.* **2004**, *13*, 1063–1071. [CrossRef]

19. Dong, S.K.; Lee, H.S.; Lee, B.K.; Sang, S.Y.; Tai, H.K.; Lee, S.S. Replications and analysis of microlens array fabricated by a modified LIGA process. *Poly. Eng. Sci.* **2006**, *46*, 416–425. [CrossRef]
20. Lan, S.; Lee, H.J.; Kim, E.; Ni, J.; Lee, S.H.; Lai, X.; Song, J.H.; Lee, N.K.; Lee, M.G. A parameter study on the micro hot-embossing process of glassy polymer for pattern replication. *Microelectron Eng.* **2009**, *86*, 2369–2374. [CrossRef]
21. Youn, S.W.; Okuyama, C.; Takahashi, M.; Maeda, R. A study on fabrication of silicon mold for polymer hot-embossing using focused ion beam milling. *J Mater. Process. Technol.* **2008**, *201*, 548–553. [CrossRef]
22. Deng, Y.; Yi, P.; Peng, L.; Lai, X.; Lin, Z. Experimental investigation on the large-area fabrication of micro-pyramid arrays by roll-to-roll hot embossing on PVC film. *J. Micromech. Microeng.* **2014**, *24*, 45023–45034. [CrossRef]
23. Yeo, L.P.; Ng, S.H.; Wang, Z.F.; Xia, H.M.; Wang, Z.P.; Thang, V.S.; Zhong, Z.W.; De Rooij, N.F. Investigation of hot roller embossing for microfluidic devices. *J. Micromech. Microeng.* **2009**, *20*, 837–854. [CrossRef]
24. Schulze, J.; Ehrfeld, W.; Loewe, H.; Michel, A.; Picard, A. Contactless embossing of microlenses: A new technology for manufacturing refractive microlenses. *Proc. SPIE* **1997**, 89–98. [CrossRef]
25. Schulze, J.; Ehrfeld, W.; Mueller, H.; Picard, A. Compact self-aligning assemblies with refractive microlens arrays made by contactless embossing. *Proc. SPIE* **1998**. [CrossRef]
26. Xie, D.; Chang, X.; Shu, X.; Wang, Y.; Ding, H.; Liu, Y. Rapid fabrication of thermoplastic polymer refractive microlens array using contactless hot embossing technology. *Opt. Express* **2015**, *23*, 5154–5166. [CrossRef] [PubMed]
27. Moore, S.; Gomez, J.; Lek, D.; You, B.H.; Kim, N.; Song, I.H. Experimental study of polymer microlens fabrication using partial-filling hot embossing technique. *Microelectron. Eng.* **2016**, *162*, 57–62. [CrossRef]
28. Ho, K.H.; Newman, S.T. State of the art electrical discharge machining (EDM). *Int. J. Mach. Tool Manuf.* **2003**, *43*, 1287–1300. [CrossRef]
29. Abbas, N.M.; Solomon, D.G.; Bahari, M.F. A review on current research trends in electrical discharge machining (EDM). *Int. J. Mach. Tool Manuf.* **2006**, *47*, 1214–1228. [CrossRef]
30. Worgull, M.; Xe, H.; Tu, J.F.; Kabanemi, K.K.; Heckele, M. Hot embossing of microstructures: Characterization of friction during demolding. *Microsyst. Technol.* **2008**, *14*, 767–773. [CrossRef]
31. Guo, Y.; Liu, G.; Zhu, X.; Tian, Y. Analysis of the demolding forces during hot embossing. *Microsyst. Technol.* **2007**, *13*, 411–415. [CrossRef]
32. Zhao, F.; Li, H.; Ji, L.; Wang, Y.; Zhou, H.; Chen, J. Ti-DLC films with superior friction performance. *Diam. Relat. Mater.* **2010**, *19*, 342–349. [CrossRef]
33. Male, A.T.; Depierre, V. The Validity of Mathematical Solutions for Determining Friction from the Ring Compression Test. *J. Tribol.* **1970**, *92*, 389. [CrossRef]
34. Yao, Z.; Mei, D.; Chen, Z. A Friction Evaluation Method Based on Barrel Compression Test. *Tribol. Lett.* **2013**, *51*, 525–535. [CrossRef]
35. Zhu, Y.; Zeng, W.; Ma, X.; Tai, Q.; Li, Z.; Li, X. Determination of the friction factor of Ti-6Al-4V titanium alloy in hot forging by means of ring-compression test using FEM. *Tribol. Int.* **2011**, *44*, 2074–2080. [CrossRef]
36. Sarhadi, A.; Hattel, J.H.; Hansen, H.N. Evaluation of the viscoelastic behaviour and glass/mold interface friction coefficient in the wafer based precision glass molding. *J. Mater. Process. Technol.* **2014**, *214*, 1427–1435. [CrossRef]
37. Xu, B.; Wu, X.Y.; Lei, J.G.; Cheng, R.; Ruan, S.C.; Wang, Z.L. Laminated fabrication of 3D queue micro-electrode and its application in micro-EDM. *Int. J. Adv. Manuf. Technol.* **2015**, *80*, 1701–1711. [CrossRef]
38. Xu, B.; Wu, X.Y.; Lei, J.G.; Zhao, H.; Liang, X.; Cheng, R.; Guo, D.J. Elimination of 3D micro-electrode's step effect and applying it in micro-EDM. *Int. J. Adv. Manuf. Technol.* **2018**, *96*, 1–10. [CrossRef]
39. Williams, D.; Landel, R.F.; Ferry, J.D. The temperature dependance of relaxation mechanisms in amorphous polymers and other glass form liquids. *J. Am. Chem. Soc.* **1955**, *77*, 3701–3707. [CrossRef]
40. Cheng, E.; Yin, Z.; Zou, H.; Jurčíček, P. Experimental and numerical study on deformation behavior of polyethylene terephthalate two-dimensional nanochannels during hot embossing process. *J. Micromech. Microeng.* **2013**, *24*, 5004. [CrossRef]

41. Omar, F.; Brousseau, E.; Elkaseer, A.; Kolew, A.; Prokopovich, P.; Dimov, S. Development and experimental validation of an analytical model to predict the demolding force in hot embossing. *J. Micromech. Microeng.* **2014**, *24*, 55007. [CrossRef]

42. He, Y.; Fu, J.Z.; Chen, Z.C. Research on optimization of the hot embossing process. *J. Micromech. Microeng.* **2007**, *17*, 2420. [CrossRef]

Article

Chatter Identification of Three-Dimensional Elliptical vibration Cutting Process Based on Empirical Mode Decomposition and Feature Extraction

Mingming Lu [1], Bin Chen [1], Dongpo Zhao [1], Jiakang Zhou [1], Jieqiong Lin [1,*], Allen Yi [2] and Hao Wang [1]

[1] Key Laboratory of Micro-Nano and Ultra-precision Manufacturing of Jilin Province,
School of Mechatronic Engineering, Changchun University of Technology, Changchun 130012, China;
lumm@ccut.edu.cn (M.L.); chen0202130612@163.com (B.C.); dpzhao1@163.com (D.Z.);
zhoujiakang07@163.com (J.Z.); wanghao9439@163.com (H.W.)

[2] Department of Industrial, Welding and Systems Engineering, Ohio State University,
210 Baker Systems Building, 1971 Neil Avenue, Columbus, OH 43210, USA; yi.71@osu.edu

* Correspondence: linjieqiong@ccut.edu.cn; Tel.: +86-13756067918

Received: 13 November 2018; Accepted: 14 December 2018; Published: 21 December 2018

Abstract: Three-dimensional elliptical vibration cutting (3D-EVC) is one of the machining methods with the most potential in ultra-precision machining; its unique characteristics of intermittent cutting, friction reversal, and ease of chip removal can improve the machinability of materials in the cutting processes. However, there is still not much research about the chattering phenomenon in the 3D-EVC process. Therefore, based on the empirical mode decomposition (EMD) technique and feature extraction, a chatter identification method for 3D-EVC is proposed. In 3D-EVC operations, the vibration signal is collected by the displacement sensors and converted to frequency domain signal by fast Fourier transform (FFT). To identify tool cutting state using the vibration frequency signal, the vibration signals are decomposed using empirical mode decomposition (EMD), a series of intrinsic mode functions (IMFs), so the instantaneous frequency can be reflected by the vibration signals at any point. Then, selecting the primary IMFs which contain rich chatter information as the object in feature extraction identification, and two identification indexes, that is, the mean square frequency and self-correlation coefficient, are calculated for the primary IMFs by MATLAB software, to judge the chatter phenomenon. The experimental results showed that the mean square frequency and self-correlation coefficient of the three cutting states increase with the increase in the instability of the cutting state. The effectiveness of the improved chatter recognition method in 3D-EVC machining is verified.

Keywords: chatter identification; three-dimensional elliptical vibration cutting; empirical mode decomposition; intrinsic mode function; feature extraction

---

1. Introduction

Three-dimensional elliptical vibration cutting (3D-EVC) is an ultra-precision cutting technology with significant development potential. Its unique characteristics of intermittent cutting, friction reversal, and ease of chip removal can improve the machinability of difficult-to-cut materials in the cutting process [1–3]. Since 3D-EVC technology was first introduced in 2005, the research field has mainly focused on the mechanism of cutting, the design of 3D-EVC apparatus, path planning of the tool, specific applications, and so on. [4,5]. At present, how to study the chatter phenomenon in the process of non-resonance 3D-EVC machining is an important problem. In the ultra-precision cutting technology processes, the main reasons that the effect of unstable vibration could be reduced is by the increase of the spindle speed, small cutting depth, and great performance of the diamond tool. But the damage accompanying chatter is mainly reflected in reducing the accuracy of the work-piece of the optical parts, increasing the surface roughness of the work-piece, speeding up the wear on the cutting tools, and causing damage to the spindle of the machine tool [6,7].

EVC is an intermittent cutting technology, which has the characteristics of friction reversal, tool wear suppression, and easy chip removal. However, there is no specific literature on the identification of chatter in the 3D-EVC cutting processes [8]. In 2011, Ma et al. proposed that the two-dimensional EVC technology had a certain inhibitory effect in chatter suppression compared to the traditional cutting method, and the effectiveness of the effect of chatter suppression was verified by experiment, but this was limited to two-dimensional elliptical vibration cutting [9]. In 2016, Jung et al. collected the real-time signal from the 2D-EVC process and analyzed the chatter phenomenon according to the frequency domain analysis and surface of the work-piece [10]. They obtained the friction chatter mechanism of the plough force during the initial period of each ellipse movement cycle and suppressed the chatter vibration by changing the guiding angle in 2D-EVC process. It was also the earliest research on identification and suppression of the chatter phenomenon in the 2D-EVC cutting processes. Therefore, exploring a new method to identify the chattering phenomenon applied to the 3D-EVC processes is critical to improving the machining accuracy of the 3D-EVC technology.

In the field of ultra-precision cutting, research on chatter is focused on the influence of undesired vibration, the chatter identification and the modeling method of chatter mechanism, and so on. Chatter is a poor vibration phenomenon in metal processing, which has effects on the machined work-piece surface quality, reduced cutting efficiency and productivity, as well as producing a lot of noise, and so on. The mechanical vibration has three forms: free vibration, forced vibration, and self-excited vibration. The free vibration is caused by impact during the machining process. The forced vibration is mainly caused by the unbalance effect of the machine tool components. The free vibration and forced vibration are relatively easy to identify and eliminate. However, the self-excited vibration is a significant factor affecting cutting stability. Zhang et al. proposed the physical modes of vibration, in ultra-precision machining, and the effect on the generated surface is related to the many factors, including the properties of materials, cutting conditions and the relative vibration between tool and work-piece [11]. Chen et al. presented the mathematical models for modeling and analyzing the vibration and surface roughness in precision turning, and also established a relationship model between tool vibration and surface profile and roughness in the precision end-milling process [11,12]. A simple method for the detection of milling chatter was presented by Zhang. The assessment of milling process stability by recursive drawing method combined with Hilbert Huang transform was proposed by Rafal [13]. In addition, with the development of the sensor and the modern signal processing technology, chatter identification technology has been greatly improved. Various signals have been applied in chatter identification, such as cutting force [14–16], sound [17] and the acceleration [18,19]. Cao et al. presented an effective chatter identification method based on the two advanced signals processing techniques, wavelet package transform (WT) and Hilbert-Huang transform (HHT), and the experimental results proved that the method can identify the chatter effectively [20]. In 2015, Cao et al. adopted a self-adaptive analysis method named ensemble empirical mode decomposition (EEMD), extracted two nonlinear indices as the indices of chatter symptoms, and presented a method for identifying the chattering phenomenon based on the two points in the end milling process. The results have shown that the mean value and standard deviation increase with the cutting state's instability [21,22].

The vibration signals belong to unsteady and nonlinear signals in the 3D-EVC process [23], traditional time-frequency analysis methods can analyze the global mean of the signal in the time-frequency domain, but it cannot analyze the feature information about the local area. In order to solve the lack of self-adaptability or poor adaptability in traditional time-frequency analysis methods, Huang et al. proposed a novel and adaptive decomposition method, which was named empirical mode decomposition (EMD), in 1998 [24]. Liu et al. decomposed the motor current signal into intrinsic mode functions (IMFs) and extracted the energy index and kurtosis index based on those IMFs for chatter detection [21]. Compared with the traditional time-frequency analysis, EMD is more appropriate for analyzing the non-stationary and nonlinear signals, as the EMD method can not only yield clear distribution for instantaneous frequency and amplitude, but also reflect the time-variation of the total energies, which is significant for the non-stationary and nonlinear signals. In recent years, the EMD method has been applied to signal processing, fault diagnosis [25] and pattern recognition [25,26].

Based on the above research, this paper presents chatter identification based on the empirical mode decomposition (EMD) technique and feature extraction. The method of chatter identification developed is divided into two parts: initial identification and feature extraction identification. In 3D-EVC operations, the vibration signals are collected by displacement sensors and converted to frequency domain signals by fast Fourier transform (FFT), and the cutting state of the whole machining system is determined according to the change of frequency. The vibration signals are decomposed by EMD, a series of intrinsic mode functions (IMFs), so the instantaneous frequency can be reflected by the vibration signals at any point. Selecting the primary IMFs which contain rich chatter information as the object in feature extraction identification, two identification indexes, that is, the mean square frequency (MSF) and self-correlation coefficient (ρ_1), were calculated from the primary IMFs using MATLAB software to judge the chatter phenomenon. In addition, chatter identification is organized as follows. First step is the fast Fourier transform (FFT) analysis of the time domain signal to observe whether the frequency changes during the cutting process. If there is a frequency domain mutation, it indicated that chatter occurs during the cutting process. The second step is eigenvalue identification, which serves two purposes. One needs to test the chatter identification result of the initial step by using two distinct eigenvalue changes during the cutting process. More importantly, the eigenvalue can reflect the state change of the cutting process to a certain extent, which is used as a theoretical basis for the online identification of chatter.

This paper is structured as follows. Section 2 briefly introduces the signal processing method empirical mode decomposition (EMD), and describes the theory and strategy of the chatter identification. Then experimental setup and the cutting parameter setting are described in Section 3. The results and discussions of the chatter identification method developed are given in Section 4. Finally, the conclusions are given in Section 5.

2. Theory of Chatter Identification

2.1. Empirical Mode Decomposition

With the rapid development of the time-frequency analysis, some signal processing technologies (such as short time Fourier transform, Wigner-Ville distribution and Wavelet transform) have the ability to analyze non-stationary and nonlinear signals. However, the methods of traditional time-frequency analysis have some limitations during analysis of non-stationary and nonlinear signals, such as a lack of adaptability. Compared with the methods of traditional time-frequency analysis, empirical mode decomposition (EMD) is more appropriate for analyzing the non-stationary and nonlinear signals. The EMD method can obtain clear instantaneous frequency and amplitude distribution and can reflect the time variation of total energy. This is very important for non-stationary and nonlinear signals. During the process of EMD, the vibration signal is processed smoothly, the fluctuations under different scales and the trend of the change are decomposed step by step, a series of data columns are generated with different characteristic scales, named the intrinsic mode functions (IMFs). The IMF obtained by the EMD is the approximate single-frequency signal which exists only at one frequency in every moment. It contains the real physical process in the vibration signals. In this method, the chatter phenomenon can be identified correctly by extracting the sensitive characteristic undesired vibration signal in each IMF.

In the process of EMD, if the local mean value is set to zero and has locally symmetric properties for a function or signal, then the essential condition of the mean instantaneous frequency is the same number of extreme points and zero-crossing points. Therefore, Huang et al. proposed the compact of the intrinsic mode function (IMF). These mode functions must satisfy the following two conditions:

(1) The number of extreme points and the number of zero-crossing points must be either equal or differ at most by one in the whole dataset.
(2) The local upper and lower envelope at any data point is symmetrical, which means the envelope of the local minimum value and the maximum value is zero.

The IMF reflects the inherent oscillatory of non-stationary and nonlinear signals, and there is only first -order oscillation mode in each of cycles. The phenomenon of mode mixing is nothingness. A typical of the IMF have the same extreme points and the number of zero-crossing points, at the same time, the upper and lower envelope is symmetrical about the time-axis, just exist a single frequency component at any time, as shown in Figure 1.

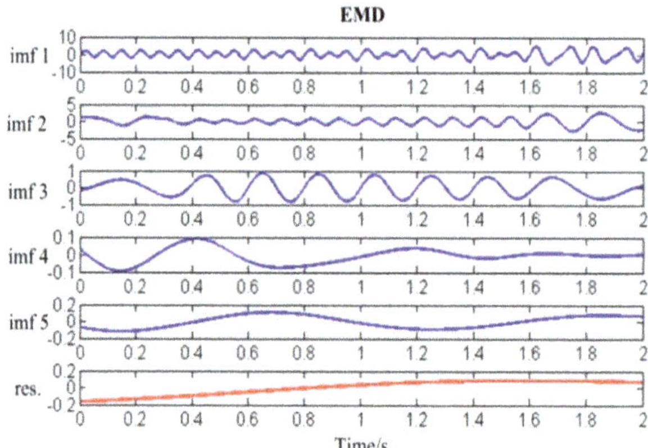

Figure 1. A series of intrinsic mode functions (IMF) obtained using empirical mode decomposition (EMD).

In addition, the EMD method which decomposes the signal is based on the following three assumptions:

(1) The signal has two extreme points at least—a maximum point and a minimum point.
(2) The characteristic time scale is defined as the time interval of the adjacent extreme points.
(3) If the signal does not have the extreme point but an inflection point exists, then derivation is performed one or more times to obtain the extreme points before decomposing the signal, and the results can be obtained by integrating the corresponding component.

The following is the brief description of the EMD algorithm:

First, calculating all local extreme point for a given signal $X(t)$, and forming the upper envelope by adopting the line cubic spline curve, and the lower envelope line also formed by the same way. The value of difference between $X(t)$ and the mean m_1 of upper (E1) and lower (E2) envelope is h_1. Then the new data h_1 are obtained as shown as follows:

$$h_1 = X(t) - m_1 \tag{1}$$

Regard h_1 as the new signal $X(t)$, and repeat the above steps until h_i meets the two required conditions of IMFs, where C_1 represents the first order IMF sifting from the original signal. In general, the first order IMF contains the highest frequency component of the signal.

Then obtain a difference signal r_1 which removes the high frequency component after separated C_1 from the original $X(t)$:

$$r_1 = X(t) - C_1 \tag{2}$$

where r_1 is regarded as the new signal, and repeat the sifting process until the mean line between the upper and lower envelope line is close to zero at any point, denoted by r_n:

$$r_n = r_{n-1} - C_n \tag{3}$$

Finally, $X(t)$ could be represented as the sum of IMFs and a residual function:

$$X(t) = \sum_{j=1}^{n} C_j(t) + r_n(t) \tag{4}$$

where $r_n(t)$ is a residual function which represents the average trend of the signal. The IMF component $C_j(t)$ represents the signal components from high to low in different frequency, respectively. The flowchart of the EMD algorithm as follows in Figure 2.

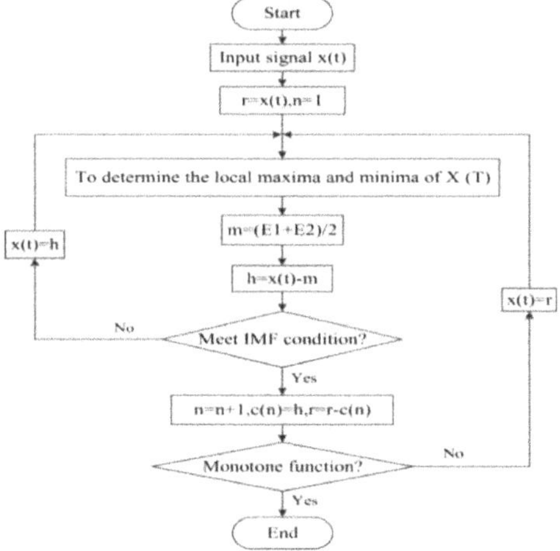

Figure 2. Flowchart of EMD algorithm.

2.2. Principle of Chatter Identification

Because of the inevitability of vibration in the cutting process, the displacement signals collected in the 3D-EVC processes are actually nonlinear and non-stationary signals. In the process of 3D-EVC, the selection of the chatter signal is significant for correctly identifying instability phenomena. In the process of the specific chatter identification experiment, the signal selected during the 3D-EVC process should follow the following two points: (1) the change of the state in the cutting process can be reflected from two aspects of the time domain and the frequency domain; and (2) the assembly of the sensor cannot affect the stability of the whole cutting system.

In the traditional machining process, the common signals applied in chatter identification are mainly the tool tip vibration displacement signal, the cutting force signal, the sound signal and the cutting acceleration signal. However, 3D-EVC technology is an ultra-precision manufacturing technology, and the cutting process has high precision as the small cutting rate, high cutting speed, and low cutting depth. In addition, the displacement sensor is easy to assemble with the tool without affecting the whole cutting system characteristics with high resolution. Therefore, the displacement sensor is used to measure the vibration and displacement signals of the tool tip as the main signal for chatter identification in the process of 3D-EVC.

In the process of 3D-EVC, the vibration frequency under a stable machining state is the frequency of the tool tip vibration set in the 3D-EVC apparatus. In order to avoid the phenomenon of over cutting or not cutting during the process, the vibration frequency of tool ellipse is relatively low. In this paper, the vibration frequency of the tool tip in the three directions is set as 40 Hz, so the chatter frequency is higher than the original vibration frequency of the tool. When the cutting process becomes unstable during processing, the phenomenon of frequency shift will appear and the unstable frequency of chatter phenomenon during the cutting process will exceed the elliptical vibration frequency applied to the diamond tool in the non-resonant 3D-EVC apparatus. On the other hand, another important factor in the chatter identification strategy is the selection of chatter characteristics. There are many methods for extracting the features of the chatter state in the time domain and frequency domain. The mean square frequency (MSF) is an ideal characteristic value which could reflect the changing situation of the cutting signal in the frequency domain. MSF represents the weighted mean of the square of the vibration frequency, and the weight is the amplitude of the power spectrum. In the cutting process, the machining frequency is not a constant value due to the chatter appearance as a result of the change of cutting parameters, tool wear, the properties of the work-piece material and the processing mechanism under different processing methods, and so on. The chatter phenomenon will appear and the frequency of chatter gradually increase and exceed the elliptical vibration frequency applied to the diamond tool in the non-resonant 3D-EVC apparatus with the change of the cutting parameter. The formula of MSF is shown as the following:

$$MSF = \sum_{i=2}^{N} x_i^2 / 4\pi^2 \sum_{i=1}^{N} x_i^2 \tag{5}$$

In addition, the other parameter used for chatter identification is the one step self-correlation coefficient ρ_1. In general, ρ_1 has a consanguineous relationship with the variety of energy in vibration signals, the value of ρ_1 will reduce suddenly when the energy concentrates around the chatter frequencies when the amplitude of vibration signals becomes uneven with the chatter frequency. The value of ρ_1 can be expressed as:

$$\rho_1 = \cos 2\pi f_i \Delta \tag{6}$$

where Δ is the sampling interval, the value of ρ_{1i} decreased with the increase of f_i.

If the signal contains a variety of harmonic components, then the value of ρ_1 can be expressed as:

$$\rho_1 = \frac{\sum_{i=1}^{n} a_i^2 \cos 2\pi f_i \Delta}{\sum_{i=1}^{n} a_i^2} = \frac{\sum_{i=1}^{n} a_i^2 \rho_{1i}}{\sum_{i=1}^{n} a_i^2} \tag{7}$$

In general, the conventional fast calculation method is shown as the following:

$$MSF = \frac{D}{4\pi^2 B} \tag{8}$$

$$\rho_1 = \frac{C - A^2}{B - A^2} \tag{9}$$

where, $A = \sum_{i=0}^{N} x_i$, $B = \sum_{i=0}^{N} x_i^2$, $C = \sum_{i=0}^{N} x_i x_{i-1}$, $D = \sum_{i=0}^{N} \dot{x}_i^2$, respectively. $x_i (i = 0, 1, 2, ..., N)$ represents the sample data, $\dot{x}_i = \frac{x_i - x_{i-1}}{\Delta} (i = 1, 2, ..., N)$ is one order difference, $N + 1$ and Δ represents the sampling point and the sampling interval, respectively.

Therefore, synthesizing two characteristic parameters of MSF and ρ_1, which represents the characteristics of vibration signal in the frequency domain, the changes of vibration states can be reflected accurately.

2.3. The Flow Chart of Chatter Identification

The flow chart of chatter identification in the 3D-EVC process is illustrated in Figure 3. Firstly, the driving signal is generated by Power PMAC controller (Delta Tau Data Systems, Inc., USA), and used for driving the piezoelectric stack in the cutting tool with the elliptical trajectory in three-dimensional space. The different vibration states can be obtained by changing the cutting parameters (the depth of cut, spindle speed or the feed rate). The vibration signal of the diamond tool in the processing of 3D-EVC was collected by the displacement sensors, and converted to frequency domain signal by fast Fourier transform (FFT), and the cutting state is determined (stable cutting, transient state and chatter state) according to the variation of frequency. On the other hand, the method of EMD is used for the vibration signals under various vibration states, selecting the sensitive IMFs which including the rich chatter information and calculate the value of MSF and ρ_1 corresponding to three cutting states. Finally, the chatter phenomenon in the 3D-EVC process can be identified correctly based on the two aspects of change characteristics.

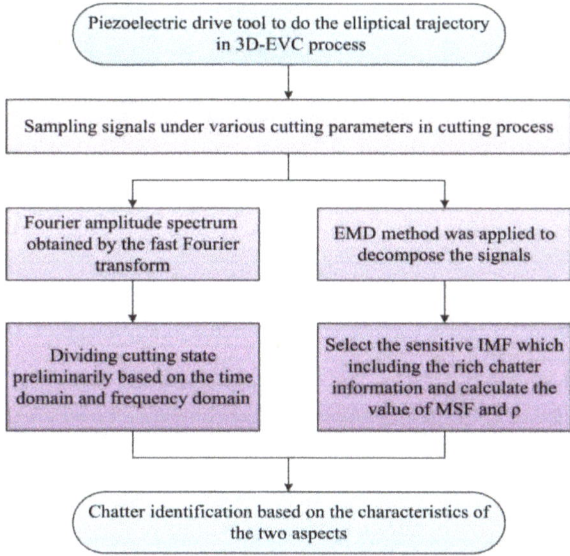

Figure 3. Flow chart of chatter identification in the three-dimensional elliptical vibration cutting (3D-EVC) process.

3. Experimental Setup

The proposed chatter identification method has been carried out in an ultra-precision machining machine, as shown in Figure 4.

A copper bar is chosen as the work-piece and installed on the air spindle of the ultra-precision machining machine. The capacitive micro-displacement sensors with four measurement channels (Micro-sense DE 5300-013) are mounted on the handle of the apparatus to measure the vibration signals in three directions during cutting, the vibration signals are collected and then transmitted to the computer, which is used for data storage and signal processing. In addition, the power amplifier (PI, E-500), which has a criterion amplification factor 1060.1, is employed to amplify the driving signal. The Power PMAC controller generates the driving signals to drive the piezoelectric stack in the apparatus. The apparatus of 3D-EVC is driven by the piezoelectric hybrid, which, given the sinusoidal excitation signal, can be expressed as:

$$\begin{cases} x = A_1 sin(\omega_1 t + \varphi_1) \\ y = A_2 sin(\omega_2 t + \varphi_2) \\ z = A_3 sin(\omega_3 t + \varphi_3) \end{cases} \tag{10}$$

where, x, y and z are the given sinusoidal excitation signal in X, Y and Z directions of tool tip corresponding to the coordinate system in machine tools, respectively. A_1, A_2 and A_3 are the amplitudes of driving singles in three directions generated by three piezoelectric stacks, respectively. t represents the time. In addition, φ_1, φ_2 and φ_3 are the angular frequency imposed on the piezoelectric stacks, and ω_1, ω_2 and ω_3 are the phase of driving signal along X, Y and Z directions, respectively.

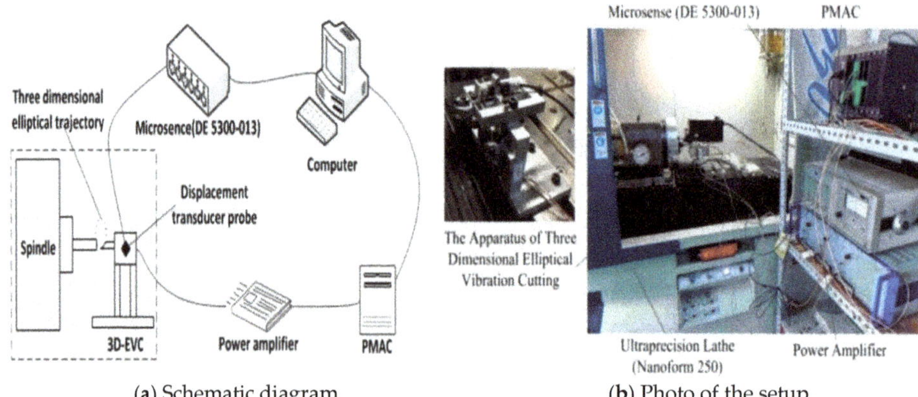

(**a**) Schematic diagram   (**b**) Photo of the setup

Figure 4. Experimental setup.

In order to compare the different cutting states during the cutting process, the value of spindle speed and the feed rate is fixed (90 rpm/min and 10 mm/min, respectively), the depth of cut is set to 5 μm, 10 μm and 20 μm, respectively. The cutting process was slotting with the oil mist cutting fluid. To avoid the phenomenon of overcut or the larger cutting marks between the two periods, the amplitude of piezoelectric driving signal is 6 μm. The detailed experimental conditions are listed in Table 1, and the main parameters of the material are listed in Table 2.

Table 1. Experimental conditions.

	Parameter	Value
Work-piece	Material	Copper (H62)
	Length (mm)	50
	Diameter (mm)	12.7
Diamond tool	Nose radius (mm)	0.2
	Nominal rake angle (°)	0
	Nominal clearance angle (°)	7
Cutting parameter	Depth of cut (μm)	5, 10, 20
	Spindle speed (rpm/min)	90
	Feed rate (mm/min)	10
	vibration amplitude (μm)	6

Table 2. Material properties.

Material	Tensile Strength Rm (N/mm^2)	Elongation to Failure/(>%)	Vickers Hardness	Density (g/cm^3)
Copper	335–370	24	85–128	8.27

4. Results and Discussions

4.1. Initial Identification

In the cutting process, there are two main characteristics to identify undesired vibrations: (1) the vibration amplitude increases gradually in the time domain as a result of the vibration energy increases; and (2) the frequency shift phenomenon will appear with the change of cutting the state in the frequency domain. This is one of the most significant signs to determine the occurrence of undesired vibration.

The time domain signal of three directions in the cutting process is collected by the displacement sensor, when the depth of cut is set to 5 μm, as shown in Figure 5. It can be seen from the time domain signals of three directions on the cutting tool that the vibration amplitudes along the Z direction have greater values compared to the other two directions. This is because that the oscillation generated on the tool tip of 3D-EVC apparatus in Z direction contains more capacity. Consequently, vibration signals in Z direction are considered as the research object and further analysis in this paper.

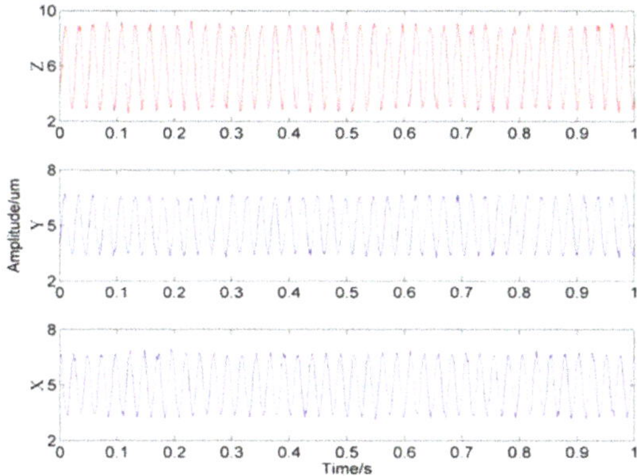

Figure 5. The time domain signal of three directions in 3D-EVC process (spindle speed was 90 rpm/min, and the feed rate was 10 mm/min, slotting.

In order to obtain the different cutting states in 3D-EVC process, the original signals of vibration in Z direction are obtained under various cutting conditions (the depth of cut was 5 μm, 10 μm and 20 μm, respectively). The time domain signals in three kinds of cutting depth are shown in Figure 6a–c, the amplitude of vibration signals in frequency domain obtained by FFT are shown in Figure 7a–c, respectively.

Three typical states (stable cutting, transient state, and chatter state) in the cutting process of 3D-EVC can be obtain based on the variation of frequency with the varied of cutting depth. In the stable condition of the cutting process, the amplitude of vibration signal is relatively small and there is no larger peak or trough in Figure 6a; the frequency domain spectrum obtained by FFT is shown in Figure 7a. It is observed that the characteristics of the peaks appear at the frequency about 40 Hz corresponding to the natural frequency of 3D-EVC apparatus. In the condition of a transient state, some smaller fluctuations appeared in the time domain, and the amplitude of vibration grows, with a small change in Figure 6b, while the frequency domain spectrum has a significant change which appears as two peaks in 150 Hz and 200 Hz, as shown in Figure 7b. But the amplitude is relatively small and there is no obvious change in the time domain. In the condition of a chatter state, the amplitude of vibration increases obviously, the change of amplitude has an effect on the three-dimensional elliptic trajectory of the tool-tip in Figure 6c. Moreover, the frequencies of the characteristic peaks at 150 Hz and 200 Hz are formed by the major components in the frequency spectrum, as shown in Figure 7c. In this stage, the cutting process is accompanied by harsh noise, and the work-piece appears obvious vibration mark. This is caused by the extrusion pressure between the tool-tip and the work-piece increasing with the rise of cutting depth.

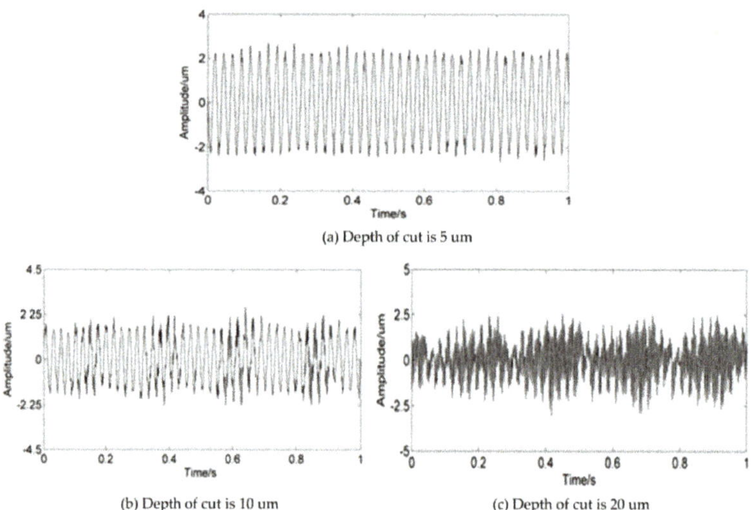

Figure 6. Vibration displacement signal obtained in 3D-EVC process (spindle speed was 90 rpm/min, the feed rate was 10 mm/min, misting).

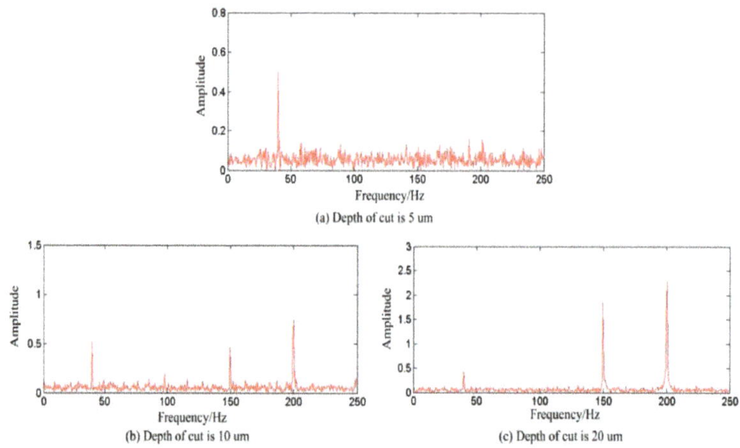

Figure 7. Frequency obtained by fast Fourier transform (FFT).

On the other hand, the cutting condition also affects the quality of the finished surface. Figure 8 shows quality of the finished surface under the three different conditions during the 3D-EVC process, respectively. The measurements of the surface roughness for the finished surface are implemented by using the ZYGO-Newview 8200, a white light interferometer, and the surface roughness R_a of the finished surface is used to estimate the three cutting states under various cutting conditions. In the condition of stable cutting, as shown in Figure 8a, the finished surface has a clear and regular elliptical trajectory, owing to the small cutting parameters in 3D-EVC process, and the value of the surface roughness R_a = 0.046 μm. In the condition of a transient state, with the increase of cutting depth, the cycle of each elliptical motion starts to appear slight chatter marks, owing to the additional components corresponding to undesired vibration frequencies (150 Hz and 200 Hz) as shown in Figure 8b. Meanwhile, the value of surface roughness also increased with the augment of cutting depth R_a = 0.245 μm. In the condition of a chatter state, the elliptical motion trajectory becomes irregular in the finished surface, as shown in Figure 8c, the undesired vibration causes the discontinuous contact between the tool tip and the work-piece in the cycle of each elliptical motion, and the intermittent chopping-elliptical-vibration cutting is formed in the process of 3D-EVC, which corresponds to the large value of surface roughness R_a = 0.853 μm.

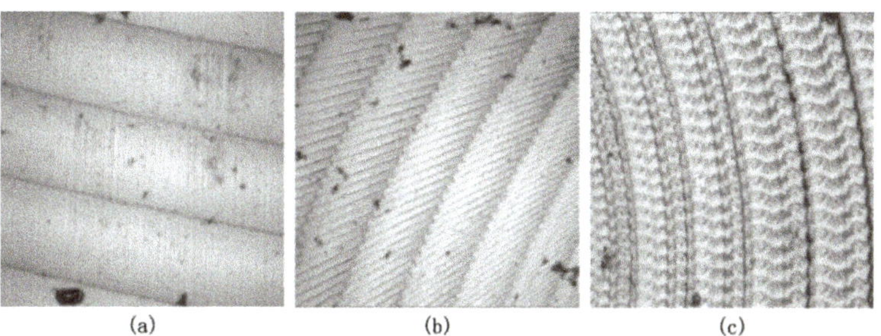

(a)   (b)   (c)

Figure 8. The quality of finished surface (**a**) Stable cutting (**b**) Transient state (**c**) Chatter state.

4.2. Feature Extraction Identification

In order to identify the chatter phenomenon accurately in the 3D-EVC process, feature extraction identification, or extracting the features of the chatter symptoms from the various cutting parameters, was proposed. Firstly, three groups of the original signals are decomposed with the EMD and a series of IMFs in stable cutting, transient and chatter state, are obtained. Then, the mean square frequency (MSF) and one step self-correlation coefficient ρ_1 of each IMF, which processes the primary energy in the 3D-EVC process, is calculated as the criteria for chatter identification. In this paper, the first six IMFs which contain mainly vibration information are selected, as shown in Figures 9–11, respectively.

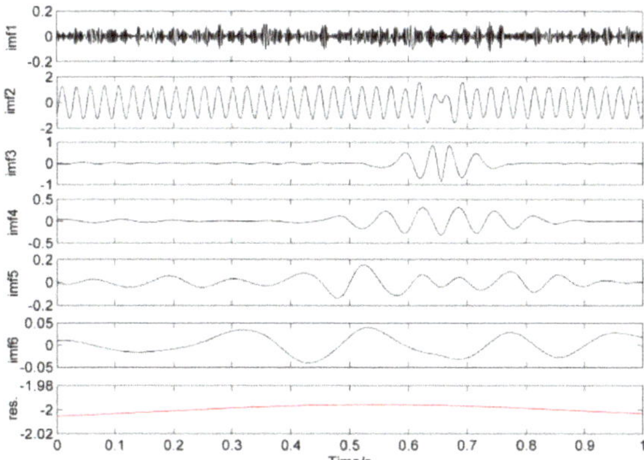

Figure 9. The result of EMD in stable cutting (spindle speed is 90 rpm/min, the feed rate is 10 mm/min, and the depth of cut is 5 μm).

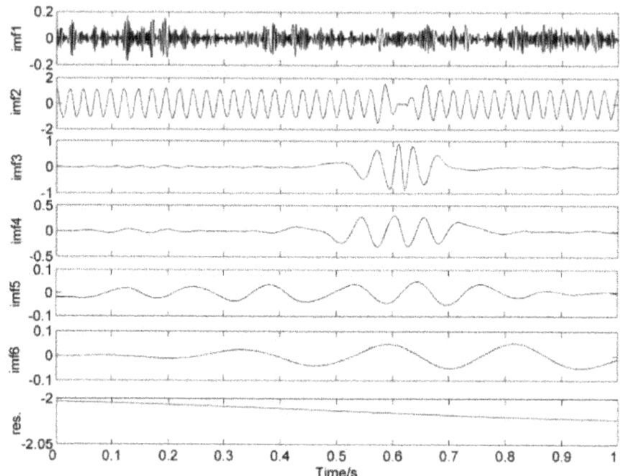

Figure 10. The result of EMD in transient state (spindle speed is 90 rpm/min, the feed rate is 10 mm/min, and the depth of cut is 10 μm).

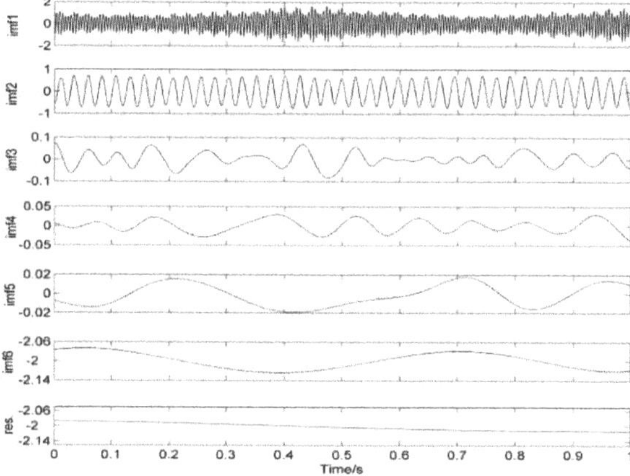

Figure 11. The result of EMD in chatter state (spindle speed is 90 rpm/min, the feed rate is 10 mm/min, and the depth of cut is 20 μm).

The values of MSF and ρ_1 are calculated to judge the development trend of the vibration signals, and regarded as the index of the criterion for the undesired vibration in the 3D-EVC process, as listed in Table 3. As mentioned in the third section, MSF represents the weighted mean of the square value of the vibration frequency. When the undesired vibration happens suddenly, the frequency-shifting phenomenon will appear and the chatter frequency gradually increases. The value of MSF also increased with the cutting state growing fluctuant. In addition, ρ_1 also has consanguineous relationship with the variety of the energy in vibration signals. The value of ρ_1 will reduce suddenly when the energy concentrates around the chatter frequencies, as a result of the uneven amplitude of vibration signals. In a stable cutting process, the values of MSF and ρ_1 are 210 and 0.25, this value represents a relatively stable state in the cutting process. For the transient state, the value of MSF increases to 270, moreover, the value of ρ_1 is reduced to 0.57. In the chatter state, the value of MSF increases to 6800, this increased value is explained by the fact that the chatter frequency has turned into a significant component in the 3D-EVC process. However, such as in the above theoretical study, the value of ρ_1 is reduced to 0.29, as the correlation between the vibration signals had reduced. Therefore, the numerical variation of MSF and ρ_1 can be applied to an effective criterion, which identifies the change of cutting state in the 3D-EVC process.

Table 3. The value of mean square frequency (MSF) and ρ_1 under the three cutting conditions.

Cutting Conditions	MSF	ρ_1
Stable cutting	210	0.95
Transient state	270	0.57
Chatter state	6800	0.29

5. Conclusions

In this paper, a new chatter identification method based on signal processing was proposed to identify chatter phenomenon during the 3D-EVC process. The 3D vibration signals of the tool-tip are obtained by the displacement sensor, which is installed in the 3D-EVC apparatus, and the vibration signals were chosen as the object of decomposition, based on the visualized time and frequency domain data in the Z direction. The method of chatter identification developed is divided into two parts: initial identification and feature extraction identification. The amplitude spectrum obtained by FFT was used as a preliminary criterion for chatter identification. Then, the EMD method was employed to decompose the vibration signal including the rich undesired vibration information, the mean square frequency, and the one-step self-correlation coefficient, which were calculated as the identification criteria for identifying the chatter in the 3D-EVC process. The experimental results show:

(1) The vibration signals obtained by the displacement transducer and the amplitude spectrum obtained by FFT have reflected the trend of vibration state in the 3D-EVC process from the time and the frequency domain, respectively. The frequency shift phenomenon appeared with the vibration energy increasing gradually in the machining processes.

(2) Both the mean square frequency and the one step self-correlation coefficient have been changed suddenly in the cutting process. The value of the mean square frequency increases from 210 to 6800, in contrast, the value of the one step self-correlation coefficient reduces from 0.95 to 0.29. Therefore, these two parameters can be used as the index of the chatter phenomenon.

In future work, the effects of model parameters will be theoretically analyzed with stability lobe theory. Moreover, a chatter suppression method should be developed by combining sensorless chatter detection with an optimization algorithm.

Author Contributions: J.L., B.C. and M.L. conceived and designed the experiments; B.C., J.Z., and D.Z. performed the experiments; B.C., A.Y. and H.W. analyzed the data; J.L. and M.L. contributed reagents/materials/analysis tools; B.C. wrote the paper; M.L., B.C., J.Z. and D.Z. reviewed and revised the paper.

Funding: This research was funded by [Ministry of Science and Technology State Key Support Program] grant number [2016YFE0105100], [Micro-Nano and Ultra-Precision Key Laboratory of Jilin Province] grant number [20140622008JC], [Science and Technology Development Projects of Jilin Province] grant number [20180101034JC, 20180201052GX], [Education Department Scientific Research Planning Project of Jilin Provincial] grant number [JJKH20181038KJ] And the APC was funded by [20180101034JC].

Acknowledgments: Special thanks to the experimental installation provided by the Key Laboratory of Micro-Nano and Ultra-precision Manufacturing of Jilin Province.

Conflicts of Interest: The authors declare no conflict of interest.

References

1. Shamoto, E.; Moriwaki, T. Study on elliptical vibration cutting. *CIRP Ann. Manuf. Technol.* **1994**, *43*, 35–38. [CrossRef]
2. Shamoto, E.; Moriwaki, T. Ultaprecision diamond cutting of hardened steel by applying elliptical vibration cutting. *CIRP Ann. Manuf. Technol.* **1999**, *48*, 441–444. [CrossRef]

3. Lu, M.; Zhou, J.; Lin, J.; Gu, Y.; Han, J.; Zhao, D. Study on Ti-6Aql-4V alloy machining applying the non-resonant three-dimensional elliptical vibration cutting. *Micromachines* **2017**, *8*, 306. [CrossRef] [PubMed]
4. Shamoto, E.; Suzuki, N.; Tsuchiya, E.; Hori, Y.; Inagaki, H.; Yoshino, K. Development of 3 DOF ultrasonic vibration tool for elliptical vibration cutting of sculptured surfaces. *CIRP Ann. Manuf. Technol.* **2005**, *54*, 321–324. [CrossRef]
5. Lu, M.; Zhao, D.; Lin, J.; Zhou, X.; Zhou, J.; Chen, B.; Wang, H. Design and analysis of a novel piezoelectrically actuated vibration assisted rotation cutting system. *Smart Mater. Struct.* **2018**, *27*, 095020. [CrossRef]
6. Quintana, G.; Ciurana, J. Chatter in machining processes: A review. *Int. J. Mach. Tools Manuf.* **2011**, *51*, 363–376. [CrossRef]
7. Siddhpura, M.; Paurobally, R. A review of chatter vibration research in turning. *Int. J. Mach. Tools Manuf.* **2012**, *61*, 27–47. [CrossRef]
8. Shamoto, E.; Suzuki, N.; Hino, R. Analysis of 3D elliptical vibration cutting with thin shear plane model. *CIRP Ann. Manuf. Technol.* **2008**, *57*, 57–60. [CrossRef]
9. Ma, C.; Ma, J.; Shamoto, E.; Moriwaki, T. Analysis of regenerative chatter suppression with adding the ultrasonic elliptical vibration on the cutting tool. *Precis. Eng.* **2011**, *35*, 329–338. [CrossRef]
10. Zhang, S.J.; To, S.; Zhang, G.Q.; Zhu, Z.W. A review of machine-tool vibration and its influence upon surface generation in ultra-precision machining. *Int. J. Mach. Tools Manuf.* **2015**, *91*, 34–42. [CrossRef]
11. Chen, C.C.; Liu, N.M.; Chiang, K.T.; Chen, H.L. Experimental investigation of tool vibration and surface roughness in the precision end-milling process using the singular spectrum analysis. *Int. J. Adv. Manuf. Technol.* **2012**, *63*, 797–815. [CrossRef]
12. Chen, C.C.; Chiang, K.T.; Chou, C.C.; Liao, Y.C. The use of D-optimal design for modeling and analyzing the vibration and surface roughness in the precision turning with a diamond cutting tool. *Int. J. Adv. Manuf. Technol.* **2011**, *54*, 465–478. [CrossRef]
13. Liu, C.; Zhu, L.; Ni, C. The chatter identification in end milling based on combining EMD and W PD. *Int. J. Adv. Manuf. Technol.* **2017**, *91*, 3339–3348. [CrossRef]
14. Lamraoui, M.; Thomas, M.; El Badaoui, M.; Girardin, F. Indicators for monitoring chatter in milling based on instantaneous angular speeds. *Mech. Syst. Signal Process.* **2014**, *44*, 72–85. [CrossRef]
15. Tangjitsitcharoen, S.; Saksri, T.; Ratanakuakangwan, S. Advance in chatter detection in ball end milling process by utilizing wavelet transform. *J. Intell. Manuf.* **2015**, *26*, 485–499. [CrossRef]
16. Rafal, R.; Pawel, L.; Krzysztof, K.; Bogdan, K.; Jerzy, W. Chatter identification methods on the basis of time series measured during titanium superalloy milling. *Int. J. Mech. Sci.* **2015**, *99*, 196–207. [CrossRef]
17. Thaler, T.; Potočnik, P.; Bric, I.; Govekar, E. Chatter detection in band sawing based on discriminant analysis of sound features. *Appl. Acoust.* **2014**, *77*, 114–121. [CrossRef]
18. Hynynen, K.M.; Ratava, J.; Lindh, T.; Rikkonen, M.; Ryynänen, V.; Lohtander, M.; Varis, J. Chatter detection in turning processes using coherence of acceleration and audio signals. *J. Manuf. Sci. Eng.* **2014**, *136*, 044503. [CrossRef]
19. Lamraoui, M.; Thomas, M.; El Badaoui, M. Cyclostationarity approach for monitoring chatter and tool wear in high speed milling. *Mech. Syst. Signal Process.* **2014**, *44*, 177–198. [CrossRef]
20. Cao, H.; Lei, Y.; He, Z. Chatter identification in end milling process using wavelet packets and Hilbert–Huang transform. *Int. J. Mach. Tools Manuf.* **2013**, *69*, 11–19. [CrossRef]
21. Cao, H.; Zhou, K.; Chen, X. Chatter identification in end milling process based on EEMD and nonlinear dimensionless indicators. *Int. J. Mach. Tools Manuf.* **2015**, *92*, 52–59. [CrossRef]
22. Jung, H.; Hayasaka, T.; Shamoto, E. Mechanism and suppression of frictional chatter in high-efficiency elliptical vibration cutting. *CIRP Ann.* **2016**, *65*, 369–372. [CrossRef]
23. Vela-Martínez, L.; Jáuregui-Correa, J.C.; Rubio-Cerda, E.; Herrera-Ruiz, G.; Lozano-Guzmán, A. Analysis of compliance between the cutting tool and the workpiece on the stability of a turning process. *Int. J. Mach. Tools Manuf.* **2008**, *48*, 1054–1062. [CrossRef]
24. Huang, N.E.; Shen, Z.; Long, S.R.; Wu, M.C.; Shih, H.H.; Zheng, Q.; Yen, N.C.; Tung, C.C.; Liu, H.H. The empirical mode decomposition and the Hilbert spectrum for nonlinear and non-stationary time series analysis. *R. Soc.* **1998**, *454*, 903–995. [CrossRef]

25. Saidi, L.; Ali, J.B.; Fnaiech, F. Bi-spectrum based-EMD applied to the non-stationary vibration signals for bearing faults diagnosis. *ISA Trans.* **2014**, *53*, 1650–1660. [CrossRef] [PubMed]
26. Li, X.; Mei, D.Q.; Chen, Z.C. Feature extraction of chatter for precision holes boring processing based on EMD and HHT. *Opt. Precis. Eng.* **2011**, *19*, 1291–1297.

 *applied sciences*

Article

Theoretical Study of Path Adaptability Based on Surface Form Error Distribution in Fluid Jet Polishing

Yanjun Han [1,2], Lei Zhang [3,*], Cheng Fan [3], Wule Zhu [2] and Anthony Beaucamp [2]

1   School of Mechanical and Aerospace Engineering, Jilin University, Changchun 130025, China; hanyj14@mails.jlu.edu.cn
2   Department of Micro-Engineering, Kyoto University, Kyoto 606-8501, Japan; wule5033@gmail.com (W.Z.); beaucamp@me.kyoto-u.ac.jp (A.B.)
3   Robotics and Micro-Systems Research Center, Soochow University, Suzhou 215021, China; chfan@suda.edu.cn
*   Correspondence: zhanglei@jlu.edu.cn; Tel.: +86-186-2667-7495

Received: 28 August 2018; Accepted: 28 September 2018; Published: 3 October 2018

Abstract: In the technology of computer-controlled optical surfacing (CCOS), the convergence of surface form error has a close relationship with the distribution of surface form error, the calculation of dwell time, tool influence function (TIF) and path planning. The distribution of surface form error directly reflects the difference in bulk material removal depth across a to-be-polished surface in subsequent corrective polishing. In this paper, the effect of path spacing and bulk material removal depth on the residual error have been deeply investigated based on basic simulation experiments excluding the interference factors in the actual polishing process. With the relationship among the critical evaluation parameters of the residual error (root-mean-square (RMS) and peak-to-valley (PV)), the path spacing and bulk material removal depth are mathematically characterized by the proposed RMS and PV maps, respectively. Moreover, a variable pitch path self-planning strategy based on the distribution of surface form error is proposed to optimize the residual error distribution. In the proposed strategy, the influence of different bulk material removal depths caused by the distribution of surface form error on residual error is compensated by fine adjustment of the path spacing according to the obtained path spacing optimization models. The simulated experimental results demonstrate that the residual error optimization strategy proposed in this paper can significantly optimize the overall residual error distribution without compromising the convergence speed. The optimized residual error distribution obtained in sub-regions of the polished surface is more uniform than that without optimization and is almost unaffected by the distribution of parent surface form error.

Keywords: fluid jet polishing; deterministic polishing; variable pitch path; residual error optimization; path adaptability

---

1. Introduction

Whether it is in the high-end fields such as the defense industry, aviation and aerospace, or in the civilian fields of medical devices, mobile phones, cameras, etc., high-precision optical components are playing an increasingly important role. At the same time, there are increasingly higher requirements for surface form error, medium frequency error, surface roughness and subsurface damage of optical components [1]. With the introduction of aspherical and free form surface design methods, precision devices composed of optical components have been unified in terms of portability and functionality. However, this also poses a great challenge to the ultra-precision machining of optical components. Ultra-precision grinding and deterministic polishing are the main means of processing ultra-precision optical components [2]. Traditional manual polishing has the disadvantages of high labor intensity, poor working environment, being time consuming, high dependence on workers' experience, difficulty in controlling the consistency of quality, etc. [3]. Automated polishing is inevitably a trend to replace manual polishing. Computer-controlled optical surfacing technology (CCOS), first proposed by Itek. Inc. in 1970, has been widely used and developed in optical surfacing with high form accuracy [4,5].

In recent years, with the development of a series of advanced non-traditional sub-aperture CCOS polishing technologies such as bonnet polishing [6], fluid jet polishing [7], magnetorheological polishing [8], ion beam polishing [9], laser polishing [10], etc., the controllability of deterministic polishing has taken a new step, which provides more possibilities for the improvement of the correction effect of the surface form error. Among them, fluid jet polishing is an ultra-precise finishing technology that combines fluid mechanics, precision manufacturing and surface engineering. In this polishing technology, the fluid is used as a carrier for the abrasive particles suspended within, forming a flexible polishing head that acts on the workpiece surface with a good fit and gets rid of tool wear issues altogether. Therefore, this technology has a strong adaptability to complex free form shapes and is especially suitable for the ultra-precision polishing of optical lenses, mirrors and molds made of hard and brittle materials. In addition, fluid jet polishing in the general submerged polishing case and some non-submerged polishing cases with narrow sapphire nozzles (with a diameter less than 0.3 mm) can generate tool influence function (TIF) with a small radius and a Gaussian or near Gaussian distribution. These advantages make fluid jet polishing a strong technology to achieve precise correction of the surface form error.

In the technology of CCOS, the convergence of surface form error has a close relationship with the distribution of surface form error, calculation of dwell time, tool influence function (TIF) and path planning. In practice, after measurement of the workpiece to be polished, the resulting surface form error can be generally expressed as a continuously changing free form surface. The optimal distribution of residual error after deterministic polishing should be as close to an ideal plane as possible and ensure sufficient convergence accuracy. However, the distribution of surface form error directly brings a difference in bulk material removal depth across a to-be-polished surface in subsequent corrective polishing. After the surface has been polished with a traditional path with a fixed pitch, it can be seen that the residual error distribution after deterministic polishing is largely affected by the distribution of parent surface form error. In this paper, this phenomenon is identified as a genetic phenomenon of error distribution. This makes it difficult to ensure the consistency of the accuracy in the entire region of the workpiece after ultra-precision polishing, and it also brings difficulties for subsequent iterations of corrective polishing. Therefore, on the premise of not affecting the degree of convergence, after every corrective polishing iteration in the pre-polishing stage, the residual error distribution that is more uniform and almost free from the influence of the parent surface form error distribution should be obtained as much as possible, which is of great theoretical significance and engineering value.

The mid-spatial-frequency (MSF) error, i.e., the ripple error of optical components, has an important impact on the performance of laser systems and high quality imaging systems. At present, some scholars mainly have been focusing on the suppression of ripple error. However, there is not much research on the genetic phenomenon of the error distribution and its relationship with ripple error. When using jet polishing to correct the surface with a large difference in surface form error, Li et al. proposed a selective region path planning strategy to optimize the residual error distribution and successfully avoided the introduction of path error in the region where the surface form accuracy meets the requirements [11]. The influencing factors of ripple error mainly include initial surface error, TIF and polishing path. In terms of the polishing path, the ripple error can be suppressed by changing the randomness of the path to optimize the period of the ripple error and choosing a reasonable path spacing to optimize the height of the ripple error [12]. Dunn et al. [13], Wang et al. [14] and Takizawa et al. [15] proposed the six-direction, four-direction and circular pseudo-random paths, respectively. Their experimental results showed that the random variability of the path direction can effectively suppress the MSF error. In the deterministic surface correction process, a small path spacing should generally be preferred. Too large a path spacing is not conducive to correcting the surface form error and suppressing the ripple error. However, due to changes in material removal and the speed limitation of the axes of the polishing machine, polishing along an overly-dense path will introduce additional dwell time and cause excessive material removal, especially in regions where material removal is low, which is unfavorable for the convergence of surface accuracy. Wang et al. proposed to use the inflection point of the approximately reversed "L"-shaped curve between path spacing and ripple error to determine the optimal path spacing, which provides a fast and efficient method for optimal selection of path spacing [16]. Hou et al. proposed to divide the surface into several regions from the perspective of suppressing the MSF error and plan polishing paths with different pitches in different form error regions [17]. In their implementation process, the transition of polishing paths in different regions and the avoidance of the path edge effect are complicated. Hu et al. also proposed a polishing path with random spacing based on the distribution of surface form error [18]. Their choice of path spacing is based on empirical formulas, and the intrinsic relationship between form error and path spacing has not been studied in depth.

In this paper, the influence law of path spacing and bulk material removal depth on residual error is deeply studied. It is not difficult to understand that the influence law of the distribution of surface form error on residual error can be revealed by studying the effects of different bulk material removal depths on residual error. On this basis, mathematical modeling is used to demonstrate the feasibility of the novel idea consisting of the normalizing residual error distribution by optimizing the path spacing under the condition that other polishing conditions including TIF are unchanged. This idea considers the fine adjustment of the path spacing to compensate for the effect of the surface form error on the residual error. A complete residual error optimization strategy based on the form error distribution is proposed. During the implementation of this optimization strategy, according to the established path spacing optimization model and the actual surface form error distribution of the workpiece to be polished, the corresponding variable pitch polishing path with seamless transition is planned for polishing, thereby optimizing the residual error distribution and greatly reducing the influence of the parent form error distribution on the child residual error distribution.

The contents of this paper are organized as follows. Section 2 describes the convolution removal principle in CCOS and the fast solution algorithm for the dwell time based on linear equations. Section 3 deeply studies and analyzes the influence of path spacing and bulk material removal depth on residual error. In Section 4, the residual error optimization strategy based on root-mean-square (RMS) and the peak-to-valley (PV) map is proposed. In this strategy, the corresponding path spacing optimization model and variable pitch path planning method are described, respectively. Section 5 makes a comparative study of optimization strategies, and Section 6 concludes the full text.

2. Theoretical Background

2.1. Convolution Removal Principle in Computer-Controlled Optical Surfacing

The deterministic polishing material removal process can be viewed as a two-dimensional convolution of the TIF and dwell time. The surface form error can be obtained by comparing the actual measured surface data with the ideal surface data and using filtering and other technical means to eliminate the middle-to-high-spatial frequency component and retain the low-spatial frequency component. The deterministic polishing process can precisely control the dwell time of the polishing tool at each dwelling point according to the obtained surface form error and the planned polishing path, thereby realizing different amounts of material removal at different positions, which can not only improve the accuracy of the surface, but also enhance the surface quality. In general, the above process is implemented by converting calculated dwell time to feed rate of the polishing tool, which is an iterative process. The convolution material removal and resulting residual error for each iteration can be expressed by Equation (1) as follows:

$$\begin{cases} Z(x,y) = R(x,y) \otimes D(x,y) \\ E(x,y) = Z_0(x,y) - Z(x,y) \end{cases} \tag{1}$$

where $Z(x,y)$ represents the convolution material removal; $R(x,y)$ represents the material removal function per unit time; $D(x,y)$ stands for the dwell time function; $E(x,y)$ represents the residual error after deconvolution removal; $Z_0(x,y)$ represents the surface form error that is expected to be removed.

2.2. Dwell Time Solution Based on Linear Equations

According to Equation (1), given the material removal map and material removal function, the solution of the dwell time can be regarded as a deconvolution process, and its solution algorithm has an important effect on the deterministic polishing process and results. Existing dwell time algorithms are mainly split into two categories: discrete convolution model and linear equation model. The former has a fast computation speed, but usually low convergence accuracy. The latter has high accuracy, but the solution speed is slow. With the introduction of regularization and sparse matrix operation, the solution algorithm based on the linear equation model is becoming more stable and fast, and it can be applied to a flexible and changeable planning path, which makes it more feasible and flexible in practical application [19]. It is assumed that there are n dwell points along the planned path on the to-be-polished surface; the surface is divided into m control points; and correction of the surface is simplified to reduce the form error at these control points. When the TIF dwells at a dwell point, the material removal at all control points within a unit of time can be written in the form of $m \times 1$, and then, the TIF removal matrix $A_{m \times n}$ for the n dwell points can be obtained, which is usually a sparse matrix with m rows and n columns. The form error to be removed at the m control points is represented as a column $b_m \times 1$. Similarly, the dwell time of each dwell point to be calculated is also written as a column vector form, i.e., $x_n \times 1$. In the linear model solution, the dwell time corresponding to each dwell point can be estimated by solving the equation $A_{m \times n} x_{n \times 1} = b_{m \times 1}$. However, because it is an indefinite equation and the large condition number means that rounding error and other errors will seriously affect the solution of the problem, the regularization method using the smart damping factor is used here to ensure stability of the solution, as shown in Equation (2) [20].

$$\begin{cases} \begin{pmatrix} A_{m \times n} \\ \gamma \times E_{m-n} \end{pmatrix} x_{n \times 1} = \begin{pmatrix} b_{m \times 1} \\ 0 \times I_{(m-n) \times 1} \end{pmatrix} \\ \gamma = 3 \times TRP \end{cases} \tag{2}$$

where γ represents the regularized damping coefficient, E_{m-n} represents a unit matrix of $(m-n)$ order and $I_{(m-n) \times 1}$ represents a column vector containing $(m-n)$ elements. TRP represents the maximum removal depth of the TIF per unit time, in λ/min ($\lambda = 632.8$ nm).

By using the least squares method and sparse matrix method to solve the objective function of Equation (3), the stable and reliable solution of dwell time can be obtained.

$$\min \left\{ \|Ax - b\|^2 + \|\gamma x\|^2 \right\}, x \geq 0 \tag{3}$$

3. Research on the Effect of Path Spacing and Bulk Material Removal Depth on Residual Error

The actual polishing experiment is more or less affected by various interference factors such as clamping error and stability of the polishing liquid, which make it difficult to extract separately the influence of path spacing and bulk material removal depth on residual error. This section proposes the use of simulation experiments to investigate the laws of influence among them. A similar polishing simulation method can effectively eliminate the interference of other factors and greatly saves time and money.

3.1. Simulation Experiment Design

In submerged jet polishing (for narrow sapphire nozzles with a diameter less than 0.3 mm) and other deterministic polishing, the TIF usually presents a Gaussian distribution with a maximum removal at the center. In the simulation study of this paper, the TIF established by the Gaussian mathematical model given in Equation (4) is used, where R_x and R_y represent the material removal area radius (mm) of the TIF in the x direction and the y direction, respectively. a represents the maximum material removal depth per unit time (μm/min). Figure 1a shows the rotationally-symmetric three-dimensional TIF profile with a removal radius of 1 mm and a maximum removal depth of 0.1 μm/min. Table 1 gives the parameter values of the TIF used in this paper.

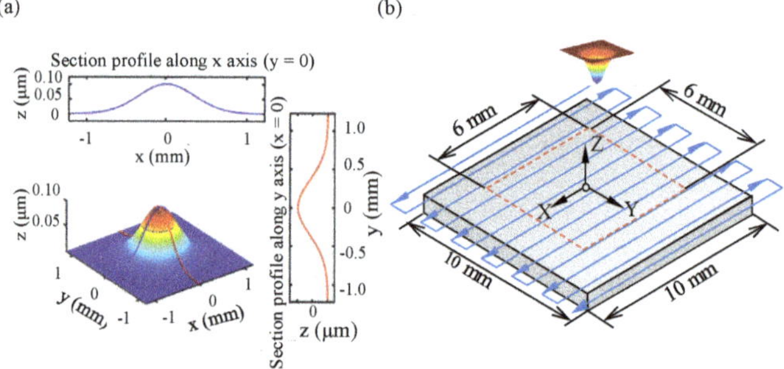

Figure 1. (a) Tool influence function (TIF) with Gaussian shape; (b) tool path design.

$$\begin{cases} sigmax = R_x/2\sqrt{2},\, sigmay = R_y/2\sqrt{2} \\ z = a \times e^{-\left(x^2/(2sigmax)+y^2/(2sigmay)\right)} \\ TRP = a \times 10^3 \big/ \lambda,\, \lambda = 632.8\ \text{nm} \end{cases} \tag{4}$$

Table 1. The parameter values of tool influence function (TIF).

Parms	a (μ/min)	R_x (mm)	R_y (mm)
Values	0.1	1.0	1.0

Without loss of generality, a scanning path was planned on a planar specimen with a length of 10 mm and a width of 10 mm, and the deterministic polishing simulation was implemented according to the given TIF, as shown in Figure 1b. Since it is difficult for the computer to achieve continuous removal in simulations, in order to make the discrete removal simulation more closely approximate the continuous removal commonly used in actual polishing, the number of dwell points along the planned path was increased. Here, the distance between adjacent dwell points along the path was set to 0.05 mm. The simulation results in our test show that the actual continuous polishing removal can be well approximated. The RMS and PV values of the residual error obtained after correction were used as evaluation indexes for the deterministic polishing convergence accuracy and the removal effect. In addition, the power spectral density (PSD) method was used to analyze the MSF error component in the residual error and study the effect of path spacing and surface form error on the component. In order to avoid the influence of edge effect in the simulation process, the central region of the specimen with a length of 6 mm and width of 6 mm served as a residual error evaluation region.

In order to deeply study the influence of path spacing and bulk material removal depth on the residual error obtained after polishing, three groups of basic simulation experiments were carefully designed. The first set of experiments studied the effect of path spacing on residual error under fixed bulk material removal depth. The second set of experiments investigated the effect of bulk material removal depth on the residual error in the case of fixed path spacing. The third set of experiments investigated the effect of path spacing variations on the residual error resulting from different bulk material removal depths. Table 2 gives details of the experimental conditions for the three simulation experiments.

Table 2. Simulation experiment conditions.

Experiment Set	Path Pitch (mm)	Bulk Material Removal Depth (μm)
Exp. 1	0.1 , 0.2, 0.3, ... , 2.5	0.5
Exp. 2	0.2	0, 0.1, 0.2, ... , 1.0
Exp. 3	0.1, 0.2, 0.3, ... , 2.5	0, 0.1, 0.2, ... , 1.0

3.2. Simulation Results and Analysis

3.2.1. Influence of Path Spacing on Residual Error

By introducing regularization and sparse matrix operation methods, the dwell time solution algorithm based on the linear equations can effectively control the negative ratio (number of dwell points with a negative dwell time calculated/total number of dwell points) of dwell time calculated to be less than 1%, as shown in Figure 2a. Note that the influence of path spacing on the RMS and PV values of the residual error is almost consistent when bulk material removal depth is constant. As the path spacing increases, the RMS and PV values of the residual error first increase slowly and approximately linearly within the range of 0.1 mm to 0.6 mm, then rapidly increase within the range of 0.6 mm to 2.0 mm and finally stabilize in the range of 2.0 mm to 2.5 mm. This is because the deterministic polishing removal is caused by the convolution effect formed by the overlap of tool influence functions. When the distance between the paths is small, i.e., less than 3/10 of the diameter of the TIF, convergence of the deterministic polishing removal process can be effectively ensured. From the residual error profile in Figure 2a, it can be seen that as the path spacing increases, it becomes increasingly difficult to ensure convergence of the regions between adjacent paths. When the distance between the paths is larger than the diameter of the TIF, the material between the adjacent paths cannot be removed. In this case, the PV values of the residual error will be slightly larger than the original surface form error and remain basically constant. However, it should be pointed out that the RMS curve is fluctuating more at these later stages than the PV curve. This may be because the RMS value of the residual error is more susceptible to dramatic changes in convergence of the surface form error.

Figure 2b shows that as path spacing increases, the total length of the path is drastically shortened, but the total polishing time is only slightly reduced within the path range of 0.1 mm to 1.0 mm. The polishing time for deterministic polishing is mainly determined by the TIF and volume of material removal. More polishing time means more volume removal. The error of a certain control point basically determines the total dwell time at that point. As the path spacing decreases, although the path length increases sharply, the number of dwell points increases at the same time, and the average dwell time allocated to each dwelling point will be reduced accordingly, so that the total polishing time remains basically unchanged. In addition, considering that as the path spacing increases, the degree of convergence of deterministic polishing decreases, the overall material removal will be reduced, and thus, the polishing time will tend to decrease. When the path spacing is greater than 1.0 mm, the degree of convergence of the polishing removal sharply changes and shows a downward trend, so that the total polishing time is drastically reduced with a certain degree of fluctuation.

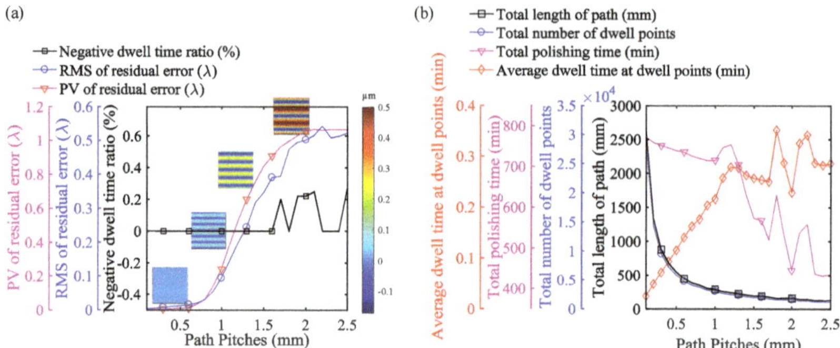

Figure 2. Effect of path pitch on (**a**) residual error and (**b**) other path parameters (bulk material removal depth = 0.5 μm).

Figure 3a shows the power spectral density curve of the residual error calculated along the *y*-direction, which is the perpendicular direction of the planned path when the path distance is 0.1 mm, 0.4 mm, 0.7 mm and 1.0 mm, respectively. It can be seen that as the path spacing increases, the MSF error becomes larger and larger, which is mainly due to waviness error caused by the tool path. It can be concluded that for deterministic polishing removal, a small path spacing should be preferred, which can effectively reduce the RMS and PV values of the residual error and effectively suppress the MSF.

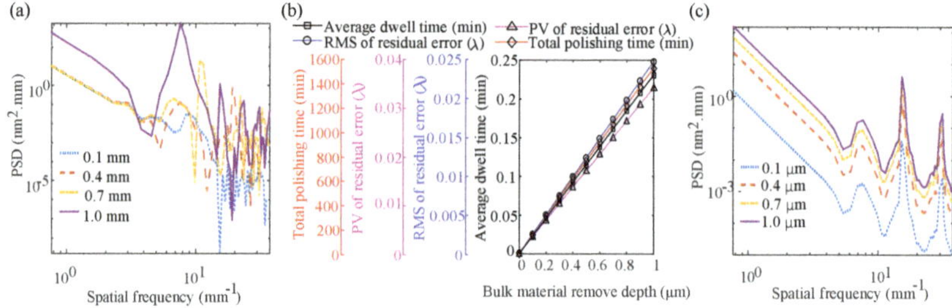

Figure 3. (**a**) *y*-direction power spectral density (PSD) results for different path pitches; (**b**) influence of the bulk material removal depth on residual error (path pitch = 0.2 mm); (**c**) *y*-direction PSD results for different path pitches and bulk material removal depths.

3.2.2. Influence of Bulk Material Removal Depth on Residual Error

Figure 3b shows the influence of the bulk material removal depth on the residual error when the path spacing is constant. With the increase of bulk material removal depth, RMS and PV values of the residual error showed a linear growth trend. This also reflects the linear decreasing trend of convergence of deterministic polishing with the increase of the bulk material removal depth. With the increase of the bulk material removal depth, the linear increase of the polishing time indicates that polishing time has a positive correlation with the amount of material removal. Since the number of pre-planned dwell points remains constant, the average dwell time allocated to each dwell point increases accordingly. Figure 3c shows the power spectral density calculation results for the residual error along the vertical direction of the planned path for different bulk material removal depths (0.1 μm, 0.4 μm, 0.7 μm, 1.0 μm) when the path pitch is 0.2 mm. It can be seen that the profiles of the power spectral density curve corresponding to different bulk material removal depths are similar, but as the bulk material removal depth increases, the power density amplitude corresponding to each spatial frequency shows a significant increase. This may be because with the increase of the bulk material removal depth, the dwell time of each dwell point is prolonged. The increase in the dwell time of the polishing tool at a certain point will make the path effect more pronounced, thereby introducing deeper ripple errors caused by the tool path. Therefore, for a large form error, a small path spacing should be preferred, which can effectively improve the convergence of deterministic removal, reduce the RMS and PV value of the residual error and effectively suppress the MSF error component in the residual error.

3.2.3. Effect of Path Spacing Variations on the Residual Error Resulting from Different Bulk Material Removal Depths

As can be seen from Figure 4a,b, the influence law of path spacing variations on the RMS and PV values of residual error resulting from different bulk material removal depth is consistent. For deterministic polishing removal, a reasonable polishing path spacing value should be less than 3/10 of the diameter of the TIF. When the value is within this range, drastic variation of RMS and PV values of the residual error can be effectively avoided. Therefore, in this simulation, the RMS and PV curves with a path pitch of less than 0.6 mm are the parts of interest. As can be seen from the partially enlarged view, shown in Figure 4c,d, the slope of the curve increases significantly with the increase of the form error. This indicates that the larger the bulk material removal depth, the greater the effect of the change in the path spacing on the residual error, which makes the subsequent adjustment of the path spacing more useful, especially for regions with large bulk material removal depth.

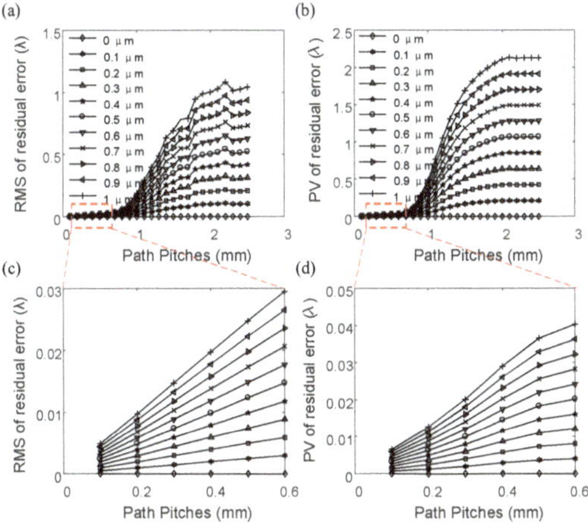

Figure 4. The influence of path spacing variations on the (**a**,**c**) root-mean-square (RMS) and (**b**,**d**) peak-to-valley (PV) of the residual error resulting from different bulk material removal depths.

3.3. Summary and Discussion of Simulation Results

Through the above three sets of simulation experiments, the influence rules of path spacing and bulk material removal depth on the residual error obtained after deterministic polishing were thoroughly studied and analyzed. When the path spacing was fixed, the larger the bulk material removal depth, the larger the RMS and PV values of residual error. The distribution of surface form error directly reflected the difference in bulk material removal depth across a to-be-polished surface in subsequent corrective polishing. This is a good explanation of the genetic phenomenon of the error distribution during the surface correction process. In the case where the bulk material removal depth was constant, the path spacing also affected the RMS and PV values of residual error approximately linearly. Once the surface form error distribution of the actual workpiece was determined, we could get the variation in bulk material removal depth across the surface and could consider the use of fine-tuning of the path spacing to compensate for the effect of surface form error on the residual error, as well as achieve the purpose of optimizing the residual error distribution. That is to say, small path spacing was used where surface form error was large, and a large path spacing was used where surface form error was small, so that the path spacing adapted to the variation of the surface form error, which was expected to make the optimized residual error distribution obtained more uniform and almost unaffected by the distribution of parent surface form error.

4. Residual Error Optimization Strategy Based on Root-Mean-Square and Peak-to-Valley Maps

Based on results of the simulation experiment in Section 3, it can be seen that the RMS and PV values of the residual error after deterministic polishing are influenced by both the bulk material removal depth and the path spacing. This section aims to establish their relationship models with the path spacing and bulk material removal depth, respectively. To achieve this goal, this paper introduces the concept of RMS and PV maps for the first time and proposes a path spacing optimization model based on the above two maps in order to optimize the corresponding RMS and PV values of the residual error. The detailed flow for variable pitch spiral path planning in residual error optimization strategy is also given.

4.1. Derivation of the RMS and PV Maps

The RMS (PV) map characterizes the relationship of the RMS (PV) value of the residual error obtained after theoretical deterministic polishing with the path spacing and the surface form error. From the results of the third set of simulation experiments in Section 3, the experimental data in the range of reasonable path spacing (0.1 mm to 0.6 mm) were extracted and collated, and then, the initial RMS and PV maps were obtained, as shown in Figure 5a,c. Using surface fitting techniques, we obtained the final smoothed RMS and PV maps, as shown in Figure 5b,d. The proposed RMS and PV maps provide a theoretical model reference for predicting the RMS and PV values of the residual error after practical deterministic polishing.

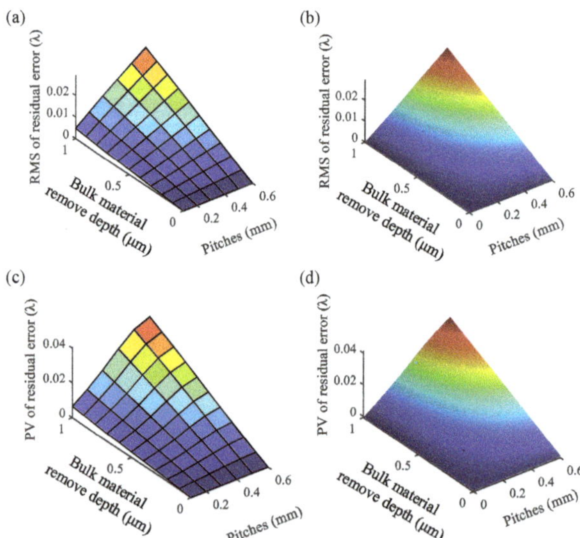

Figure 5. (a) Original RMS map; (b) fitted map of RMS; (c) original PV map; (d) fitted map of PV.

4.2. Path Spacing Optimization Model Based on Surface Form Error

Based on the RMS map of Figure 5b, the RMS contour of the residual error shown in Figure 6a is obtained. On the RMS contour, all points with the same RMS value form continuous curves in sequence. The curve reflects the correspondence between the bulk material removal depth to be corrected and path spacing for which the RMS value of residual error is controlled to be a constant value. Although corrective polishing should be preferably performed with a small path pitch, the path pitch in actual polishing cannot be set infinitely small. The variation range of expected bulk material removal depth in one polishing iteration is constant. As shown in Figure 6a, once the minimum path spacing is set, for the minimum path used in the region with largest bulk material removal depth, the theoretical optimum RMS value of the residual error after polishing is determined. Conversely, given the RMS value of the residual error we expect, we can get the corresponding minimum path spacing. Moreover, the smaller the RMS value of the desired residual error, the smaller the corresponding minimum path spacing. In the simulation experiments in this paper, the bulk material removal depth varies from 0.5 μm to 1.0 μm. We specify the minimum path spacing as $p_{min} = 0.2$ mm and the maximum path spacing is $p_{max} = 0.6$ mm. The P1 point determined by the minimum pitch gives the best RMS value of the residual error corresponding to key point P2, and then, the contour line of the best RMS value is located. The contour curve determines the point P3 at the position of the maximum path spacing. P3 determines the critical value of bulk material removal depth corresponding to point P4. When the bulk material removal depth of the region to be corrected is smaller than the critical value, the path spacing corresponding to the region should be set to the maximum path spacing. Finally, the relationship between the bulk material removal depth and the path spacing based on the RMS map is obtained, as shown in Figure 6b.

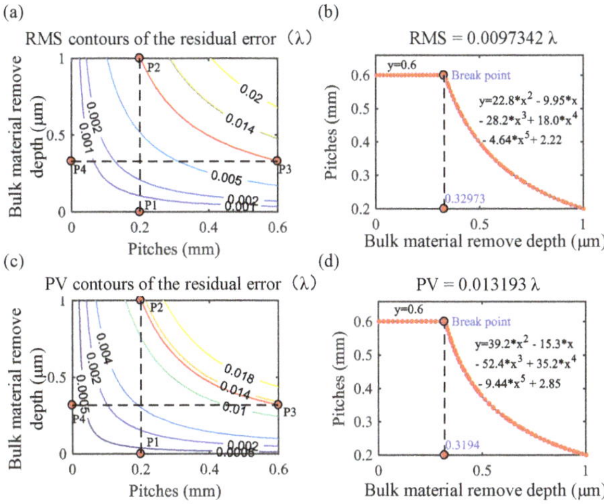

Figure 6. (**a**) RMS contours of the residual error; (**b**) relationship curve between bulk material removal depth and the path pitch based on the RMS map; (**c**) PV contours of the residual error; (**d**) relationship curve between bulk material removal depth and the path pitch based on the PV map.

Similarly, the PV contour of residual error and the relationship between the bulk material removal depth and the path spacing based on the PV map are obtained, as shown in Figure 6c,d. A comparison of Figure 6b,d shows that the trend of the two curves is similar, which is attributed to the similarity of the effects of bulk material removal depth, and path spacing on the RMS and PV values of the residual error. A piecewise polynomial fitting method was used to fit the two curves, respectively, and the path spacing optimization models based on the RMS map (Equation (5)) and the PV map (Equation (6)) were obtained, respectively.

$$p_{rms}(z_h) = \begin{cases} 0.6 & , z_h < 0.3297 \\ -4.46z_h{}^5 + 18.0z_h{}^4 - 28.2z_h{}^3 + 22.8z_h{}^2 - 9.95z_h + 2.22 & , z_h \geq 0.3297 \end{cases} \tag{5}$$

$$p_{pv}(z_h) = \begin{cases} 0.6 & , z_h < 0.3194 \\ -9.44z_h{}^5 + 35.2z_h{}^4 - 52.4z_h{}^3 + 39.2z_h{}^2 - 15.3z_h + 2.85 & , z_h \geq 0.3194 \end{cases} \tag{6}$$

in which z_h stands for the bulk material removal depth (μm); p_{rms}, p_{pv} represent the resulting path spacing (mm) based on the RMS-map optimization model and the PV-map optimization model, respectively.

It is important to point out that several simulation experiments have shown that when the fluctuation range of the parent surface form error is large, expectations for the degree of residual error optimization should be reduced in one iteration, and the minimum path spacing should be selected to be larger, thereby increasing the critical value of bulk material removal depth determined by point P4 and reducing the variation range of the polishing path pitch. This is more conducive to reducing the impact of the distribution of the parent surface form error on the child residual error distribution, while avoiding distortion of the planned variable pitch path and affecting the final polishing effect. Equation (7) gives the empirical formula for the choice of minimum and maximum path spacing under the same parameters of the optimization strategy proposed in this paper.

$$\begin{cases} p_{max} = 3R_{tif}\big/5 \\ p_{min} = 2p_{max} \times [\max(Z_0(x,y)) - \min(Z_0(x,y))]/(3\max(Z_0(x,y))) \end{cases} \tag{7}$$

where R_{tif} represents the radius of the TIF; $\max(Z_0(x,y))$ and $\min(Z_0(x,y))$ represent the largest and the smallest form error, respectively.

4.3. Variable Pitch Spiral Path Planning Method

In order to realize the optimization process of the spiral polishing path efficiently and quickly, this paper proposes to plan the variable pitch spiral polishing path on a two-dimensional plane. Moreover, the path pitch along the radial direction of the spiral path is iteratively optimized according to the proposed path spacing optimization model in Section 4.2. At the same time, when the variable pitch 2D path is projected onto a 3D target surface, in order to avoid as much as possible the impact of slope change on the pitch of the projected path, a path spacing compensation factor based on the gradient change distribution of the target surface is introduced. After several iterative optimizations, the final variable pitch spiral-polishing path with seamless transition will be obtained. Figure 7a gives the complete planning flowchart for the optimized path with variable pitch.

Although practical fluid jet polishing is hardly affected by the polishing edge effect [21], the edge-up phenomenon occurs when performing polishing removal simulations, which is a common problem that is difficult to avoid when solving the dwell time [22,23]. In this algorithm, to avoid the influence of this phenomenon on the solution in the region of interest, the polished surface and its form error will extend outward by a specified distance, and so does the planned polishing path, as shown in Figure 7b. Through several simulation experiments, it was found that when the extension distance is specified as $1.5R_{tif}$, the influence of the edge-up phenomenon on the region of interest can be effectively avoided and a better removal effect obtained.

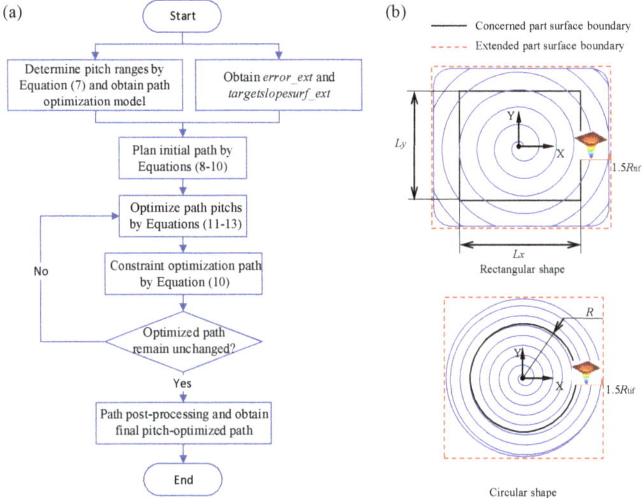

Figure 7. (**a**) The generation flowchart of the optimized path with variable pitch; (**b**) control strategy for edge-warping effect.

In this algorithm, we first determine the range of path spacing changes according to Equation (7) and give the corresponding path optimization model specified by Equations (5) or (6). Then, the initial planned Archimedes spiral path can be given by Equation (8) as follows:

$$\begin{cases} r_0 = 0.075R_{tif}; p_0 = P_{\min}; \theta_m = m\Delta\theta, m = 0, \ldots, M \\ M = \text{INT}\left(\frac{2\pi R_{\max}}{\Delta\theta P_0}\right); r_m = r_0 + p_0 \times \frac{\theta_m}{2\pi} \end{cases} \tag{8}$$

where γ_0 represents the starting spiral radius; p_0 represents the pitch of the Archimedean spiral path; $\Delta\theta$ represents the angular coordinate increment of adjacent path points in the initial planned path; $R_{\max}$ represents the limit of the radius of the circular area of the initial planned Archimedes spiral path; INT represents the operation of rounding down. For a rectangular workpiece and a circular workpiece, Equation (9) gives the calculation method of $R_{\max}$ in two cases. The initial Archimedes spiral path bounded by the maximum radius calculated by Equation (9) can ensure fast convergence of the iterative optimization process. θ_m, r_m represent the angular coordinate and radial coordinate of the m-th path point of the initially planned path in polar coordinates, respectively.

$$\begin{cases} R_{\max} = \sqrt{\left(\frac{L_x + 3R_{tif}}{2}\right)^2 + \left(\frac{L_y + 3R_{tif}}{2}\right)^2}, \text{for rectangular shape} \\ R_{\max} = \frac{2R + 3R_{tif}}{2}, \text{for circular shape} \end{cases} \tag{9}$$

in which L_x, L_y represent the length and width of the original rectangular workpiece, respectively, and R represents the radius of the original circular workpiece.

The algorithm extracts the boundary curve of the projection of the target surface on the *xoy* plane, and the new curve obtained by offsetting the distance of $1.5R_{tif}$ outward is used as the constraint boundary of the subsequent path planning. Assume that the boundary can be expressed in polar coordinates as $r_b = Boundary\left(\theta_b\right), 0 \leq \theta_b \leq 2\pi$, then the radial coordinates of all the path points beyond the constraint boundary are set as the radial coordinates of the corresponding boundary points, and the angular coordinates remain unchanged. Then, Equation (10) can be used to calculate the constrained path.

$$r_b = Boundary\left(\mathrm{mod}\left(\theta, 2\pi\right)\right); r = r_b, r \geq r_b \tag{10}$$

We need to extend the target surface and calculate the slope distribution surface corresponding to the extended surface. Then, according to Equation (11), the path points are transformed from the polar coordinate system to the Cartesian coordinate system and are projected onto the calculated slope distribution surface and the extended form error surface, respectively. The z coordinate of each projection point obtained by projecting on the slope distribution surface and the form error surface respectively correspond to the slope k_{slope} and the form error z_h. According to Equation (12), we can calculate the change of path spacing after one iterative optimization.

$$\begin{cases} x = x_0 + r\cos\theta; y = y_0 + r\sin\theta \\ z_h = \Gamma(error\_ext, x, y); k_{slope} = \Gamma(targetslopesurf\_ext, x, y) \end{cases} \tag{11}$$

in which $\Gamma(error\_ext, x, y)$ and $\Gamma(targetslopesurf\_ext, x, y)$ represent projection operations on the extended form error surface $error\_ext$ and the slope distribution surface $targetslopesurf\_ext$ corresponding to the extended target surface, respectively.

$$p = \begin{cases} p_{rms}\left(z_h\right) \Big/ \sqrt{k^2_{slope} + 1}, \mathrm{RMS} - \text{based optimization} \\ p_{pv}\left(z_h\right) \Big/ \sqrt{k^2_{slope} + 1}, \mathrm{PV} - \text{based optimization} \end{cases} \tag{12}$$

Finally, an unconstrained variable pitch spiral path after one iterative optimization is obtained according to Equation (13). Then, using Equation (10) to calculate the variable pitch spiral path with boundary constraints and eliminate the repeated path with the edge at the end of the path, a variable pitch spiral path with boundary constraints after one iteration optimization is obtained. After the iterative optimization is converged, the final variable pitch path is derived by resampling $\Delta\theta$ making point-to-point distance constant along the path.

$$\begin{cases} N = \mathrm{INT}\left(2\pi / \Delta\theta\right) \\ \theta_k = k\Delta\theta; r_k = r_0 + \frac{p_1\theta_k}{2\pi}, 0 \leq k \leq N \\ \theta_k = \theta_{k-N} + 2\pi; r_k = r_{k-N} + p_{k-N}, k > N \end{cases} \tag{13}$$

5. Comparative Study of Optimization Strategies

In order to verify the validity and feasibility of the proposed residual error optimization strategy based on the distribution of surface form error, the comparative correction polishing studies were performed in this section for one planar test piece and one aspherical test piece. The comparison results of the residual error distribution before and after optimization both verify the validity and feasibility of the proposed strategy.

5.1. Case 1: Planar Workpiece

In this case, a planar test piece with a length of 10 mm and width of 10 mm is again selected. Figure 8 shows the initial typical form error distribution and the extended form error distribution, respectively. It can be seen that the distribution of form error is a typical convex surface form with a height variation range of 0.5 μm to 1 μm. The TIF specified in Section 3 is again used in this simulation. Therefore, we can directly use the RMS-map-based (Equation (5)) and PV-map-based path-pitch model (Equation (6)) obtained in Section 3 to optimize the initially planned polishing paths. In addition, since it is a planar specimen, the slope compensation factor in Equation (11) is null, i.e., $k_{slope} = 0$. Table 3 shows the key parameters for planning the spiral-polishing path with optimized variable pitch.

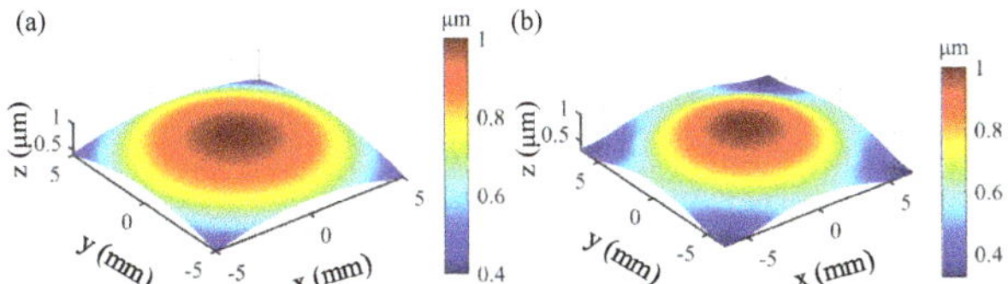

Figure 8. (**a**) Original form error distribution; (**b**) extended form error distribution.

Table 3. Parameters for planning the initial spiral-polishing path.

p_{min} (mm)	p_{max} (mm)	p_0 (mm)	$\Delta\theta$ (deg)	r_0 (mm)
0.2	0.6	0.2	$\pi/6$	0.01

According to the kind of optimization model, we get three kinds of polishing paths, namely the unoptimized path, RMS-map-based optimized path and PV-map-based optimized path. Figure 9a shows the initially planned spiral-polishing path with boundary constraints. Using the two proposed optimization models respectively to optimize the initial path given in Figure 9a, the corresponding optimized path distributions are obtained, as shown in Figure 9b,c. It can be seen that the polishing path pitch before optimization is always 0.2 mm and is evenly distributed. Although there are subtle differences in paths optimized by different optimization models, their path pitches are continuously changed with the variation of the surface form error in the range of 0.2 mm to 0.6 mm.

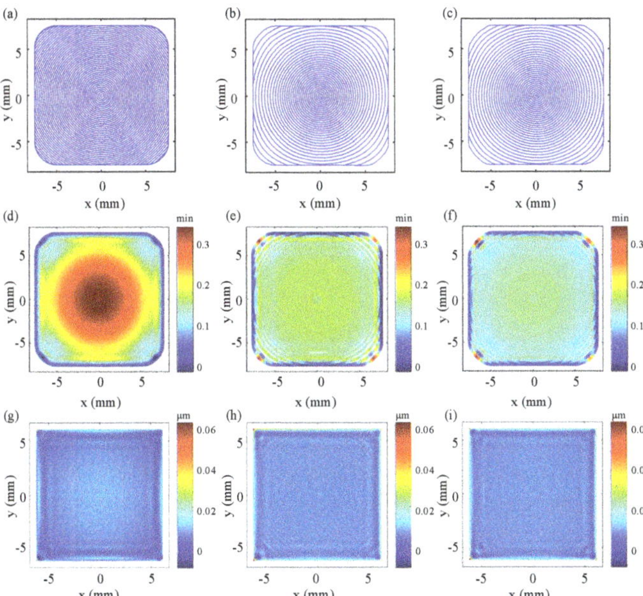

Figure 9. (**a**) Boundary-defined unoptimized path with constant pitch; (**b**) RMS-map-based optimized path with varying pitch; (**c**) PV-map-based optimized path with varying pitch; (**d**) dwell time map along the unoptimized path; (**e**) dwell time map along the RMS-map-based optimized path; (**f**) dwell time map along the PV-map-based optimized path; (**g**) residual error corresponding to the unoptimized path; (**h**) residual error corresponding to the RMS-map-based optimized path; (**i**) residual error corresponding to the PV-map-based optimized path.

By solving the dwell times corresponding to the three polishing paths, the dwell time distribution maps corresponding to the three polishing paths are obtained, as shown in Figure 9d–f. It can be seen from Figure 9d that the dwell time distribution corresponding to the initial equally-spaced spiral polishing path is consistent with the initial form error distribution. It further verifies the experimental results obtained in Section 3.2.2, that is, when the path spacing is constant, there is a positive linear relationship between the dwell time of each dwell point and the amount of form error. For the optimized paths, especially in the regions we care about, the dwell time of each dwell point does not change very much. From Figure 3b in Section 3.2.1, it can be seen that there is also a positive linear relationship between the path spacing within a certain range (0.2 mm–0.6 mm) and the dwell time of each point when bulk material removal depth is constant. It is under the combined influence of bulk material removal depth and path spacing that the dwell time at each point on the optimized path remains mostly the same. To some extent, the optimized path is a good way to improve dwell time distribution. Since spacing of the planned dwell points along the path is basically the same, the optimized polishing path will run under a more stable polishing speed, which is crucial for obtaining a stable polishing process [24]. Figure 9g–i respectively shows the residual error distribution obtained by figuring the extended form error using the three polishing paths, respectively. It can be seen that the strategy of extending the path and the surface form error effectively prevents material removal of the region of interest from being affected by the edge-warping phenomenon. Compared to the residual error distribution before optimization, the imprint of the initial surface form the error distribution can hardly be found visually in the optimized residual error distributions.

As shown in Figure 10a,c,e, in order to further clearly compare and quantify the difference of the residual error distribution before and after optimization and the effects of two optimization models on the residual error distribution, the residual error distribution of the region of interest is extracted. At the same time, two cross-sectional profiles passing through the center origin and parallel to the x-axis and y-axis are respectively given for each residual error distribution. From Figure 10a, it can be seen that the residual error distribution corresponding to the unoptimized polishing path is highly similar to the distribution of the parent surface form error shown in Figure 8a, that is the error shows a decreasing trend from the center to the periphery. However, the residual error distribution after optimization shown in Figure 10c,e, respectively, is more flattened and is almost unaffected by the distribution of parent form error.

Considering the use of RMS and PV values of the overall residual error to evaluate the residual error distribution, there are two disadvantages: (i) the optimization effect of the proposed residual error optimization strategy on residual error distribution cannot be evaluated intuitively and comprehensively; (ii) it is difficult to compare subtle differences in the residual error distribution obtained by the two optimization models proposed. In this paper, a method of sub-domain evaluation is proposed. Firstly, the residual error of the region of interest is evenly divided into 25 sub-regions, as shown in the Figure 10b,d,f. Then, the RMS and PV values of the residual error corresponding to each sub-region are calculated, and the corresponding RMS and PV curves are plotted. Figure 11a,b shows the RMS and PV curves obtained from the simulation results, respectively. It can be seen that, using the sub-domain evaluation method, the RMS curves obtained by the three polishing paths all show both obvious and inconspicuous peaks, and the occurrence frequency of the peaks has a certain regularity. The curve peaks appear in the ranges A1–A5, A6–A10, A11–A15, A16–A20 and A21–A25, respectively. For the range A1–A5, if the average value of the form errors corresponding to each sub-area is plotted as a curve, a similar peak curve is obtained. From the results obtained in Section 3, we know that the RMS value of the residual error increases with the material removal depth; therefore, the resulting RMS curve also exhibits the peak. This can also explain the formation of the corresponding peak in other ranges. In addition, the average amplitude and corresponding amplitude fluctuation range of the RMS curve with optimization are both smaller than that without optimization. This is because adjustment of the path spacing effectively compensates for the influence of different material removal depths caused by the surface form error distribution on the residual error, which greatly reduces the influence of the distribution of parent surface form error on the child residual error distribution.

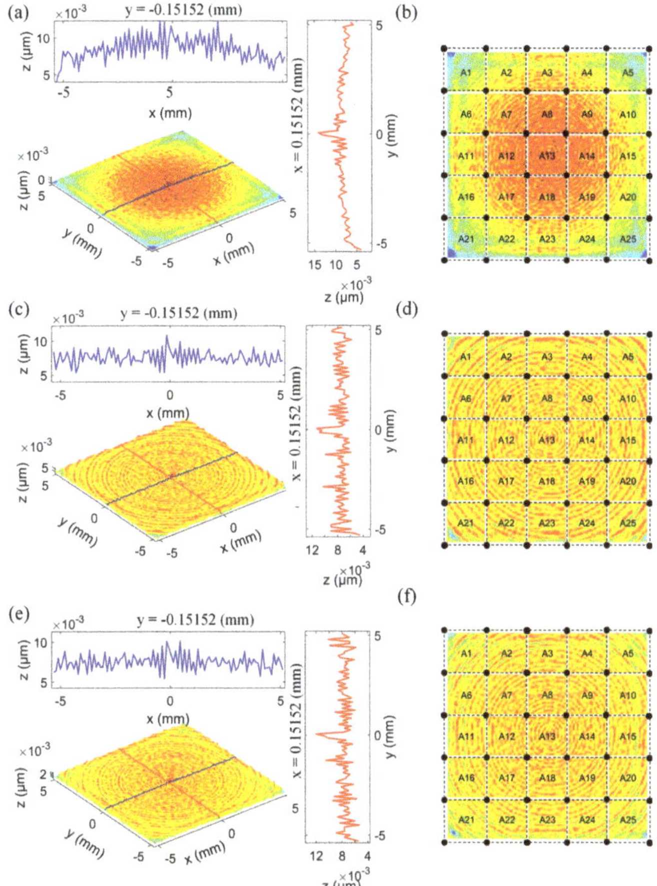

Figure 10. Residual error of the concerned surface region polished along: (**a**,**b**) the unoptimized path; (**c**,**d**) the RMS-map-based optimized path; and (**e**,**f**) the PV-map-based optimized path.

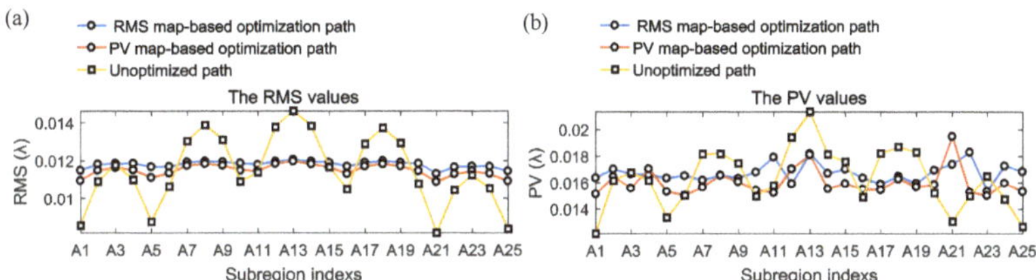

Figure 11. (**a**) RMS and (**b**) PV curves for sub-regional evaluation.

There are multiple outlier points on the PV curve given in Figure 11b, which appear at the positions of A1, A5, A13, A21 and A25, respectively. Comparing the residual error distribution shown in Figure 10, we can see that these outlier points appear at the center of the residual error region and at the four edge corners, respectively. The reason for the abrupt change of PV value in the residual error region corresponding to A13 is related to the center effect produced by the spiral path polishing. The sudden change in PV value at the four edge corners may be related to a 90 degree change in the polishing direction of the planned polishing path at the four corner points. Regardless of the above outlier points, it can be seen that the residual errors obtained by using the PV-map-based optimization model have smaller and closer PV values. In addition, it can be seen that the RMS-map-based optimization model improves the spread of PV values over the sub-regions of residual error.

As shown in Figure 6b,d, the ideal RMS and PV values for the two optimized models are RMS = 0.0097342 λ and PV = 0.013193 λ, respectively. However, we found that RMS and PV values obtained after optimization are about 20% larger than the target values. This may be related to the fact that the bulk material removal depth varies in the adjacent path interval caused by varying surface form error, unlike the constant bulk material removal depth in the experiment in Section 3.

Power spectral density analysis was performed on the residual errors obtained by the above three polishing paths, and the results along the x-direction and y-direction are, respectively, shown in Figure 12a,b. It can be seen from the PSD curves that whether the path optimization model is based on the RMS or PV map, the amplitude variation of the power spectral density of the mid-spatial frequency components in the residual error obtained by the optimized path is more stable. It can be concluded that the proposed optimization strategy can effectively smooth the distribution of MSF error components in the residual error.

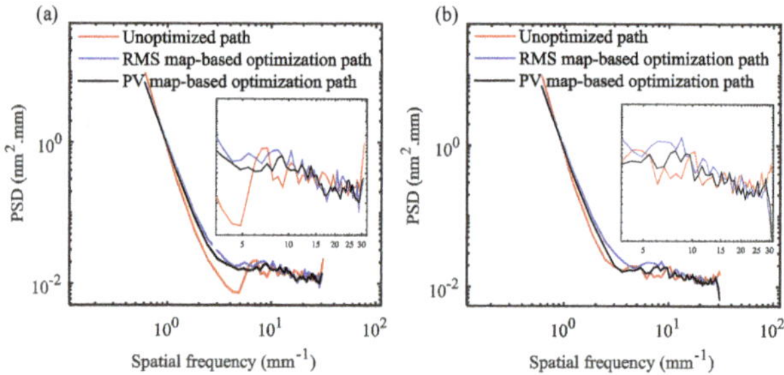

Figure 12. PSD analysis results of residual error in the (**a**) x-direction and (**b**) y-direction.

5.2. Case 2: Aspheric Workpiece

To further verify the feasibility of the proposed optimization strategy for corrective polishing of complex surfaces, in this case, we selected the aspherical surface shown in Figure 13a as the test sample. Table 4 gives the specific parameters of the aspheric surface, which is expressed by Equation (14) as follows:

$$z = \frac{c\left(x^2 + y^2\right)}{1 + \sqrt{1 - c^2\left(1 + k\right)\left(x^2 + y^2\right)}} + a_2\left(x^2 + y^2\right)^2 + a_3\left(x^2 + y^2\right)^3 + a_4\left(x^2 + y^2\right)^4 + \cdots \tag{14}$$

where R_0 represents the radius of curvature of the aspherical vertex; c represents the curvature of the aspherical vertex, with $c = 1/R_0$; D represents the aspherical aperture; k represents the quadratic constant; a_2, a_3, a_4 represent the coefficients of higher order terms in the aspheric equation, respectively.

Table 4. Detailed parameters of the aspheric surface.

R_0 (mm)	c (mm^{-1})	D (mm)	k	a_2	a_3	a_4
20	$1/R_0$	20	-0.32	-1.5×10^{-14}	-1.5×10^{-14}	-3.0×10^{-16}

Figure 13d shows the slope distribution corresponding to the aspherical surface. It can be seen that the slope of the surface is rotationally symmetric and gradually increases from the center to the periphery. Figure 13b,c,e,f respectively shows the isometric and top view of the aspherical surface form error and of the extended surface form error obtained after extending a distance of $1.5R_{tif}$. Since the material removal depth variation in all regions in this case is small (about 0.75 μm to 1 μm), the chosen minimum path in Equation (7) and its variation range are narrow. The path obtained in the Cartesian space is too dense to observe the effect of pitch change by the naked eye; thus, Figure 13g,h gives the RMS-map-based optimization path in polar coordinate and its pitch-variation along the path.

Similar to Case 1, Figure 14 shows the distribution of residual error and dwell time, respectively. It can be seen that the residual error distribution obtained by the optimization strategy is hardly affected by the distribution of parent form error, and the approximately uniform distribution of dwell time indicates that the polishing speed for the whole process is maintained at a stable level, such that polishing dynamics are optimized. Figure 15a–c shows the profile changes of the two section lines corresponding to three residual errors, respectively. The imprint of the parent form error is almost invisible in the optimized section profile curve, which further verifies the usefulness of the proposed optimization strategy. For the circular workpiece, the sub-domain evaluation strategy proposed in Section 5.1 is also used to obtain the RMS and the PV curves in sub-regions of the residual error, as shown in Figure 15d,e. It can be seen from the variation of the curve that the optimization strategy can significantly improve the spread of RMS and PV values across the residual error and that different optimization models result in different optimization effects. It can be concluded that for the corrective polishing of the surface, the proposed optimization strategies are both effective, and the distribution of residual error is significantly improved.

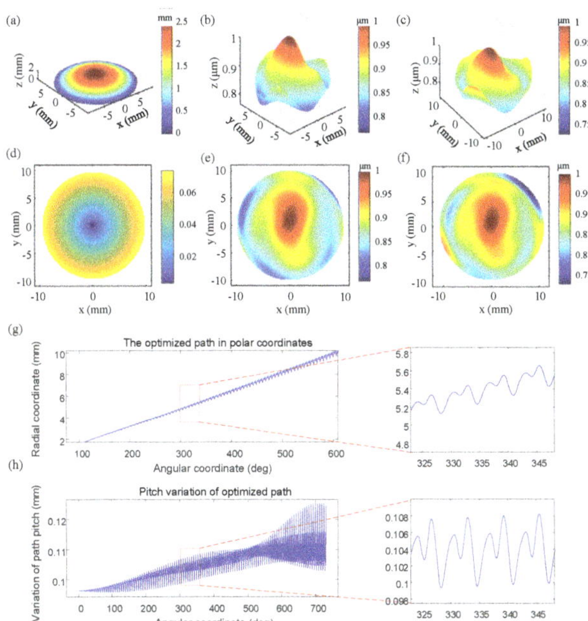

Figure 13. (**a**) Aspheric surface; (**b**) isometric view of the original form error distribution; (**c**) isometric view of the extended form error distribution; (**d**) slope distribution of the aspheric surface; (**e**) top view of the original form error distribution; (**f**) top view of the extended form error distribution; (**g**) RMS-map optimization path in polar coordinates; (**h**) variation of path pitches of the RMS-map optimization path.

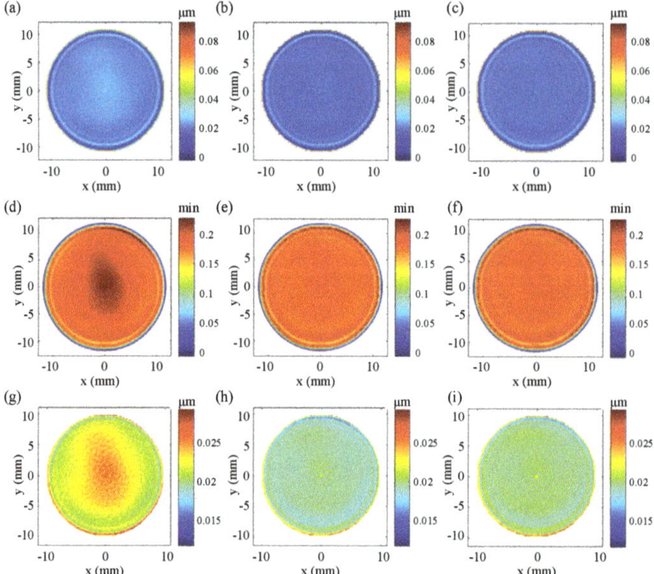

Figure 14. Residual error corresponding to: (**a**) the unoptimized path; (**b**) the RMS-map-based optimized path; and (**c**) the PV-map-based optimized path. Dwell time map along: (**d**) the unoptimized path; (**e**) the RMS-map-based optimized path; and (**f**) the PV-map-based optimized path. Residual error of the concerned surface region polishing along: (**g**) the unoptimized path; (**h**) the RMS-map-based optimized path; and the (**i**) PV-map-based optimized path.

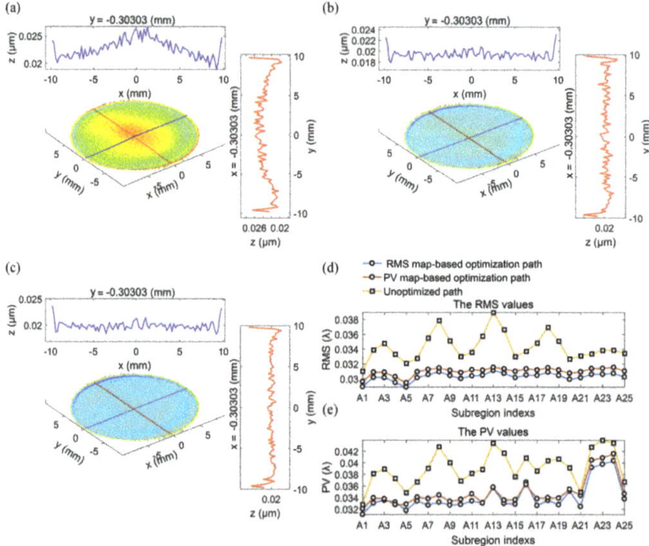

Figure 15. Residual error of the concerned surface region polished along: (**a**) the unoptimized path; (**b**) the RMS-map-based optimized path; and (**c**) the PV-map-based optimized path; (**d**) RMS and (**e**) PV curves for sub-regional evaluation.

6. Conclusions

This paper deeply studied and analyzed the influence law of bulk material removal depth and path spacing on residual error. It was found that with the increase in surface form error, the RMS and PV values of the residual error also increase. For deterministic polishing, a reasonable polishing path spacing should be less than 3/10 of the diameter of the TIF. In this range, as the path spacing increases, the corresponding RMS and PV values of the residual error increase linearly, and the larger the bulk material removal depth, the greater the influence of the path spacing on the residual error. This paper further proposed a residual error optimization strategy based on surface form error distribution. In this strategy, the mapping relationship between bulk material removal depth, path spacing and residual error was established and integrated into the optimization goal. The influence of varying bulk material removal depths resulting from the form error distribution on the residual error was compensated by adjusting the path spacing, and a specific path optimization model was obtained. In the implementation of this process, a variable-pitch polishing path is planned in accordance with the path optimization model and the actual surface form error distribution of the workpiece to be polished. Polishing with this optimized path can effectively optimize the residual error distribution and greatly reduce the influence of parent form error distribution on outcome RMS and PV values in sub-regions of the residual error. Furthermore, this optimization drastically reduces variations in tool feed across the workpiece, which facilitates dynamic control of the process.

Author Contributions: Y.H. wrote the manuscript and analyzed the data. L.Z., C.F. and W.Z. provided valuable suggestions for the manuscript. A.B. improved the theory of the manuscript and fully polished the language. All authors discussed the results and commented on the manuscript.

Funding: The authors would like to express sincere thanks for the financial support provided by the Young Creative Leading Talent and Team Program of Jilin Province (Grant No. 20150519005JH), the Chinese National Natural Science Foundation (Grant No. 51505312) and the Basic Research Program of Jiangsu Province (Grant No. BK20150330).

Acknowledgments: The authors also acknowledge support from China Scholarship Council (CSC) for funding Y.H. as a joint training doctoral student at Kyoto to conduct this research.

Conflicts of Interest: The authors declare no conflict of interest.

References

1. Guo, Z.; Jin, T.; Ping, L.; Lu, A.; Qu, M. Analysis on a deformed removal profile in FJP under high removal rates to achieve deterministic form figuring. *Precis. Eng. J. Int. Soc. Precis. Eng. Nanotechnol.* **2018**, *51*, 160–168, doi:10.1016/j.precisioneng.2017.08.006. [CrossRef]

2. Brinksmeier, E.; Mutlugünes, Y.; Klocke, F.; Aurich, J.C.; Shore, P.; Ohmori, H. Ultra-precision grinding. *CIRP Ann.* **2010**, *59*, 652–671, doi:10.1016/j.cirp.2010.05.001. [CrossRef]

3. Zhang, L.; Han, Y.J.; Fan, C.; Tang, Y.; Song, X.P. Polishing path planning for physically uniform overlap of polishing ribbons on free form surface. *Int. J. Adv. Manuf. Technol.* **2017**, *92*, 4525–4541, doi:10.1007/s00170-017-0466-z. [CrossRef]

4. Jones, R.A. Optimization of computer controlled polishing. *Appl. Opt.* **1977**, *16*, 218–224, doi:10.1364/ao.16.000218. [CrossRef] [PubMed]

5. Wan, S.L.; Zhang, X.C.; Zhang, H.; Xu, M.; Jiang, X.Q. Modeling and analysis of sub-aperture tool influence functions for polishing curved surfaces. *Precis. Eng. J. Int. Soc. Precis. Eng. Nanotechnol.* **2018**, *51*, 415–425, doi:10.1016/j.precisioneng.2017.09.013. [CrossRef]

6. Cao, Z.C.; Cheung, C.F.; Liu, M.Y. Model-based self-optimization method for form correction in the computer controlled bonnet polishing of optical free form surfaces. *Opt. Express* **2018**, *26*, 2065–2078, doi:10.1364/oe.26.002065. [CrossRef] [PubMed]

7. Beaucamp, A.; Namba, Y. Super-smooth finishing of diamond turned hard X-ray molding dies by combined fluid jet and bonnet polishing. *CIRP Ann. Manuf. Technol.* **2013**, *62*, 315–318, doi:10.1016/j.cirp.2013.03.010. [CrossRef]

8. Temple, P.A.; Lowdermilk, W.H.; Milam, D. Carbon dioxide laser polishing of fused silica surfaces for increased laser-damage resistance at 1064 nm. *Appl. Opt.* **1982**, *21*, 3249–3255, doi:10.1364/ao.21.003249. [CrossRef] [PubMed]

9. Ghadikolaei, A.D.; Vahdati, M. Experimental study on the effect of finishing parameters on surface roughness in magneto-rheological abrasive flow finishing process. *Proc. Inst. Mech. Eng. Part B. J. Eng. Manuf.* **2014**, *229*, 1517–1524, doi:10.1177/0954405414539488. [CrossRef]

10. Stognij, A.I.; Novitskii, N.N.; Stukalov, O.M. Nanoscale ion beam polishing of optical materials. *Tech. Phys. Lett.* **2002**, *28*, 17–20, doi:10.1134/1.1448630. [CrossRef]

11. Li, W.; Bin, F.; Shi, C.; Jia, W.; Bin, Z. A path planning method used in fluid jet polishing eliminating lightweight mirror imprinting effect. *Proc. SPIE* **2014**, *9281*, doi:10.1117/12.2070967. [CrossRef]

12. Wang, T.; Cheng, H.; Zhang, W.; Yang, H.; Wu, W. Restraint of path effect on optical surface in magnetorheological jet polishing. *Appl. Opt.* **2016**, *55*, 935–942, doi:10.1364/ao.55.000935. [CrossRef] [PubMed]

13. Dunn, C.; Walker, D. Pseudo-random tool paths for CNC sub-aperture polishing and other applications. *Opt. Express* **2008**, *16*, 18942–18949, doi:10.1364/OE.16.018942. [CrossRef] [PubMed]

14. Wang, C.; Wang, Z.; Xu, Q. Unicursal random maze tool path for computer-controlled optical surfacing. *Appl. Opt.* **2015**, *54*, 10128–10136, doi:10.1364/AO.54.010128. [CrossRef] [PubMed]

15. Takizawa, K.; Beaucamp, A. Comparison of tool feed influence in CNC polishing between a novel circular-random path and other pseudo-random paths. *Opt. Express* **2017**, *25*, 22411–22424, doi:10.1364/OE.25.022411. [CrossRef] [PubMed]

16. Wang, C.; Yang, W.; Ye, S.; Wang, Z.; Yang, P.; Peng, Y.; Guo, Y.; Xu, Q. Restraint of tool path ripple based on the optimization of tool step size for sub-aperture deterministic polishing. *Int. J. Adv. Manuf. Technol.* **2014**, *75*, 1431–1438, doi:10.1007/s00170-014-6223-7. [CrossRef]

17. Hou, J.; Liao, D.; Wang, H. Development of multi-pitch tool path in computer-controlled optical surfacing processes. *J. Eur. Opt. Soc.-Rapid Publ.* **2017**, *13*, 22, doi:10.1186/s41476-017-0050-z. [CrossRef]

18. Hu, H.; Dai, Y.; Peng, X. Restraint of tool path ripple based on surface error distribution and process parameters in deterministic finishing. *Opt. Express* **2010**, *18*, 22973–22981, doi:10.1364/oe.18.022973. [CrossRef] [PubMed]

19. Dong, Z.; Cheng, H. Toward the complete practicability for the linear-equation dwell time model in subaperture polishing. *Appl. Opt.* **2015**, *54*, 8884–8890, doi:10.1364/ao.54.008884. [CrossRef] [PubMed]

20. Dong, Z.; Cheng, H.; Tam, H.Y. Robust linear equation dwell time model compatible with large scale discrete surface error matrix. *Appl. Opt.* **2015**, *54*, 2747–2756, doi:10.1364/ao.54.002747. [CrossRef] [PubMed]

21. Guo, P.; Fang, H.; Yu, J. Edge effect in fluid jet polishing. *Appl. Opt.* **2006**, *45*, 6729–6735, doi:10.1364/AO.45.006729. [CrossRef] [PubMed]

22. Li, H.; Zhang, W.; Yu, G. Study of weighted space deconvolution algorithm in computer controlled optical surfacing formation. *Chin. Opt. Lett.* **2009**, *7*, 627–631, doi:10.3788/col20090707.0627. [CrossRef]

23. Wang, C.; Yang, W.; Wang, Z.; Yang, X.; Hu, C.; Zhong, B.; Guo, Y.; Xu, Q. Dwell-time algorithm for polishing large optics. *Appl. Opt.* **2014**, *53*, 4752–4760, doi:10.1364/AO.53.004752. [CrossRef] [PubMed]

24. Guillerna, A.B.; Axinte, D.; Billingham, J. The linear inverse problem in energy beam processing with an application to abrasive waterjet machining. *Int. J. Mach. Tools Manuf.* **2015**, *99*, 34–42, doi:10.1016/j.ijmachtools.2015.09.006. [CrossRef]

MDPI
St. Alban-Anlage 66
4052 Basel
Switzerland
Tel. +41 61 683 77 34
Fax +41 61 302 89 18
www.mdpi.com

Applied Sciences Editorial Office
E-mail: applsci@mdpi.com
www.mdpi.com/journal/applsci

www.ingramcontent.com/pod-product-compliance
Lightning Source LLC
LaVergne TN
LVHW071517180726
843512LV00014B/1099